Animal Nutrition and Metabolism

Animal Nutrition and Metabolism

Editor: Jasper Mitchell

Callisto Reference
www.callistoreference.com

Callisto Reference,
118-35 Queens Blvd., Suite 400,
Forest Hills, NY 11375, USA

Visit us on the World Wide Web at:
www.callistoreference.com

ISBN: 978-1-64116-773-4 (Hardback)

Cataloging-in-Publication Data

Animal nutrition and metabolism / edited by Jasper Mitchell.
p. cm.
Includes bibliographical references and index.
ISBN 978-1-64116-773-4
1.Animal nutrition. 2. Metabolism. 3. Nutrition. 4. Livestock--Metabolism. I. Mitchell, Jasper.
SF95 .A55 2023
636.085 2--dc23

Table of Contents

Preface

The science and practice of animal nutrition focuses on the dietary needs of food-producing, companion, or service animals. It is concerned with the preparation or formulation of feed for boosting the growth, health and performance of these animals. The main components of dietary food of an animal are carbohydrates, lipids, proteins, vitamins and minerals. Animal nutrition aims to analyze the feed and the needs of the animal for optimal utilization of nutrition in the diet. Metabolism refers to all the chemical reactions involved in maintaining the living state of the cells and the organism. For understanding metabolic disorders in animals, an analysis of the relationship among energy metabolism, endocrinology, and immunology must be carried out. The topics covered in this extensive book deal with the core aspects of animal nutrition and metabolism. It will also prove to be immensely beneficial to agricultural advisers, animal nutritionists, and animal feed manufacturers.

In the end, I would like to express my deep sense of gratitude to all the authors for meeting the set deadlines in completing and submitting their research chapters. I would also like to thank the publisher for the support offered to us throughout the course of the book. Finally, I extend my sincere thanks to my family for being a constant source of inspiration and encouragement.

Editor

1

Effects of Dietary Crude Protein Levels on Fecal Crude Protein, Amino Acids Flow Amount, Fecal and Ileal Microbial Amino Acids Composition and Amino Acid Digestibility in Growing Pigs

Zhenguo Yang *,†, Tianle He †, Gifty Ziema Bumbie †, Hong Hu, Qingju Chen, Changwen Lu and Zhiru Tang *

Laboratory for Bio-feed and Molecular Nutrition, College of Animal Science and Technology, Southwest University, Chongqing 400715, China; 18894335705@163.com (T.H.); giftyziema@gmail.com (G.Z.B.); huhong1020@163.com (H.H.); cqj9517@163.com (Q.C.); luchangwen678@163.com (C.L.)

* Correspondence: guoguo00002@163.com (Z.Y.); tangzhiru2326@sina.com (Z.T.)

† These authors contributed equally to this work.

Simple Summary: The purpose of this experiment was to evaluate a low protein corn-soybean meal-based diet with the same Lys, Met + Cys, Thr and Trp level as a high protein diet on fecal crude protein (CP), amino acid (AA) flow amount, AA digestibility and fecal and ileal microbial AA composition in growing pigs. Eighteen pigs with an initial body weight of (30 ± 1.35) kg were randomly divided into three groups, with six replicates in each group, and fed a corn-soybean meal-based diets with 12%, 15% and 18% CP levels, respectively. Our aim was to explain whether the addition of four crystalline essential AAs (EAA) to a low diet affected the digestibility of protein-bound NEAA (non-essential amino acid) and EAA and the composition of microbial AA in ileum and feces.

Abstract: The purpose of this experiment was to evaluate the effects of low protein corn-soybean meal-based diets on fecal CP, amino acid (AA) flow amount, AA digestibility and fecal and ileal microbial AA composition in growing pigs. Eighteen pigs (initial body weight = 30 ± 1.35) were randomly divided into three groups and fed with basal diets with CP levels of 12%, 15% and 18%, respectively. The Lys, Met + Cys, Thr and Trp level in the 12% CP and 15% CP groups is the same as 18% CP group by the addition of four crystalline Lys, Met + Cys, Thr and Trp to the diet. The results showed that with the decrease of dietary CP level from 18% to 12%, the fecal total nitrogen (N), CP and total AA (TAA) flow amount decreased linearly ($p < 0.05$). Dry matter (DM) digestibility, CP digestibility, TAA digestibility, essential amino acid (EAA) digestibility and non-essential amino acid (NEAA) digestibility increased linearly with the decrease of dietary CP concentration from 18% to 12%. Compared with 18% CP group, the flow amount of Asp, Ser, Glu, Gly, Tyr, Val, Leu and Phe in feces of pigs in the 15% CP group and 12% CP group decreased significantly, while the flow amount of Arg in the 15% CP group was lower than that in the 18% CP group and 12% CP group. The fecal microbial N and AA of the 15% CP group were higher than those of the 18% CP and 12% CP groups. Fecal TAA flow amount decreased linearly with the decrease of the dietary CP levels from 18% to 12%. Fecal TAA and NEAA flow amount also decreased linearly with the decrease of dietary CP level from 18% to 12%. Except for Glu, Gly, Met, Tyr, Thr and Phe, there were significant differences among the three groups in the composition of 17 kinds of AAs in fecal microorganisms. Among the 17 AA compositions of ileal microorganisms, except Tyr and Lys, the other AAs were significantly different among the three groups ($p < 0.05$)

Keywords: dietary crude protein levels; essential amino acids; growing pigs; apparent digestibility; composition of fecal and ileal microbial amino acids

1. Introduction

In recent years, the environmental pollution caused by nitrogen emission has become more and more serious with the continuous expansion of livestock breeding scales and intensive degrees. At present, the nitrogen emission from livestock and poultry in China is about 500–600 million tons, of which monogastric animals (mainly pigs) account for about 60% of the total nitrogen emissions. At the same time, China is seriously short of protein resources. In 2019, China's domestic soybeans were about 11 million tons, while imported soybeans were 95.53 million tons. Therefore, to improve protein utilization efficiency and reduce nitrogen emissions has become an urgent scientific problem to be solved.

Nitrogen emission reduction has been studied for a long time, and so far the following nitrogen emission reduction technologies have been formed: preparing dietary protein with ideal AA model [1]; low protein diet technology [2]; increasing the proportion of fermentable carbohydrates in diet [3]; adding feed additives such as enzymes, probiotics and organic acids [4]. Among them, the nitrogen emission reduction effect of low protein diet technology is the most obvious. According to the theory of protein-amino acid nutrition balance, a low protein diet is a diet that can reduce dietary protein levels and nitrogen emissions by adding appropriate kinds and quantities of synthetic AAs without affecting animal production performance and product quality. The practical significance of a low-protein diet is very significant. Studies have shown that for every 1% reduction of CP in pig diet, nitrogen emissions can be reduced by 8% to 10%, and ammonia concentration in pig houses can be reduced by more than 10% [5]. In the case of supplementation of important EAA, a reduction of 2% to 3% in dietary CP content did not affect the nitrogen deposition or growth performance of pigs [6,7], but the growth performance of pigs was usually negatively affected when the dietary CP content was more than 3% [5,8]. The increasingly serious environmental pollution forces researchers to fundamentally solve the protein and amino acid nutrition of pigs while reducing the nitrogen excretion of pig feces. It is pointed out that reducing the dietary CP of growing pigs can effectively reduce the nitrogen excretion of pig manure [9]. Reducing N excretion in swine manure can be effectively accomplished by reducing dietary CP intake. However, reducing the protein concentration of diets, with no supplemental crystalline AA, decreases growth performance and meat and carcass quality [9] and inhibits the activity of the mammalian target of rapamycin (mTOR) and s6 kinase 1 (S6K1), leading to the restriction of protein synthesis [10]. In low-protein diets, the absorption of protein-bound amino acids may be reduced, resulting in impaired animal growth performance if the dietary protein content cannot be further reduced, although the environmental benefits of low-protein diets are significant, the economic benefits are not obvious. Large-scale farms usually take economic benefits as the primary assessment index. Therefore, although the research history of low-protein diet has a history of more than 100 years, it has not been fully promoted in production [11]. In view of this, it is necessary to carry out in-depth study on the regulation mechanism of nitrogen emission reduction of livestock, identify the regulation targets, and develop a low-protein diet with a more efficient effect of nitrogen emission reduction.

The purpose of this experiment was to determine the effects of a low protein corn-soybean-based diet with the same Lys, Met + Cys, Thr and Trp level as high protein diet on fecal CP, AA flow, fecal CP, AA digestibility and fecal and ileal microbial AA composition in 30–60 kg growing pigs. Four kinds of crystalline AAs were added to the corn-soybean meal-based diet to reduce the dietary CP concentration to meet the needs of essential AA. In addition, the EAA/NEAA ratio (EAA/NEAA) decreased with the increase of dietary CP concentration. Our aim was to explain whether the addition of four crystalline EAAs to a low-CP diet affected the digestibility of protein-bound NEAA and EAA and the composition of microbial AA in ileum and feces.

2. Materials and Methods

2.1. Experimental Animals, Design and Diets

All experimental procedures in this study were approved by Sciences Animal Ethics Committee of Chinese Academy of Sciences (Hunan, China; Ethic approval number: SYXK 2014-0002). Eighteen crossbred barrows (Duroc × Landrace × Yorkshire, weighing on average 30 ± 1.35 kg) were randomly allotted to 3 experimental groups in accordance with dietary CP levels of corn-soybean meal-based diets: 12%, 15% and 18%, with six replicates each group and one piglet each replicate. Ingredients and nutrient contents of diets were given in Table 1 and the Lys, Met + Cys, Thr and Trp level in the 12% CP and 15% CP groups is the same as 18% CP group by the addition of four crystalline Lys, Met + Cys, Thr and Trp to diet. The trial lasted 30 days. All pigs were kept individually in stainless steel metabolism crates (1.5 m × 0.5 m × 0.8 m) in an ambient indoor temperature maintained approximately at 24 °C. Natural lighting was used during the experiment. All pig feeds are prepared according to various nutrient requirement recommended by the government (NRC, 2012). Feeding and watering was *ad libitum* throughout the trial.

2.2. Measurements and Sampling

On day 25 to day 30, six pigs each group were chosen to collect feces. Then the pigs were bled and anesthetized by intramuscularly injecting pentobarbital sodium (50 mg/kg·BW). Samples of terminal ileal digesta were collected and put into polythene bags for microbial or chemical analysis. The samples for microbe separation were stored at 4°C and for chemical analysis were stored at −20 °C.

2.3. Chemical Analysis

DM was determined by drying to a constant mass in a forced air oven at temperature maintained 95. The previous research methods were used to separate other digested samples by differential centrifugation [12,13]. Firstly, pooled digesta were centrifuged immediately (250× *g* for 15 min at 4 °C), to give a fraction expected to obtain porcine cells and food particles. Secondly, fractions were centrifuged at 14,500× *g* for 30 min at 4 °C, giving precipitates expected to contain microbial cells. N and ash in feed, feces and microbials were measured by the Leco total combustion method [14]. The AA in feed, feces and microorganisms was measured by a Hitachi Lmur8800 AA automatic analyzer (Hitachi, Ltd., Tokyo, Japan) [15]. Fecal and microbial protein N (PN) and non-protein N (NPN) were measured by the method of Warren et al. [13].

2.4. Data Treatment and Statistical Analyses

Acid insoluble ash (AIA), a kind of endogenous indicator was measured to estimate the apparent digestibility of DM, CP and TAA in feces. The following is the relevant formula: 1-bc/ad. where a denotes the content of DM, CP or AA in diets (%); b denotes the content of DM, CP or AA in feces (%); c denotes the content of AIA in diet (%); d denotes the content of AIA in feces (%). AA compositions (%) appertained the content of single AA in TAA.

All results were expressed as mean ± standard error of mean (SEM). All data were subjected to a one-way analysis of variance using the general linear model (GLM) procedure of SAS 8.2 statistical software (SAS Institute, Inc. Cary, NC, USA) [16] according to a completely randomized one-factorial design. Differences among experimental groups were identified via the Student–Newman–Keuls (SNK) test. A difference was considered to be statistically significant when $p < 0.05$.

Table 1. Ingredients and nutrient contents of diets in 30–60 kg growing pigs (air-dry basis).

Items	18% CP	15% CP	12% CP	Nutrition Levels	18% CP	15% CP	12% CP
Corn	58.60	67.50	77.60	DE (MJ/kg)	14.20	14.20	14.20
Soy meal	29.00	19.50	10.00	CP, %	18.27	15.16	12.35
Wheat bran	7.80	6.94	5.06	Lys, %	0.97	0.97	0.94
Soy oil	1.55	2.38	3.00	Met + Cys, %	0.57	0.56	0.55
Monocalcium phosphate	0.69	0.78	0.90	Thr, %	0.61	0.61	0.60
Stone dust	0.87	0.89	0.90	Arg, %	1.08	0.82	0.57
NaCl	0.30	0.30	0.30	Trp, %	0.17	0.17	0.17
Vitamin-mineral Premix	1.00	1.00	1.00	His, %	0.41	0.33	0.25
L-Lysine HCl (Lys)	0.18	0.46	0.74	Phe, %	0.77	0.62	0.46
DL-Methionine (Met)	0.00	0.09	0.17	Val, %	0.66	0.56	0.44
L-Threonine (Thr)	0.01	0.14	0.26	Ile, %	0.64	0.49	0.35
L-Tryptophan (Trp)	0.00	0.02	0.07	Leu, %	1.35	1.14	0.94
Total	100.00	100.00	100.00	Available *p*, %	0.51	0.48	0.45
				EAA, %	7.00	6.04	5.03
				NEAA, %	9.12	7.50	5.83
				EAA/NEAA	0.76	0.81	0.86

3. Results

3.1. Fecal N and CP Flow and Digestibility

Feces flow amount and apparent digestibility of CP and N of growing pigs were illustrated in Table 2. Feces CP, total N (TN) and NPN flow amounts in growing pigs were affected by the experimental treatments ($p < 0.01$), i.e., greatest in the 18% CP group, greater in the 15% CP group and lowest in the 12% CP group ($p < 0.05$). Fecal PN flow amounts was a highly significant difference among the three experimental groups ($p < 0.01$), i.e., lower in the 12% CP group than in the 15% and18% CP group ($p < 0.05$), but there were no significant differences in feces PN flow amount between the 15% CP group and the 18% CP group ($p > 0.05$). Feces Total microbial N and microbial PN flow amount were significantly different among the three experimental groups ($p < 0.01$), i.e., lowest in the 18% CP group, greater in the 12% CP group and greatest in the 15% CP group ($p < 0.05$). Feces apparent digestibility of CP, DM, TN, NPN and PN was also different significantly among the three experimental groups ($p < 0.01$), i.e., greatest in the 12% CP group, greater in the 15% CP group and lowest in the 18% CP group ($p < 0.05$). In addition, there were no significant discrepancies in fecal microbial NPN among the three experimental groups ($p > 0.05$).

Table 2. Feces flow amount and apparent digestibility of crude protein (CP) and N of growing pigs fed low protein diets.

Items	18% CP	15% CP	12% CP	SEM	*p*-Values
Fecal CP and N flow amount (mg/g DM)					
CP	245.80 [a]	228.60 [b]	201.50 [c]	2.70	<0.01
TN	39.40 [a]	36.50 [b]	32.20 [c]	0.43	<0.01
NPN	4.80 [a]	3.25 [b]	3.20 [b]	0.28	<0.01
PN	34.60 [a]	33.30 [a]	29.10 [b]	0.50	<0.01
Total microbial N	5.68 [c]	6.60 [a]	5.99 b	0.07	<0.01
Microbial NPN	0.64	0.70	0.64	0.02	0.14
Microbial PN	5.05 [c]	5.90 [a]	5.35 [b]	0.06	<0.01
Feces apparent digestibility (%)					
CP	85.30 [c]	88.10 [b]	91.00 [a]	0.52	<0.01
DM	89.50 [c]	92.00 [b]	94.60 [a]	0.41	<0.01
TN	85.30 [c]	88.10 [b]	91.00 [a]	0.52	<0.01
NPN	83.10 [c]	87.40 [b]	90.60 [a]	0.93	<0.01
PN	97.90 [c]	98.30 [b]	98.70 [a]	0.77	<0.01

[a,b,c] Values in the same row with different letter superscripts mean significant differences ($p < 0.05$).

3.2. Feces Total AA (TAA), Essential (EAA) Amino Acids and Non-Essential (NEAA) Flow Amount and Apparent Total Amino Acid Number (TAA) with or without Microbia

Feces TAA, EAA, NEAA flow amount and apparent digestibility with or without microbia of growing pigs were illustrated in Table 3. Feces TAA, EAA and NEAA flow amounts with or without microbia in growing pigs were affected by the experimental treatments ($p < 0.01$), i.e., greater in the 18% CP group than in the 12% CP and 15% CP groups ($p < 0.05$), but there were no significant differences in feces TAA, EAA and NEAA flow amounts with or without microbia in growing pigs between the 12% CP group and the 15% CP group ($p > 0.05$). Digestibility of TAA, EAA and NEAA in feces with microbia and digestibility of EAA in feces without microbial were affected by the experimental treatments ($p < 0.01$), i.e., greater in the 12% CP group than in the 15% CP and 18% CP groups ($p < 0.05$), but there were no significant differences in digestibility of TAA, EAA and NEAA in feces with microbia and digestibility of EAA in feces without microbia in growing pigs between the 12% CP group and the 15% CP group ($p > 0.05$). Digestibility of TAA and NEAA in feces without microbia in growing pigs was affected by the experimental treatments ($p < 0.01$), i.e., greatest in the 12% CP group, greater in the 15% CP group and lowest in the 18% CP group ($p < 0.05$). Feces EAA/NEAA flows with or without microbia in growing pigs were affected by the treatments ($p < 0.01$), i.e., greater in the 15% CP group than in the 12% CP group and the 18% CP group ($p < 0.05$). There were no significant differences in feces EAA/NEAA flow in growing pigs between the 12% CP group and the 18% CP group ($p > 0.05$).

Table 3. Feces Total AA, essentials AA, nonessential AA flow amount (mg/g DM) and apparent digestibility (%) with or without microbia of growing pigs fed low protein diets.

Items	18% CP	15% CP	12% CP	SEM	*p*-Values
Feces with microbes					
NEAA flow amount	90.50 ^{a}	73.90 ^{b}	76.20 ^{b}	1.46	<0.01
EAA flow amount	86.20 ^{a}	75.00 ^{b}	72.40 ^{b}	1.44	<0.01
TAA flow amount	176.80 ^{a}	148.80 ^{b}	148.60 ^{b}	2.57	<0.01
NEAA digestibility	92.50 ^{b}	93.00 ^{b}	94.60 ^{a}	0.29	<0.01
EAA digestibility	90.70 ^{b}	90.80 ^{b}	92.90 ^{a}	0.36	<0.01
TAA digestibility	91.70 ^{b}	92.10 ^{b}	93.90 ^{a}	0.32	<0.01
EAA/NEAA flow amount	0.95 ^{b}	1.01 ^{a}	0.95 ^{b}	0.02	0.03
Feces without microbes					
NEAA flow amount	73.50 ^{a}	54.40 ^{b}	58.50 ^{b}	1.49	<0.01
EAA flow amount	71.70 ^{a}	57.80 ^{b}	56.70 ^{b}	1.40	<0.01
TAA flow amount	145.20 ^{a}	111.90 ^{b}	115.20 ^{b}	2.55	<0.01
NEAA digestibility	93.80 ^{c}	94.70 ^{b}	95.70 ^{a}	0.24	<0.01
EAA digestibility	92.40 ^{b}	93.12 ^{b}	94.55 ^{a}	0.31	<0.01
TAA digestibility	93.20 ^{c}	94.00 ^{b}	95.20 ^{a}	0.27	<0.01
EAA/NEAA flow amount	0.98 ^{b}	1.06 ^{a}	0.97 ^{b}	0.02	0.03

^{a,b,c} Values in the same row with different letter superscripts mean significant differences ($p < 0.05$).

3.3. Feces Amino Acids Flow Amount

Feces Ser, Glu, Ala, Gly, Tyr, Asp, Thr, Cys, Leu and Phe flow amounts were different among the three experimental groups ($p < 0.01$) as illustrated in Table 4, i.e., compared with the 15% CP and 12% CP group, it was higher in the 18% CP group. However, compared with the 15% CP group, it was higher in the 12% CP group ($p < 0.05$). There were no significant differences in feces Ser, Glu, Gly, Tyr, Asp, Thr, Leu and Phe flow amounts in growing pigs between the 15% CP and 12% CP groups ($p > 0.05$). There were no significant differences in feces Pro, Met and His flow amounts in growing pigs among the three treatments ($p > 0.05$). Feces Lys flow amounts were highly significant in their difference among the three experimental groups ($p < 0.01$), i.e., greater in the 15% CP group and the 18% CP group than in the 12% CP group ($p < 0.01$). Feces Leu flow amounts were different among three treatments ($p < 0.05$), i.e., greater in the 18% CP group than in the 12% CP group and the 15% CP group ($p < 0.05$). There were no significant differences in feces Lys flow amounts in growing pigs between the 15% CP group and the 18% CP group ($p > 0.05$). Feces Ala and Val flow amounts were also very significantly different among the three experimental groups ($p < 0.01$), i.e., greatest in the 18% CP group, greater in 12% CP group and lowest in 15% CP, ($p < 0.05$). Feces Arg flow amounts were different among the three experimental groups ($p < 0.01$), i.e., greater in the 12% CP group and the 18%

CP group than in the 15% CP group ($p < 0.05$). There were no significant differences in feces Arg flow amounts in growing pigs between the 18% CP group and 12% CP group ($p > 0.05$).

Table 4. Feces amino acids flow amount (mg/g DM) of growing pigs fed low protein diets.

Items	18% CP	15% CP	12% CP	SEM	p-Values
Ser	3.30 ^{a}	2.80 ^{b}	2.80 ^{b}	0.11	<0.01
Glu	24.00 ^{a}	20.40 ^{b}	20.30 ^{b}	0.62	<0.01
Ala	19.80 ^{a}	15.00 ^{c}	17.20 ^{b}	0.51	<0.01
Gly	15.80 ^{a}	12.80 ^{b}	11.90 ^{b}	0.42	<0.01
Pro	9.80	8.90	10.10	0.39	0.13
Tyr	5.90 ^{a}	5.10 ^{b}	4.90 ^{b}	0.21	0.01
Asp	17.90 ^{a}	14.00 ^{b}	13.10 ^{b}	0.42	<0.01
Thr	7.20 ^{a}	5.60 ^{b}	4.90 ^{b}	0.26	<0.01
Val	12.60 ^{a}	9.20 ^{c}	10.30 ^{b}	0.30	<0.01
Cys	2.60 ^{a}	2.40 ^{b}	2.50 ^{b}	0.05	0.01
Met	3.00	2.90	2.70	0.12	0.18
Ile	10.50 ^{a}	9.70 ^{ab}	8.80 ^{b}	0.31	<0.01
Leu	19.00 ^{a}	16.90 ^{b}	17.10 ^{b}	0.34	<0.01
Phe	8.70 ^{a}	7.70 ^{b}	7.60 ^{b}	0.27	0.02
Lys	8.20 ^{a}	7.70 ^{a}	6.60 ^{b}	0.33	0.01
His	4.10	4.30	3.70	0.18	0.09
Arg	4.50 ^{a}	3.60 ^{b}	4.30 ^{a}	0.13	<0.01

^{a,b,c} Values in the same row with different letter superscripts mean significant differences ($p < 0.05$).

3.4. AAs Digestibility

The AAs digestibility were illustrated in Table 5. Feces Ser, Glu, Gly, Pro, Tyr, Thr, Ile, Leu, Phe and Lys digestibility were very significantly different among the three experimental groups ($p < 0.01$), i.e., greater in the 12% CP group than in the 15% CP and 18% CP groups ($p < 0.05$), but there were no significant differences in feces Ser, Glu, Gly, Pro, Tyr, Thr, Ile, Leu, Phe and Lys digestibility in growing pigs between the 15% CP and 18% CP groups ($p > 0.05$). There were no significant differences in feces Arg flow in growing pigs among the three experimental groups ($p > 0.05$). Feces Ala and Val digestibility of pigs in the 15% CP group and the 12% CP group are higher than in the 18% CP group ($p < 0.05$), while there were no significant differences in feces Ala and Val digestibility in growing pigs between the 15% CP group and the 12% CP group ($p > 0.05$). Feces Cys and His digestibility of pigs in the 12% CP group and the 18% CP group are higher than in the 15% CP group ($p < 0.05$), but there were no significant differences in feces Cys and His digestibility in growing pigs between the 12% CP and 18% CP groups ($p > 0.05$). Feces Met digestibility was very significantly different among these three experimental groups ($p < 0.01$), i.e., greatest in 12% CP group, greater in 18% CP group and lowest in the 15% CP group ($p < 0.05$).

3.5. The Feces microbial AA Flow Amount

As shown in Table 6, feces microbial Ser flow amounts were different among the three experimental groups ($p < 0.01$), i.e., greatest in the 15% CP group, greater in the 18% CP group and lowest in the 12% CP group ($p < 0.05$). Feces microbial Gly, Ala, Pro, Thr, Ile, TAA and EAA flow amounts were different among the three experimental groups ($p < 0.01$), i.e., greatest in the 15% CP group, greater in 12% and lowest in the 18% CP group ($p < 0.05$). Feces microbial Tyr, Val, Met, Phe and NEAA flow amounts were different among the three experimental groups ($p < 0.01$), i.e., greater in the 15% CP group than in the 18% CP and 12% CP groups ($p < 0.05$), but there were no significant differences in feces microbial Tyr, Val, Met, Phe and NEAA flow in growing pigs between the 12% CP and 18% CP groups ($p > 0.05$). Feces microbial Asp flows were different among the three experimental groups ($p < 0.01$), i.e., greatest in the 18% CP group, greater in the 15% CP group and lowest in the 12% CP group ($p < 0.05$). Feces microbial Cys flow amounts were different among the three experimental groups ($p < 0.01$), i.e., greater in the 18% CP group than in the 15% CP and 12% CP groups ($p < 0.05$), while there were no significant differences in feces microbial Cys flow amounts in growing pigs between the 12% CP group and the 15% CP group ($p > 0.05$). Feces microbial Lys and His flow amounts were different among the three experimental groups ($p < 0.01$), i.e., greater in the 15% CP and 18% CP groups than in the 12% CP group ($p < 0.05$). There were no significant differences in feces microbial Lys and His flow in growing pigs between the 18% CP group and the 15% CP. Feces microbial Arg flow amounts were different among the three experimental groups ($p < 0.01$), i.e., greater in the 15% CP and 12% CP groups than in the 18% CP group ($p < 0.05$). There were no significant differences in feces microbial Cys flow amounts in growing pigs between the 12% CP group and the 15% CP group ($p > 0.05$). There were no significant differences in feces microbial Glu, Leu and EAA/NEAA flow amounts in growing pigs among the three experimental groups ($p > 0.05$).

Table 5. Feces apparent amino acids digestibility (%) in growing pigs low protein diets.

Items	18% CP	15% CP	12% CP	SEM	p-Values
Ser	94.80 ^{b}	94.40 ^{b}	95.50 ^{a}	0.19	<0.01
Glu	92.00 ^{b}	92.40 ^{b}	93.90 ^{a}	0.38	<0.01
Gly	85.00 ^{b}	86.80 ^{b}	89.90 ^{a}	0.59	<0.01
Ala	87.40 ^{b}	89.10 ^{a}	90.90 ^{a}	0.60	<0.01
Pro	97.10 ^{b}	96.90 ^{b}	97.60 ^{a}	0.16	0.01
Tyr	91.50 ^{b}	91.30 ^{b}	93.60 ^{a}	0.34	<0.01
Asp	90.00 ^{c}	91.50 ^{b}	93.00 ^{a}	0.39	<0.01
Thr	91.30 ^{b}	92.40 ^{b}	94.70 ^{a}	0.43	<0.01
Cys	92.20 ^{a}	90.50 ^{b}	91.20 ^{a}	0.32	<0.01
Val	88.10 ^{b}	90.20 ^{a}	90.90 ^{a}	0.55	<0.01
Met	91.70 ^{b}	89.00 ^{c}	93.20 ^{a}	0.44	<0.01
Ile	88.50 ^{b}	88.10 ^{b}	91.00 ^{a}	0.50	<0.01
Leu	90.70 ^{b}	91.05 ^{b}	93.20 ^{a}	0.41	<0.01
Phe	90.20 ^{b}	89.50 ^{b}	91.50 ^{a}	0.48	0.03
Lys	91.50 ^{b}	91.20 ^{b}	94.60 ^{a}	0.43	<0.01
His	92.40 ^{a}	90.30 ^{b}	93.30 ^{a}	0.49	<0.01
Arg	96.70	97.10	97.10	0.18	0.21

^{a,b,c} Values in the same row with different letter superscripts mean significant differences ($p < 0.05$).

3.6. The Feces Microbial AA Composition

Feces microbial Ser of AA compositions were different among the three experimental groups ($p < 0.01$), i.e., greatest in the 15% CP group, greater in the 18% CP group, lowest in the 12% CP group as illustrated in Table 7 ($p < 0.05$). Feces microbial Ala, Pro and Ile of AA compositions were different among the three experimental groups ($p < 0.01$), i.e., greater in the 12% CP and 15% CP groups than in the 18% CP ($p < 0.05$), but there were no significant differences in feces microbial Ala, Pro and Ile of AA compositions in growing pigs between the 12% CP and 15% CP groups ($p > 0.05$). Feces microbial Asp, Cys, Lys and His of AA compositions were different among the three experimental groups ($p < 0.01$), i.e., greater in the 18% CP group than in the 15% CP and 12% CP groups ($p < 0.05$), but there were no significant differences in feces microbial Asp, Cys, Lys and His of AA composition in growing pigs between the 12% CP and 15% CP groups ($p > 0.05$). Feces microbial Val of AA compositions were different among three treatments ($p < 0.01$), i.e., greater in the 15% CP group than in the 12% CP and 18% CP groups ($p < 0.05$), but there were no significant differences in microbial Val feces of AA compositions in growing pigs between 12% CP group and 18% CP group ($p > 0.05$).

Table 6. Feces microbial amino acids flow amount (mg/g DM) of growing pigs fed low protein diets.

Items	18% CP	15% CP	12% CP	SEM	*p*-Values
Ser	0.58 ^{b}	0.84 ^{a}	0.50 ^{c}	0.02	0.08
Glu	3.99	4.49	4.20	0.14	<0.01
Gly	2.43 ^{c}	2.83 ^{a}	2.62 ^{b}	0.06	<0.01
Ala	3.08 ^{c}	3.81 ^{a}	3.56 ^{b}	0.07	<0.01
Pro	1.38 ^{c}	2.11 ^{a}	1.98 ^{b}	0.04	<0.01
Tyr	1.23 ^{b}	1.41 ^{a}	1.20 ^{b}	0.04	<0.01
Asp	4.33 ^{a}	4.01 ^{b}	3.67 ^{c}	0.07	<0.01
Thr	1.30 ^{c}	1.66 ^{a}	1.46 ^{b}	0.05	<0.01
Cys	0.38 ^{a}	0.28 ^{b}	0.26 ^{b}	0.03	0.01
Val	1.74 ^{b}	2.36 ^{a}	1.81 ^{b}	0.05	<0.01
Met	0.58 ^{b}	0.69 ^{a}	0.55 ^{b}	0.02	0.02
Ile	1.38 ^{c}	2.35 ^{a}	2.03 ^{b}	0.04	<0.01
Leu	3.59	3.74	3.79	0.07	0.12
Phe	1.51 ^{b}	1.71 ^{a}	1.54 ^{b}	0.05	0.04
Lys	1.42 ^{a}	1.50 ^{a}	1.35 ^{b}	0.04	0.05
His	0.90 ^{a}	0.91 ^{a}	0.78 ^{b}	0.03	0.01
Arg	1.72 ^{b}	2.19 ^{a}	2.17 ^{a}	0.06	<0.01
TAA	31.50 ^{c}	36.90 ^{a}	33.40 ^{b}	0.36	<0.01
NEAA	17.00 ^{b}	19.50 ^{a}	17.70 ^{b}	0.26	<0.01
EAA	14.50 ^{c}	17.40 ^{a}	15.70 ^{b}	0.20	<0.01
EAA/NEAA	0.85	0.89	0.89	0.01	0.16

^{a,b,c} Values in the same row with different letter superscripts mean significant differences ($p < 0.05$).

Feces microbial Leu of AA compositions were different among the three experimental groups ($p < 0.01$), i.e., greater in the 12% CP and 18% CP groups than in the 15% CP group ($p < 0.05$), but there were no significant differences in feces microbial Leu of AA composition in growing pigs between the 12% CP and 18% CP groups ($p > 0.05$). Feces microbial Arg of AA compositions were different among the three experimental groups ($p < 0.01$), i.e., greatest in the 12% CP group, greater in the 15% CP and lowest in the 18% CP ($p < 0.05$). There were no significant differences in feces microbial Glu, Gly, Tyr, Thr, Met, Phe, NEAA, EAA and EAA/NEAA of AA composition in growing pigs among the three experimental groups ($p > 0.05$).

3.7. The Ileal Microbial AA Composition

Table 8 shows that the ileal microbial Ser, Pro, Val and Ile of AA compositions were different among the three experimental groups ($p < 0.01$), i.e., greater in the 12% CP and 15% CP group than in 18% CP group ($p < 0.05$), but there were no significant differences in feces microbial Ser, Pro, Val and Ile of AA composition in growing pigs between the 12% CP group and the 15% CP group ($p > 0.05$). The ileal microbial Glu of AA composition was different among the three experimental groups ($p < 0.01$), i.e., greater in the 18% CP group than in the 12% CP and 15% CP groups ($p < 0.05$), but there were no significant differences in feces microbial Glu of AA composition in growing pigs between the 12%

CP group and the 15% CP group ($p > 0.05$). The ileal microbial Gly of AA composition was different among the three experimental groups ($p < 0.01$), i.e., greatest in the 15% CP group, greater in the 12% CP group and lowest in the 18% CP group ($p < 0.05$). The ileal microbial Ala of AA composition was different among the three experimental groups ($p < 0.01$), i.e., greatest in the 12% CP group, greater in the 15% CP group and lowest in the 18% CP group ($p < 0.05$). The ileal microbial Asp, Thr and His of AA composition was different among the three experimental groups ($p < 0.01$), i.e., greater in the 12% CP and 18% CP groups than in the 15% CP group ($p < 0.05$). There were no significant difference in feces microbial Asp, Thr and His of AA composition in growing pigs between the 12% CP and 18% CP groups ($p > 0.05$).

Table 7. Feces microbial amino acids composition (%) of growing pigs fed low protein diets.

Items	18% CP	15% CP	12% CP	SEM	*p*-Values
Ser	1.83 ^{b}	2.28 ^{a}	1.49 ^{c}	0.06	<0.01
Glu	12.64	12.16	12.54	0.34	0.57
Gly	7.70	7.68	7.83	0.18	0.80
Ala	9.8 ^{b}	10.3 ^{a}	10.70 ^{a}	0.21	0.03
Pro	4.39 ^{b}	5.71 ^{a}	5.89 ^{a}	0.12	<0.01
Tyr	3.91	3.83	3.59	0.12	0.11
Asp	13.7 ^{a}	10.9 ^{b}	11.0 ^{b}	0.19	<0.01
Thr	4.13	4.51	4.36	0.12	0.12
Cys	1.20 ^{a}	0.77 ^{b}	0.76 ^{b}	0.08	<0.01
Val	5.53 ^{b}	6.39 ^{a}	5.44 ^{b}	0.14	<0.01
Met	1.85	1.86	1.64	0.06	0.04
Ile	4.37 ^{b}	6.38 ^{a}	6.06 ^{a}	0.13	<0.01
Leu	11.4 ^{a}	10.1 ^{b}	11.30 ^{a}	0.23	<0.01
Phe	4.78	4.63	4.61	0.15	0.66
Lys	4.50 ^{a}	4.08 ^{b}	4.04 ^{b}	0.12	0.03
His	2.83 ^{a}	2.47 ^{b}	2.32 ^{b}	0.09	<0.01
Arg	5.44 ^{c}	5.93 ^{b}	6.45 ^{a}	0.16	<0.01
TAA	100.00	100.00	100.00	0.00	1.00
NEAA	54.00	52.80	53.00	0.42	0.16
EAA	46.00	47.10	47.00	0.42	0.16
EAA/NEAA	0.85	0.89	0.89	0.01	0.16

^{a,b,c} Values in the same row with different letter superscripts mean significant differences ($p < 0.05$).

The ileal microbial Cys and Met of AA composition was different among the three experimental groups ($p < 0.01$), i.e., greatest in the 18% CP group, greater in the 12% CP group and lowest in the 15% CP group ($p < 0.05$). The ileal microbial Leu of AA composition was different among the three experimental groups ($p < 0.01$), i.e., greater in the 15% CP and 18% CP groups than in the 12% CP group ($p < 0.05$). There were significant differences in feces microbial Leu of AA composition in growing pigs between the 18% CP group and the 15% CP group ($p < 0.05$). The ileal microbial Phe, EAA and EAA/NEAA of AA composition was different among the three experimental groups ($p < 0.01$), i.e., greatest in the 18% CP group and lowest in the 12% CP group ($p < 0.05$). The ileal microbial NEAA of AA composition was different among the three experimental groups ($p < 0.01$), i.e., greater in the 12% CP group than in the 15% CP and 18% CP groups ($p < 0.05$). There were no significant differences in feces microbial NEAA of AA composition in growing pigs between the 15% CP and 18% CP groups ($p > 0.05$). There were no significant differences in feces microbial Tyr and Lys of AA composition in growing pigs among the three experimental groups ($p > 0.05$). The ileal microbial Arg of AA composition was different among the three experimental groups ($p < 0.01$), i.e., greatest in the 15% CP group, greater in the 12% CP group and lowest in the 18% CP group ($p < 0.05$).

Table 8. Ileal microbial amino acids composition (%) of growing pigs fed low protein diets.

Items	18% CP	15% CP	12% CP	SEM	*p*-Values
Ser	3.23 [b]	4.03 [a]	4.36 [a]	0.12	<0.01
Glu	12.84 [a]	9.74 [b]	9.04 [b]	0.25	<0.01
Gly	4.93 [c]	6.54 [a]	5.59 [b]	0.12	<0.01
Ala	6.00 [c]	8.28 [b]	8.96 [a]	0.18	<0.01
Pro	4.85 [b]	6.80 [a]	6.30 [a]	0.19	<0.01
Tyr	3.50	3.97	3.64	0.19	0.24
Asp	11.8 [a]	8.8 [b]	11.8 [a]	0.23	<0.01
Thr	3.26 [a]	2.26 [b]	3.01 [a]	0.12	<0.01
Cys	2.57 [a]	1.36 [c]	1.57 [b]	0.05	<0.01
Val	5.31 [b]	6.94 [a]	6.88 [a]	0.17	<0.01
Met	3.95 [a]	2.24 [c]	3.14 [b]	0.15	<0.01
Ile	5.29 [b]	6.80 [a]	6.75 [a]	0.28	<0.01
Leu	11.8 [a]	11.8 [a]	10.6 [b]	0.23	<0.01
Phe	7.75 [a]	6.91 [b]	5.77 [c]	0.12	<0.01
Lys	4.30	4.80	4.67	0.32	0.52
His	4.38 [a]	3.71 [b]	4.61 [a]	0.19	0.01
Arg	4.22 [b]	5.00 [a]	3.26 [c]	0.15	<0.01
TAA	100.00	100.00	100.00	0.00	1.00
NEAA	43.70 [b]	44.20 [b]	46.10 [a]	0.23	<0.01
EAA	52.80 [a]	51.80 [b]	50.30 [c]	0.26	<0.01
EAA/NEAA	1.21 [a]	1.17 [b]	1.09 [c]	0.01	<0.01

[a,b,c] Values in the same row with different letter superscripts mean significant differences ($p < 0.05$).

4. Discussion

4.1. Fecal N Excretion

Livestock and poultry excreta are important sources of environmental N pollution. In feeding and management, it has become an effective way to reduce nitrogen emissions by reducing the amount of protein raw materials to reduce dietary protein levels, and then by adding synthetic AA, to meet animal AA needs of low-protein diets [16]. Compared to feeding them the 18% CP diet, fecal N losses decreased by 7.4% and 18.3% when feeding the 15% and 12% (protein-free) CP diet, respectively. Our values are lower than those reported previously. Gloaguen et al. [17] pointed out that dietary protein levels decreased from 20% to 17%, with an average reduction of 1 percentage point, and total nitrogen emissions decreased by 8.04%. At the same time, with the decrease of dietary protein levels, N intake, N uptake and N deposition decreased, while N apparent utilization efficiency and biological potency increased. In addition, previous study has shown that N excretion was reduced by 10% and 40% when decreasing dietary CP levels from 15% to 6% in growing pigs [18]. In the present study, our greatest contribution to decreasing the fecal N losses when reducing CP concentration from 18% to 12% is protein N origin. We also found that feces microbial N of growing pigs had no obvious change among the 12%, 15% and 18% CP diets; therefore, microbial N in feces is not a principal factor to reduce total N-losses. However, other research suggests that Bacterial N, combined with ammonia and urea N, accounted for approximately 61% of total nitrogenous losses in pigs which were fed free protein diets or casein diets [13]. Dietary manipulation to reduce NH_3 emission seems highly effective for realistic and efficient practice to reduce NH_3 emission throughout the process from feeding the animal to spreading the fertilizer [19]. Dietary regulation to decrease NH_3 emissions seems to be a highly effective method to cut back on NH_3 emissions. We also know that reducing dietary CP levels is one effective way to decrease ammonia emitted from pigs houses in industry [20].

4.2. AA Digestibility

Added essential crystalline AA to low-protein diet can not only increase AA digestibility and reduce AA excretion, but also provide a better balance of AA absorption for the whole intestine by decreasing the competition of intestinal epithelial cells for AA transport [18]. The research showed that adding crystalline AA to diets could improve the digestible energy, metabolizable energy and net energy of growing pigs [21]. Similarly, adding crystalline AA to low-protein diets can improve performance and carcass traits of late-breeding sows [16]. Other studies have shown that adding branched chain AA to low-protein diets can improve the growth performance in pigs to a level similar to that of adequate protein diets [22]. In addition, crystalline AA can provide growth requirements for broilers in low-protein diets [23]. In conclusion, added EAA to low-protein diet can make up for the problem of protein deficiency, and even improve animal production performance and lipid metabolism.

4.3. Microbial AA Composition

The present study indicated that the effects of dietary N substrates on ileal microbial AA composition were more significant than in feces. Ileal microbial EAA and NEAA compositions were affected by dietary AA composition, while fecal microbial EAA and NEAA compositions were not affected by dietary AA composition. This is related to the different microbiome composition from anterior to posterior [24]. Up until now, very little research has addressed the microbial AA composition of pigs. Yet, the decomposition and synthesis of N and AA in intestinal microflora were very active. Previous studies have shown that NEAA added to a low-protein Thr-deficient diet can improve the utilization of these AAs for protein deposition in young pigs [25]. Branched chain AA, Lys and Phe, synthesized by swine intestinal bacteria with intestinal fistula, can be absorbed in the small intestine and large intestine [26–28]. In addition, prior research has suggested that a carbohydrate to nitrogenous compound ratio in the large intestine is an determinant factor of microbial metabolism and gut barrier function in the colon [29]. These findings will be beneficial to enhancing our understanding of the fact that microbe-related secondary bile acid metabolism may mediate the interplay between intestinal barrier functions and microbes.

5. Conclusions

In conclusion, we had direct implications for reducing dietary CP levels from 18% to 12%, including a decreased fecal N excretion, and a principal factor to reducing total N losses in feces was dietary N except for microbial N. Microbial EAA and NEAA composition in ileum was affected by the AA composition in growing pigs' diets, while dietary AA composition had no effects on the composition of microbial EAA and NEAA in feces.

Author Contributions: Conceptualization, Z.Y., and Z.T.; Data curation, Z.Y. and T.H.; Formal analysis, Q.C., H.H., and Z.Y.; Funding acquisition, Z.Y., and Z.T.; Investigation, G.Z.B. and T.H.; Methodology, Z.Y. and Z.T.; Project administration, Z.Y. and Z.T.; Resources, Z.Y. and Z.T.; Software, C.L.; writing—original draft, G.Z.B., T.H. and Z.Y.; writing—review and editing, Z.Y. and T.H. All authors have read and agreed to the published version of the manuscript.

References

1. Chung, T.K.; Baker, D.H. Ideal amino acid pattern for 10-kilogram pigs. *J. Anim. Sci.* **1992**, *10*, 3102–3111. [CrossRef] [PubMed]
2. Boujendar, S.; Reusens, B.; Merezak, S.; Ahn, M.-T.; Arany, E.; Hill, D.; Remacle, C. Taurine supplementation to a low protein diet during foetal and early postnatal life restores a normal proliferation and apoptosis of rat pancreatic islets. *Diabetologia* **2002**, *45*, 856–866. [CrossRef] [PubMed]

3. Shriver, J.A.; Carter, S.D.; Sutton, A.L.; Richert, B.T.; Senne, B.W.; Pettey, L.A. Effects of adding fiber sources to reduce d-crude protein, amino acid-supplemented diets on nitrogen excretion, growth performance, and carcass traits of finishing pigs. *J. Anim. Sci.* **2003**, *81*, 492–502. [CrossRef] [PubMed]
4. Puiman, P.; Stoll, B.; Molbak, L.; De Bruijn, A.; Schierbeek, H.; Boye, M.; Boehm, G.; Renes, I.; Van Goudoever, J.; Burrin, D. Modulation of the gut microbiota with antibiotic treatment suppresses whole body urea production in neonatal pigs. *Am. J. Physiol Cell Physiol.* **2013**, G300–G310. [CrossRef]
5. Kerr, B.J.; Mckeith, F.K.; Easter, R.A. Effect on performance and carcass characteristics of nursery to finisher pigs fed reduced crude protein, amino acid-supplemented diets. *J. Anim. Sci.* **1995**, *73*, 433–440. [CrossRef]
6. Rotz, C. Management to reduce nitrogen losses in animal production. *J. Anim. Sci.* **2004**, *82*, E119–E137. [CrossRef]
7. Jin, J.P.; Zhang, L.P.; Jia, J.L.; Chen, Q.; Yuan, Z.; Zhang, X.Y.; Sun, W.B.; Ma, C.M.; Xu, F.F.; Zhan, S.J.; et al. Effects of Maternal Low-Protein Diet on Microbiota Structure and Function in the Jejunum of Huzhu Bamei Suckling Piglets. *Animals* **2019**, *9*, 713. [CrossRef]
8. Figueroa, J.L.; Lewis, A.J.; Miller, P.S.; Fischer, R.L.; Gómez, R.S.; Diedrichsen, R.M. Nitrogen metabolism and growth performance of gilts fed standard corn-soybean meal diets or low-crude protein, amino acid-supplemented diets. *J. Anim. Sci.* **2002**, *80*, 2911–2919. [CrossRef]
9. Kumar, A.; Bhar, R.; Mandal, A.B.; Mendiratta, S.K. Effect of low protein diets and lysine supplementation on growth performance and carcass characteristics of growing pigs. *Afr. J. Biotechnol.* **2012**, *11*, 12128–12134. [CrossRef]
10. Wang, D.; Wan, X.; Peng, J.; Xiong, Q.; Niu, H.D.; Li, H.N.; Chai, J.; Jiang, S.W. The effects of reduced dietary protein level on amino acid transporters and mTOR signaling pathway in pigs. *Biochem. Bioph. Res. Co.* **2017**, *485*, 319–327. [CrossRef]
11. Chick, H.; Hume, E.M. The Production in Monkeys of Symptoms closely resembling those of Pellagra, by Prolonged Feeding on a Diet of Low Protein Content. *Biochem. J.* **1920**, *14*, 135–146. [CrossRef] [PubMed]
12. Metges, C.C.; Petzke, K.J.; El-khoury, A.E.; Henneman, L.; Young, V.R. Incorporation of urea and ammonia nitrogen into ileal and fecal microbial proteins and plasma free amino acids in normal men and ileostomates. *Am. J. Clin. Nutr.* **1999**, *70*, 1046–1058. [CrossRef] [PubMed]
13. Miner-williams, W.; Moughah, P.J.; Fuller, M.F. Endogenous Components of Digesta Protein from the Terminal Ileum of Pigs Fed a Casein-Based Diet. *Agr. Food. Chem.* **2009**, *57*, 2072–2078. [CrossRef] [PubMed]
14. Horwitz, W. *Official Methods of Analysis of Association of Official Analytical Chemists International*; Association of Official Analytical Chemists International: Rockville, MD, USA, 2000.
15. Moughan, P.J.; Hodgkinson, S.M. The enzyme hydrolysed protein method for the determination of endogenous ileal nitrogen and amino acid flows-a modification. *Anim. Feed Sci. Tech.* **2003**, *108*, 207–214. [CrossRef]
16. Zhao, L.; Guo, H.L.; Sun, H. Effects of low-protein diet supplementation with alpha etoglutarate on growth performance, nitrogen metabolism and mTOR signalling pathway of skeletal muscle in piglets. *J. Anim. Physiol. An. N.* **2020**, *104*, 300–309. [CrossRef]
17. Gloaguen, M.; Lefloc'h, N.; Corrent, E.; Primot, Y.; VanMilgen, J. The use of free amino acids allows formulating very low crude protein diets for piglets. *J. Anim. Sci.* **2014**, *92*, 637–644. [CrossRef]
18. Otto, E.R.; Yokoyama, M.; KU, P.K.; Ames, N.K.; Trottier, N.L. Nitrogen balance and ileal amino acid digestibility in growing pigs fed diets reduced in protein concentration. *J. Anim. Sci.* **2003**, *81*, 1743–1753. [CrossRef]
19. Portejoie, S.; Dourmad, J.-Y.; Martinez, J.; Lebreton, Y. Effect of lowering dietary crude protein on nitrogen excretion, manure composition and ammonia emission from fattening pigs. *Livest. Prod. Sci.* **2004**, *91*, 45–55. [CrossRef]
20. Hansen, M.J.; Rgaard, J.V.N.; Adamsen, A.P.S.; Poulsen, H.D. Effect of reduced crude protein on ammonia, methane, and chemical odorants emitted from pig houses. *Livest. Sci.* **2014**, *169*, 118–124. [CrossRef]
21. Li, Y.K.; Lv, Z.Q.; Li, Z.C.; Liu, L.; Wang, F.L.; Li, D.F.; Lai, C.H. Effects of feeding level and dietary supplementation with crystalline amino acids on digestible, metabolizable and net energy values of corn in growing pigs. *Anim. Feed Sci. Tech.* **2018**, *240*, 197–205. [CrossRef]
22. Li, Y.H.; Wei, H.K.; Li, F.G.; Duan, Y.H.; Guo, Q.P.; Yin, Y.L. Effects of Low-Protein Diets Supplemented with Branched-Chain Amino Acid on Lipid Metabolism in White Adipose Tissue of Piglets. *J. Agric. Food Chem.* **2017**, *65*, 2839–2848. [CrossRef] [PubMed]

23. Hilliar, M.; Hargreave, G.; Girish, C.K.; Barekatain, R.; Swick, R.A. Using crystalline amino acids to supplement broiler chicken requirements in reduced protein diets. *Poultry Sci.* **2020**, *99*, 1551–1563. [CrossRef] [PubMed]
24. Isaacson, R.; Kim, H.B. The intestinal microbiome of the pig. *Anim. Health Res. Rev.* **2012**, *13*, 100–109. [CrossRef] [PubMed]
25. Świech, E.; Buraczewska, L.; Tusnio, A.; Taciak, M. The effects of supplementing a low-protein threonine-deficient diet with different sources of non-essential amino acids on nitrogen retention and gut structure in young pigs. *Archiv. Für. Tierernaehrung.* **2010**, *64*, 22–35. [CrossRef] [PubMed]
26. Goudoever, V.J.; Stoll, B.; Henry, J.F.; Burrin, D.G.; Reeds, P.J. Adaptive regulation of intestinal lysine metabolism. *Proc. Natl. Acad. Sci. USA* **2000**, *97*, 10620–10625. [CrossRef]
27. Metges, C.C. Contribution of Microbial Amino Acids to Amino Acid Homeostasis of the Host. *J. Nutr.* **2000**, *130*, 1857S–1864S. [CrossRef]
28. Torrallardona, D.; Hasrris, C.I.; Fuller, M.F. Lysine synthesized by the gastrointestinal microflora of pigs is absorbed, mostly in the small intestine. *Ajp. Endocrinol. Metab.* **2003**, *284*, E1177–E1180. [CrossRef]
29. Pi, Y.; Mu, C.L.; Gao, K.; Liu, Z.; Peng, Y.; Zhu, W.Y. Increasing the Hindgut Carbohydrate/Protein Ratio by Cecal Infusion of Corn Starch or Casein Hydrolysate Drives Gut Microbiota-Related Bile Acid Metabolism to Stimulate Colonic Barrier Function. *Msystems* **2020**, *5*, e00176. [CrossRef]

2

Carbon Footprint Assessment of Spanish Dairy Cattle Farms: Effectiveness of Dietary and Farm Management Practices as a Mitigation Strategy

Ridha Ibidhi [1,2] **and Sergio Calsamiglia** [3,*]

[1] Department of Eco-Friendly Livestock Science, Institute of Green Bio Science and Technology, Seoul National University, Pyeongchang, Gangwon 08826, Korea; ridha@snu.ac.kr or ibidhi_ridha@hotmail.fr

[2] National Institute of Agronomic Recherche of Tunisia (INRAT), Laboratory of Animal Production and Forages, Université de Carthage, rue Hédi Karray, Ariana 2049, Tunisia

[3] Animal Nutrition and Welfare Service (SNiBA), Departament de Ciència Animal i dels Aliments, Universitat Autònoma de Barcelona, 08193 Bellaterra, Spain

[*] Correspondence: Sergio.Calsamiglia@uab.cat

Simple Summary: Livestock production has been identified as an important source of greenhouse gas emissions. The current study was conducted to quantify the carbon footprint of Spanish dairy farms and to evaluate the potential of nutritional and management practices for mitigating methane emissions at farm level. The carbon footprint ranged from 0.67 to 0.98 kg CO_2-eq/kg of energy corrected milk. Simulation scenarios showed that methane emissions and the carbon footprint of milk could be reduced more through management practices rather than dietary strategies. Modelling may provide policy makers, farmers and stakeholders valuable information for planning and developing strategies to reduce the carbon footprint associated with milk production.

Abstract: Greenhouse gas emissions and the carbon footprint (CF) were estimated in twelve Spanish dairy farms selected from three regions (Mediterranean, MED; Cantabric, CAN; and Central, CEN) using a partial life cycle assessment through the Integrated Farm System Model (IFSM). The functional unit was 1 kg of energy corrected milk (ECM). Methane emissions accounted for the largest contribution to the total greenhouse gas (GHG) emissions. The average CF (kg CO_2-eq/kg of ECM) was 0.84, being the highest in MED (0.98), intermediate in CEN (0.84), and the lowest in CAN (0.67). Two extreme farms were selected for further simulations: one with the highest non-enteric methane (MED1), and another with the highest enteric methane (CAN2). Changes in management scenarios (increase milk production, change manure collection systems, change manure-type storage method, change bedding type and installation of an anaerobic digester) in MED1 were evaluated with the IFSM model. Changes in feeding strategies (reduce the forage: concentrate ratio, improve forage quality, use of ionophores) in CAN2 were evaluated with the Cornell Net Carbohydrate and Protein System model. Results indicate that changes in management (up to 27.5% reduction) were more efficient than changes in dietary practices (up to 3.5% reduction) in reducing the carbon footprint.

Keywords: greenhouse gas; carbon footprint; dairy farm; methane

1. Introduction

Climate change is a worldwide concern and the carbon footprint (CF) was proposed as an indicator to tackle the greenhouse gas (GHG) emissions from anthropogenic activities to the atmosphere [1]. Agricultural production, in general, and livestock in particular is recognized as an important contributor to GHG production [2]. The global livestock population contributes about 14.5% to global GHG emissions and the dairy sector is estimated to contribute by 4% of all anthropogenic GHG emissions [3,4]. Driven by population growth and diet changes towards an increase meat and dairy consumption, the production of livestock products globally is increasing. This increase will carry significant environmental costs through the accumulation of GHG in the atmosphere [5].

The main sources and types of GHG from ruminants include methane (CH_4) production from enteric fermentation and animal manure, carbon dioxide (CO_2) from land use, and nitrous oxide (N_2O) from manure and slurry management [6]. Mathematical modelling approaches, including life cycle assessment (LCA), Intergovernmental Panel on Climate Change (IPCC) guidelines [6,7], and predictive models (empirical and mechanistic), have been used to evaluate the contribution of livestock production to GHG emissions [8,9]. Many studies have been conducted to quantify GHG emissions from individual farm sources, but few have provided estimates of total farm-level emissions [10]. Therefore, whole farm predictive models were proposed to estimate the CF and the profile of GHG emissions by production stage integrating all sources of emissions [9,11]. In addition, whole farms models can be used as an effective tool to evaluate the overall impact of management strategies to reduce farm gate GHG emissions. The Integrated Farm System Model (IFSM) is a process-based mechanistic whole-farm simulation model that incorporates soil processes, crop growth, tillage, planting and harvest operations, feed storage, feeding, herd production, manure storage, and economics [12]. The IFSM was originally used in dairy farms in the United States [13] and successfully adapted to be used in dairy farms in Australia and Canada [14,15].

The Spanish dairy sector is a major industry with about 850,000 milking cows producing 7,100,000 ton/year of raw milk [16]. Furthermore, Spain is committed to the United Nations Framework Convention on Climate Change (UNFCCC) and annually reports its national emissions inventory following guidelines promoted by the IPCC [6] to limit or reduce GHG from different sectors through national measures. Previous studies conducted in Spain used different methods to estimate the CF of milk from dairy cattle [17,18]. Del Prado et al. [17] assessed the CF of milk from 17 commercial confined dairy farms in the Basque Country (northern Spain) using a combination of LCA [19], NGAUGE [20], and $\text{LAND}_{\text{DAIRY}}$ [13,21] models. Results revealed that the CF of milk ranged from 0.84 to 2.07 kg CO_2-eq/kg energy corrected milk (ECM). Laca et al. [18] also reported that the milk CF (CO_2-eq/kg fat-protein corrected milk) was higher in the semi-confined system (1.22 kg) compared with the pasture-based system (0.99 kg) in northern Spain. Despite the significant contributions of these studies, none has yet compared results from different regions and assess the impact of feeding and management changes on CH_4 emissions and the milk CF.

The objectives of the current study were: (1) to predict GHG emissions and assess the CF from twelve dairy farms from the three most productive regions in Spain; (2) to evaluate the diet contribution on enteric CH_4 emissions; and (3) to evaluate the impact of dietary and management modifications on CH_4 emissions and the milk CF.

2. Materials and Methods

2.1. Farm Selection and Data Collection

This study was carried out in the three most important regions in dairy production in Spain and with the highest potential for future expansion: Mediterranean (MED: Catalonia, Valencia and Murcia), Cantabric (CAN: Galicia, Asturias and Cantabria) and Central (CEN: Castilla-La Mancha, Castilla-Leon, Madrid and Aragon). MED has predominantly a Mediterranean climate, CAN has an Atlantic climate and CEN has a temperate continental climate [22]. CAN is home of 52% of the total dairy cattle milk production, while MED and CEN represent 14 and 20% of the total milk production, respectively [16].

Farm data were collected through a questionnaire from twelve Holstein dairy farms in Spain, four from each region. Selected dairy farm represented a significant number of dairy farms in a region in terms of size, forage and crops grown, livestock systems, labor organization, management practices and production technology used [23]. Within each region, farms were selected in a two-step process: In the first step, technical advisors with a sound knowledge of local conditions and with good contacts with farm managers were contacted. The main characteristics of farms were discussed with the aim to describe a typical farm for each region. In the second step, selected farms data were obtained by visiting the farms and completing a structured questionnaire. The questionnaire collected information from different areas: crop and soil, grazing, machinery, tillage and planting, crop harvest, feed storage, facilities, bedding type, herd productivity, feeding strategies and manure management. Diet composition from each cattle group of each herd was also requested (type and amounts of forages and concentrates) together with nutritional management strategies (feeding groups for each herd, feeding system, grazing activity and mineral supplementation). The general characteristics of the selected dairy farms from each region are presented in Table 1.

Table 1. Characteristics of the selected dairy farms in Spain by region.

Parameters	Mediterranean Region Farms				Cantabric Region Farms				Central Region Farms			
	Mean	Max	Min	SD	Mean	Max	Min	SD	Mean	Max	Min	SD
Farm structure												
Farm size (ha)	154	330	25	127.7	56	82	25	26.5	105	180	60	52.3
Grazing area (ha)	-	-	-	-	14.5	40.0	0.0	19.00	-	-	-	-
Italian Ryegrass (ha)	13.5	26.0	0.0	11.70	14.8	26.0	4.0	12.42	-	-	-	-
Alfalfa (ha)	8.3	24.0	0.0	11.32	0.0	0.0	0.0	0.00	10.3	25.0	0.0	12.39
Maize (ha)	27.5	60.0	0.0	32.02	14.3	36.0	0.0	17.56	12.5	40.0	0.0	18.93
Sorghum (ha)	8.8	35.0	0.0	17.50	-	-	-	-	-	-	-	-
Oat (ha)	29.3	70.0	0.0	35.06	-	-	-	-	70.5	180.0	25.0	73.29
Herd and production characteristics												
Number of animals	280	440	106	154.9	161	240	64	72.6	319	400	198	88.3
Number of lactating cows	133	220	50	80.0	85	112	42	30.9	185	220	130	38.7
Number of dry cows	22	36	9	11.1	8	15	4	5.2	13	20	7	5.6
Heifers (1–2 years)	58	78	22	26.0	33	51	10	17.4	49	70	20	22.6
Calves	67	106	25	40.7	32	57	8	20.2	73	100	41	25.0
Milk yield (Tm/cow/year)	10.4	11.0	9.6	0.71	10.6	12.1	8.5	1.52	11.4	12.2	10.5	0.75
Milk protein (%)	3.6	3.9	3.4	0.23	3.4	3.8	3.1	0.30	3.3	3.5	3.2	0.15
Milk fat (%)	3.8	3.9	3.6	0.13	3.7	4.2	3.3	0.43	3.3	3.6	3.1	0.21
Feed composition and intake												
DMI [1] (kg/dairy cow/day)	24.5	25.8	22.3	1.53	23.5	25.2	21.3	1.64	25.6	27.2	23.8	1.73
% concentrates feeds	36	39	34	2.4	44	55	32	10.2	44	50	40	4.7
% forages	64	66	61	2.4	56	68	45	10.2	56	60	50	4.8
Manure and fertilizer application												
Liquid slurry application (m^3/ha)	45	50	40	4.1	27	50	0	20.7	49	60	40	8.3
NPK [2] (kg/year)	264	300	230	30.4	219	300	0	146.4	194	275	80	82.2

[1] DMI; dry matter intake; [2] NPK: nitrogen, phosphorus and potassium.

2.2. Modeling Procedure

Farm emissions were simulated using the IFSM, a mechanistic model developed and validated by Chianese et al. [12] to calculate CH_4, N_2O, CO_2 and total GHG emissions from dairy cattle production systems through a partial LCA. The IFSM was further refined into a process-based whole farm simulation including major components for soil processes, crop growth, tillage, planting, harvesting, feed storage, feeding, herd production, manure storage and economics in dairy farms (Figure 1) [24]. The functional unit considered in this study was 1 kg of ECM at the farm gate. The system boundaries contained all relevant activities from cradle to farm gate including on-farm activities, the production of farm inputs and carbon sequestration. Emissions attributable to imported feeds and its transportation were excluded. In addition, the model was adapted to be used in Spanish conditions such as soil type, nutritional characteristics of forages and concentrates, and the 10-year average daily local weather. Weather variables included annual temperature, relative humidity, wind velocity, mean precipitation and solar radiation from three meteorological stations corresponding to each region.

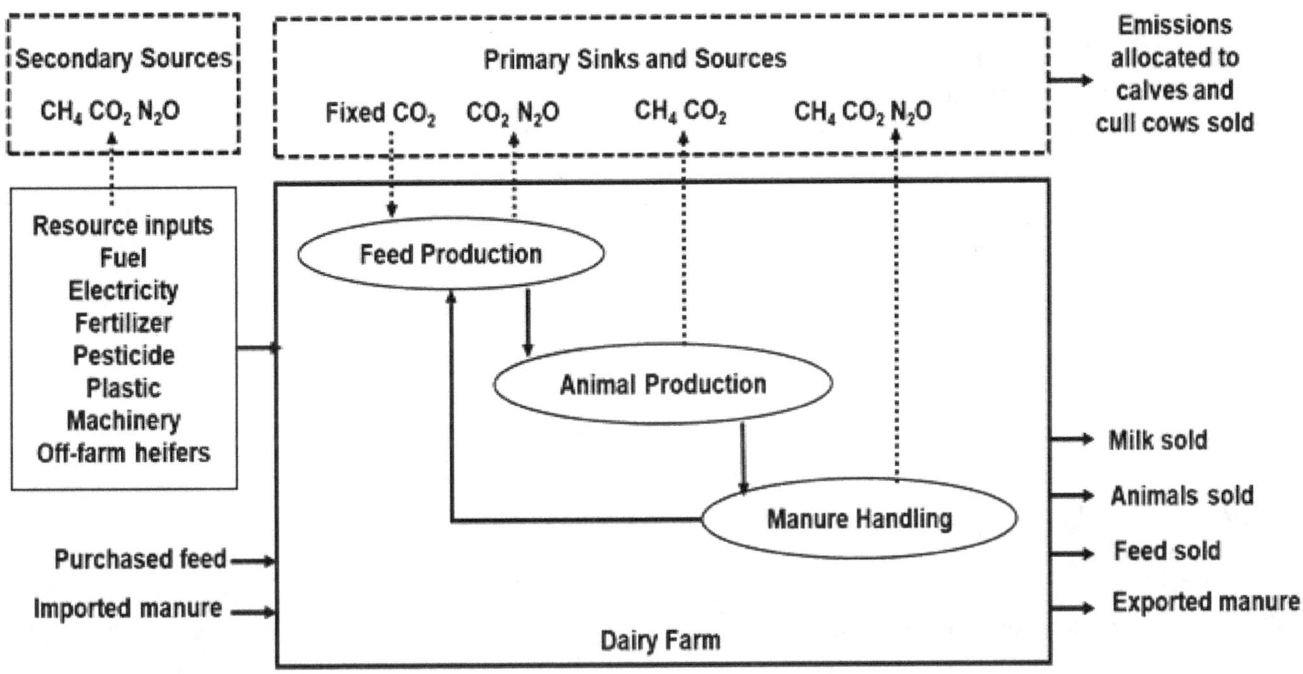

Figure 1. System boundaries, primary and secondary greenhouse gas emissions sources and sinks for a partial life cycle assessment of the carbon footprint (adopted from Rotz et al. [13]).

In addition, the Cornell Net Carbohydrate and Protein System Model (CNCPS version 6.1; [25]) was used to evaluate the contribution of diets to enteric CH_4 emissions of lactating cows on each farm and explore the impact of dietary modifications on CH_4 emissions. The CNCPS is a mechanistic mathematical model that estimates cattle requirements and nutrient supply based on animal, environment, and feed composition information and predicts nitrogen (N), phosphorus (P) and CH_4 emissions, enabling its integration in a whole-farm nutrient management plans [25,26].

2.3. Methane Mitigation Scenarios Simulation

To assess how changing dietary and management practices affects the enteric, non-enteric CH_4 emissions and milk CF, different scenarios were compared. Two extreme dairy farms were selected for further simulations: one with the largest enteric CH_4 emissions, and another with the largest non-enteric CH_4 emissions.

The CNCPS model [25] was used in the farm with the largest share of enteric CH_4 emissions to evaluate three changes in the diet composition of milking cows to determine how these changes would affect enteric CH_4 emissions:

(i) Reducing by 10% the forage to concentrate ratio;
(ii) Improving forages quality: the original ryegrass silage (9% crude protein, CP; 65% neural detergent fiber, NDF; and 8% lignin, ADL) was changed by an improved ryegrass silage (21% CP, 50% NDF and 7% ADL);
(iii) Supplying 330 mg/day of monensin.

The IFSM model was used in the farm with the largest share of non-enteric CH_4 emissions to analyze the impact of five management changes on CH_4 emissions:

(i) Changing milk production from 9565 to 11,000 (+13%) and 8000 (−15%) kg milk/cow/year;
(ii) Comparing five manure collection systems: top-loaded lined earthen basin, top-loaded lined concrete tank, bottom-loaded tank, covered tank and enclosed tank;
(iii) Comparing four types of manure storage: semi-solid (12–14% dry matter; DM), solid (20% DM), slurry (8–10% DM) and liquid-slurry (5–7% DM);
(iv) Comparing five types of bedding: straw, chopped straw, sand, sawdust and manure solids.
(v) Installing an anaerobic digester.

3. Results

3.1. Greenhouse Gas Emissions Profile and the Carbon Footprint of the Selected Dairy Farms

The estimated annual averages for CH_4, N_2O, CO_2, total GHG emissions and CF of milk (kg of CO_2-eq/kg of ECM) from dairy cattle farms by source of emissions (animal housing, manure storage, manure field application, grazing, fuel combustion and secondary emissions) from MED, CAN and CEN regions are presented in Table 2. The annual average CH_4 emissions were the highest in MED (328 ± 25.5 kg/cow), intermediate in CEN (273 ± 52.0 kg/cow) and the lowest in CAN (251 ± 48.0 kg/cow). The highest annual average N_2O emissions were in MED (6.5 ± 1.81 kg/cow), intermediate in CEN (4.0 ± 1.52 kg/cow) and the lowest in CAN (3.2 ± 1.66 kg/cow). Negative values were obtained for the annual average net CO_2 emissions on all farms as the positive contributions from animal and housing, manure storage, grazing and fuel combustion were offset by the net feed production. The average CF of milk in selected farms was 0.87 kg CO_2-eq/kg of ECM, being the highest in MED (1.0 ± 0.12), intermediate in CEN (0.8 ± 0.08) and the lowest in CAN (0.7 ± 0.07 kg of CO_2-eq/kg of ECM). The breakdown of total GHG emissions into component gases (enteric CH_4, non-enteric CH_4, N_2O and CO_2) was examined in absolute terms and relative to each other for all farms.

CH_4 emissions from enteric fermentation and manure handling (non-enteric fermentation) represented the largest share of GHG emissions, accounting for more than 70% of total farm emissions.

Table 2. The profile of greenhouse gas by source of emissions from the selected dairy farms by region.

Greenhouse Gas Emissions	Mediterranean Region Farms				Cantabric Region Farms				Central Region Farms			
	Mean	Max	Min	SD	Mean	Max	Min	SD	Mean	Max	Min	SD
CH_4 (kg/cow) [1]												
Animals and housing	202	217	180	17.2	110	194	25	93.3	187	205	168	17.0
Manure storage	126	152	96	23.8	56	91	14	40.3	86	166	33	60.8
Field-applied manure	0.2	0.4	0.1	0.14	0.1	0.2	0.0	0.08	0.3	0.6	0.1	0.24
Grazing	0.0	0.0	0.0	0.00	84.5	196.4	0.0	100.07	0.0	0.0	0.0	0.00
Total	328	352	292	25.5	251	284	180	48.0	273	335	223	52.0
N_2O (kg/cow) [2]												
Animals and housing	3.1	3.6	1.8	0.86	0.6	1.2	0.0	0.67	2.3	3.2	0.0	1.52
Manure storage	0.2	0.8	0.0	0.40	0.0	0.0	0.0	0.00	0.7	1.4	0.0	0.58
Cropland	3.2	4.8	2.4	1.13	2.7	5.7	1.0	2.17	1.0	1.6	0.6	0.43
Total	6.5	8.4	4.2	1.81	3.2	5.7	2.2	1.66	4.0	5.6	2.0	1.52
CO_2 (kg /cow) [3]												
Animals and housing	6391	6850	6107	340	3776	6721	751	334	6106	6474	5806	308
Manure storage	418	503	357	71.2	199	321	50	140.4	333	649	131	248.9
Net feed production	−10,553	−10,155	−10,776	275	−9551	−7431	−10,301	1414	−9681	−8843	−10,400	738
Fuel combustion	239	313	169	60.2	163	239	61	74.9	204	372	89	128.7
Grazing animals	0.0	0.0	0.0	0.00	2399	5541	0.0	2835.6	0.0	0.0	0.0	0.00
Net Emissions	−3506	−3075	−3909	465	−3014	−2516	−3398	367	−3038	−2787	−3162	173
Total GHG (ton CO_2-eq [4])												
Animals and housing	670	1036	288	370.2	414	579	162	186.6	849	1045	536	218.8
Manure storage	554	841	182	338.6	161	310	19	134.4	1769	5143	348	2259.8
Net Feed production	122	204	36	68.8	67	172	30	70.0	78	184	5	78.5
Fuel combustion	38	49	7	20.4	15	25	2	9.4	48	75	26	19.9
Secondary sources	549	991	213	352.0	217	310	98	91.5	562	772	167	271.2
CF (kg of CO_2-eq/kg of ECM) [5]	1.0	1.1	0.8	0.12	0.7	0.8	0.6	0.07	0.8	0.9	0.8	0.08

[1] CH_4: methane; [2] N_2O: nitrous oxide; [3] CO_2: carbon dioxide; [4] CO_2-eq: CO_2 equivalent; [5] CF: carbon footprint expressed as kg of CO_2 per kg of energy corrected milk with 3.5% fat and 3.1% protein concentrations.

3.2. Diet Evaluation with the CNCPS Model and Its Contribution to Enteric Methane Emissions

Diet evaluation of each farm with the CNCPS model and its contribution to enteric CH_4 emissions per unit of milk produced are presented in Table 3. The highest enteric CH_4 emissions (g/kg of milk) were recorded in CAN farms with an average of 13.6 and a range from 12.2 in the CAN1 to 16.2 in the CAN2 farms. Intermediate CH_4 emissions (g/kg of milk) were recorded in MED with an average of 12.5 and a range from 11.2 to 13.3 in the MED3 and MED4 farms, respectively. The lowest CH_4 emissions (g/kg of milk) were recorded in CEN farms with an average of 12.4 and a range from 10.7 to 14.4 in CEN4 and CEN1 farms, respectively.

Table 3. Enteric methane emissions evaluation from feed intake in the selected farms using the CNCPS [1] model.

Farm	Number of Milking Cows	Milk Yield (kg/day)	Enteric Methane (g/kg Milk)
Mediterranean Region (MED)			
MED1	82	32.0	12.3
MED2	59	36.5	11.2
MED3	180	34.5	13.1
MED4	220	33.0	13.3
Mean	135	34.0	12.5
Cantabric Region (CAN)			
CAN1	85	36.5	12.2
CAN2	42	28.5	16.2
CAN3	102	36.0	13.4
CAN4	112	40.0	12.4
Mean	85	35.3	13.6
Central Region (CEN)			
CEN1	220	36.5	14.4
CEN2	200	40.0	11.2
CEN3	190	34.0	13.2
CEN4	130	38.0	10.7
Mean	185	37.1	12.4

[1] CNCPS: Cornell Net Carbohydrate and Protein System Model.

3.3. Characteristics of the Two Selected Extreme Farms

Two extreme dairy farms in term of CH_4 emissions were selected for further simulations (Table 4): one with the largest non-enteric CH_4 emissions (MED1), and the other with the largest enteric CH_4 emissions (CAN2). The first farm was used to simulate the effects of changes in management on non-enteric CH_4 emissions using the IFSM model. Farm MED1 is located on a 330 ha clay loam soil in Girona that is used to provide most of the forage required for feeding cows (ryegrass and corn silage). It has 82 milking Holstein cows plus replacement heifers housed in a free stall barn and bedded pack barns, respectively. Average intake was 26.5 kg DM of a 55:45 forage to concentrate ratio, and average production was 32.0 L/cow/day. The second farm was used to simulate the effects of changes in diet composition on enteric CH_4 emissions using the CNCPS model. The CAN2 farm is on a 25 ha of loam soil in Lugo (northern Spain) that produces all required ryegrass and forage for pasture. It has 42 milking Holstein cows housed in a free stall barn with access to pasture. Average estimated intake and production of milking cows was 21.3 kg DM (68:32 forage to concentrate ratio) and 28.5 L/cow/day of milk, respectively.

Table 4. Characteristics of the selected two extreme dairy farms in term of enteric and non-enteric methane emissions.

Parameters	Selected Dairy Farms	
	Farm MED1 [1]	Farm CAN2 [2]
Farm area (ha)	330	25
Soil type	Clay loam	Loam
Number of animals	197	64
Number of lactating cows	82	42
Number of dry cows	21	4
Heifers (1–2 years)	54	10
Calves	40	8
Calving strategy	Whole year	Whole year
Housing of animals	Free stall	Free stall
Milk yield (kg/cow/year)	9565	8500
Primary feed	Corn, sorghum and ryegrass silage; and concentrates	Pasture, ryegrass silage and concentrates
Grazing time	0	Whole day
Intake (kg DM [3]/cow/day)	26.5	21.3
Type of manure	Semi-solid (12–14% DM)	Liquid-slurry (5–7% DM)
Type of storage pit	Top-loaded lined earthen basin	Covered basin
Storage period for manure	6 months	4 months
Bedding type	Straw (4 kg)	None
Enteric CH_4 [4] emissions (g/kg milk)	12.3	16.2
Enteric CH_4 emissions (kg/cow/year)	145	167
Non-enteric CH_4 emissions (kg/cow/year)	191	29

[1] Mediterranean region farm 1; [2] Cantabric region farm 2; [3] DM: dry matter; [4] CH_4: methane.

3.4. Dietary and Management Changes Scenarios to Reduce Enteric and Non-Enteric Emissions

3.4.1. Dietary Change Scenarios

The CNCPS model was used to analyze three potential changes in diet composition and quality in the CAN2 farm to determine how these changes affected their enteric CH_4 emissions. Results of the simulated scenarios are shown in Table 5.

Table 5. Potential of dietary strategies on the methane emissions change in the Cantabric region farm (CAN2).

Parameters	Baseline	Scenario 1	Scenario 2	Scenario 3
		Ratio F/C [1]	Silage Quality [2]	Ionophores [3]
Dry matter intake (kg/day/cow)	21.6	21.6	21.6	21.6
Ration forage: concentrate	68/32	58/42	68/32	68/32
Milk yield kg/cow/day	28.5	31.2	28.7	28.6
CH_4 [4](g/kg milk)	16.4	15.6	16.3	16.3
Total CH_4 production (kg/cow/year)	170	177	170	170.4
% of CH_4 change per kg of milk	0.0	−5.0	−0.5	−0.3
CF [5] (kg CO_2-eq/kg ECM)	0.59	0.57	0.59	0.59
% CF change	-	3.5	0.0	0.0

[1] F/C: 10% reduction in the forage/concentrate ratio; [2] Improved silage quality; [3] Supplement 330 mg monensin/day; [4] CH_4: methane; [5] CF: carbon footprint.

The 10% reduction in the forage: concentrate ratio reduced CH_4 emissions from 16.4 to 15.6 g/kg milk, representing a 5% reduction in CH_4 emissions, but only 3.5% of the milk CF. In contrast, change in the nutritional value of the ryegrass silage had a small impact on CH_4 emissions which decreased from 16.4 to 16.3 g/kg milk. Finally, the addition of 330 mg/day monensin also had small effects on CH_4 emissions from 16.4 to 16.3 g/kg milk. The impact of these two dietary factors on the milk CF was negligible.

3.4.2. Management Change Scenarios

The effects of management changes on CH_4 emissions and the CF in the MED1 farm are presented in Table 6. Simulation with the IFSM revealed that the increased milk production by 13% reduced slightly CH_4 emissions (−0.2%), but the impact on the milk CF was moderate (−6.4%). Change in bedding types from straw to sand had a larger effect, reducing CH_4 emissions by 10% and the milk CF by 13.8%. Farm MED1 has already a semi-solid manure management system that it is rather efficient, reducing CH_4 emissions around 32% and the milk CF by 7.4% compared with slurry or liquid-slurry systems. The utilization of a solid manure management system would further reduce the milk CF by 17.4%. The construction and use of an anaerobic digester would reduce CH_4 emissions by 29.9% and the milk CF by 24.8%. The highest reduction potential would be obtained if the top-loaded lined earthen basin is changed into an enclosed tank, with a 37.5% and 27.5% reduction in CH_4 emissions and the milk CF, respectively.

Table 6. Potential of management practices change on methane emissions in the Mediterranean region farm (MED1).

Simulation Scenario	Baseline and Changes Details		CH_4 [2] Emissions (kg/cow/year)	% CH_4 Change	CF [3] (kg CO_2-eq/kg ECM [4])	% CF Change
Change in milk production (kg/cow/year)	Baseline	9565	335.5	-	1.09	-
	Change 1	8000	335.1	+0.1	1.16	+6.4
	Change 2	11,000	334.8	−0.2	1.02	−6.4
Change in manure collection system [1]	Baseline	Semi-solid	230.4	-	1.09	-
	Change 1	Solid	226.6	−1.6	0.90	−17.4
	Change 2	Slurry	335.1	+31.2	1.17	+7.4
	Change 3	Liquid-slurry	338.5	+31.9	1.17	+7.4
Change in manure storage method	Baseline	Top-loaded lined earthen basin	334.8	-	1.09	-
	Change 1	Top-loaded lined concrete tank	332.6	−0.7	1.09	0.0
	Change 2	Bottom-loaded tank or basin	301.9	−9.8	1.01	−7.3
	Change 3	Covered tank or basin	264.7	−20.9	0.93	−14.7
	Change 4	Enclosed tank	215.3	−35.7	0.79	−27.5
Change in bedding type	Baseline	Straw	335.5	-	1.09	-
	Change 1	Chopped straw	334.7	−0.2	1.09	0.0
	Change 2	Sawdust	334.8	−0.2	1.08	−1.0
	Change 3	Sand	302.2	−9.9	0.94	−13.8
	Change 4	Manure solids	302.5	−9.8	1.00	−8.3
Anaerobic digester installation	Baseline	No anaerobic digester	335.6	-	1.09	-
	Change 1	With anaerobic digestor	235.2	−29.9	0.82	−24.8

[1] Manure collection system: Semi-solid (12–14% DM); Solid (20% DM); Slurry (8–10%DM); Liquid-slurry (5–7% DM); [2] CH_4: methane; [3] CF: carbon footprint; [4] ECM: energy corrected milk with 3.5% fat and 3.1% protein concentrations.

4. Discussion

4.1. The Carbon Footprint

The milk CF among all farms ranged from 0.6 to 1.1 kg CO_2-eq/kg ECM and was within the range reported in earlier LCA studies using the same model for conventional dairy farms [13,15]. Del Prado et al. [17] reported a slightly higher CF (0.84 to 2.07 kg CO_2-eq/kg ECM) in 17 commercial confined dairy farms in the Basque Country (northern Spain) using a combination of sub-models approaches. Laca et al. [18] also reported a higher CF of milk in semi-confined dairy cows in northern Spain (1.22 kg CO_2-eq/kg ECM). Differences compared with our results are explained by the different system boundaries delimitation used, where emissions from transport and purchased feeds were included in these previous studies but not in ours, which justifies higher values. In contrast, the milk CF in Pennsylvania and California farms were lower (0.60, 0.55 and 0.61 kg CO_2-eq/kg ECM, respectively) [13] than results reported herein. Belflower et al. [27] found that the CF in conventional United States (USA) farms was about 0.87 kg CO_2-eq/kg ECM, closer to our estimates, but variation was much larger in Holstein cows in Australia (range from 0.54 to 1.35 kg CO_2-eq/kg ECM; [15]). Results obtained in our simulations are within the range of those reported elsewhere in similar conditions.

4.2. Methane and Carbon Footprint Mitigation Strategies

CH_4 is one the main GHG produced at farm gate from enteric and non-enteric fermentation in dairy farms which mitigation may help reduce the CF [28–30]. It is very important to identify which strategies may have the highest impact on the reduction of CH_4 emissions and the milk CF, so policy-makers, farmers and stakeholders may direct their policies toward the most efficient processes. In the current study, we evaluated the effects of dietary and management changes on CH_4 emissions and milk CF using CNCPS and IFSM, respectively.

4.2.1. Dietary Change Scenarios

Dietary changes in the CAN2 farm were simulated with the CNCPS model, and reformulated to optimize production using the ingredients available in the farm. Our results suggest that CH_4 emissions originated from enteric fermentation and the CF of milk could be reduced up to 5.0% and 3.5%, respectively, when the proportion of forage was reduced by 10%. Feeding grain increases ruminal propionate while lowering acetate levels from microbial fermentation [29–31]. While fibrolytic bacteria ferment toward acetate, which releases a carbon that methanogenic bacteria reduce with metabolic hydrogens to produce CH_4, amylolytic bacteria ferment towards propionate that consumes metabolic hydrogens. Therefore, when the fermentation is shifted towards a lower acetate to propionate ratio, CH_4 production is reduced and justifies the results observed. Also, the increase in concentrate in the diet resulted in an increase in milk production that contributed to the dilution of the amount of CH_4 per kg of milk. In contrast, improving silage quality had negligible impact on reducing CH_4 emissions and the milk CF. The small impact of an improved silage quality was expected. Most dairy diets in Spain contain under 55% forage in their diets, and quality is already reasonably high. Increasing forage quality implies a better digestibility of the fiber components, but probably has a small impact on the fermentation pathways, producing similar proportions of CH_4. There are few in vivo data exploring the impact of forage quality on CH_4 emissions and the milk CF. Boadi and Wittenberg [32] reported that CH_4 emissions in dairy heifers were reduced by 23.4 and 42.5% when forage quality improved from low to medium and high quality silage, respectively. However, the high proportional impact is counterbalance with the small overall contribution of young non-producing animals to the overall farm GHG production. Finally, the impact of monensin on CH_4 emissions observed in our report was small. Monensin decreased the acetate-to-propionate ratio in rumen fluid through increasing the flow of reducing equivalents to form propionate [33]. In fact, results from a meta-analysis conducted based on 11 experiments on lactating dairy cows showed that monensin supplementation reduced CH_4 production by about 2–3% [34]. It is possible that the small changes reported herein result from an underestimation of the effects of monensin on CH_4 emissions in the CNCPS model. However, the effects of monensin, even considering in vivo evaluations, are relatively small. All dietary changes proposed herein are within reasonable changes to be implemented in a dairy farm. Therefore, the overall potential of dietary changes appears to have a relatively small impact of up to 5% in the reduction of CH_4 emissions and up to 3.5% in the CF.

4.2.2. Management Change Scenarios

On the other hand, the IFSM model was used to simulate management scenarios in the MED1 farm with the highest non-enteric CH_4 emissions. Three major strategies should be considered: increase productivity to dilute fixed GHG emissions; use bedding material that produce less emissions; and modify manure management practices.

Increasing productivity of milking cows is an effective strategy to mitigate GHG emissions, which may allow a reduction in animal numbers providing the same edible product output at a reduced CF. We estimated that an increase of around 1.500 L/cow/year would reduce the CF of milk by 6.4%. Boadi and Wittenberg [32] also reported that increasing milk production of dairy cows from 5000 to 10,000 L/cow/year in the European Union would increase total CH_4 production per animal by 23%, but the production/ECM would be reduced up to 40%. In fact, the sustained increase in productivity was the single most influential mitigation factor which reduced the CF of milk in the United States dairy industry from 1944 to 2007 [35].

Bedding material may also have a relevant impact on CH_4 emissions and the milk CF. The MED1 farm had a straw-type bedding. While differences in CF compared with other organic alternatives were small, the use of sand would have a moderate, yet relevant, impact, reducing CH_4 emissions by 9.8% and the milk CF by 13.8%. The potential impact of bedding type on the CF reported herein is relevant and within the ranges observed by Aguirre-Villegas and Larson [36].

Manure is the second largest source of GHG emissions from dairy farms [36]. Therefore, manure handling strategies can be effective strategies to reduce non-enteric CH_4 emissions and the milk CF. In the current study, the MED1 farm had a semi-solid manure collection system. The solid collection systems are the least contaminating in terms of GHG emissions compared with semi-solid, multiple and liquid systems [36,37]. Solid manure has less available water and is usually stored in stock piles that promote aeration, which reduces CH_4 emissions. In contrast, liquid manure is stored on pits that promote anaerobic conditions, increasing CH_4 emissions. Compared with slurry or liquid-slurry systems, MED1 farm has already reduced CH_4 emissions and the CF by around 32 and 7%, respectively. However, moving MED1 semi-solid manure into a solid system would have an additional important impact on the CF (17.4% reduction). The installation of an anaerobic digestor could reduce enteric CH_4 and the milk CF by 29.9 and 24.8%, respectively. Aguirre-Villegas and Larson [36] reported that anaerobic digestion can reduce GHG emissions related to manure management by more than 50%, mostly in the form of CH_4 during storage. However, the largest contribution to the reduction of the milk CF would come from shifting an uncovered storage tank to an enclosed one that would reduce non-enteric CH_4 and the CF of milk up to 35.7 and 27.5%, respectively. This large impact of an enclosed manure handling system agrees with Rotz et al. [13], who reported that enclosing manure storage with a flare to burn the escaping biogas would reduce by 39% the milk CF. Overall, manure handling has the largest potential to effectively contribute to the reduction of CH_4 and the milk CF compared with those obtained through changes in dietary strategies.

This study has several limitations. For example, the costs of the proposed mitigating strategies were not considered. Therefore, future studies aimed at estimating the cost of implementation of each strategy will help clarify the cost effectiveness of these mitigation strategies. The uncertainties and assumption in primary and secondary input data such as the average body weight, feed dry matter intake, ingredient and chemical composition of diets, fertilizers and manure application to croplands and pasture emissions factors may affect the accuracy of the results of this analysis using the IFSM. Rotz et al. [37] reported that uncertainties for IFSM predictions of GHG emissions were ±10 and ±20% for enteric and manure handling CH_4, respectively, ±50% for N_2O, and ±20% for fuel combustion and upstream emissions. In addition, the current analysis considered all relevant activities from cradle to farm gate, but excluded emissions attributable to imported feeds and its transportation. Finally, strategies to reduce CH_4 emissions could be associated with other environmental impact such as water contamination, denitrification and eutrophication. Therefore, the milk CF analysis should be integrated with other footprint indicators (water, land and energy) in order to cover other environmental impacts spectrum of dairy farming.

5. Conclusions

The IFSM was used in this study to quantify the profile of GHG emissions in twelve dairy cattle farms from three Spanish regions. This study revealed that Cantabric farms are more efficient in term of GHG emissions compared with Mediterranean and Central dairy farms. CH_4 was found to be the major contributor to the milk CF. Mitigation efforts are frequently targeted to changes in dietary strategies while management efforts are often ignored. However, simulation scenarios showed that CH_4 emissions and the CF of milk could be reduced more through management practices compared with dietary strategies. Modelling may provide policy makers, farmers and stakeholders valuable information for planning and developing strategies to reduce the CF associated with milk production.

Author Contributions: Conceptualization, R.I. and S.C.; methodology, R.I. and S.C.; software, R.I. and S.C.; formal analysis, R.I.; investigation, R.I. and S.C.; data curation, R.I.; writing—original draft preparation, R.I.; writing—review and editing, S.C; supervision, project administration, S.C.; funding acquisition, S.C. All authors have read and agreed to the published version of the manuscript.

Acknowledgments: Authors would like to thank dairy farmers for their collaboration.

References

1. Persson, U.M.; Johansson, D.J.A.; Cederberg, C.; Hedenus, F.; Bryngelsson, D. Climate metrics and the carbon footprint of livestock products: Where's the beef? *Environ. Res. Lett.* **2015**, *10*, 034005. [CrossRef]
2. Steinfeld, H.; Gerber, P.; Wassenaar, T.D.; Castel, V.; Rosales, M.; Rosales, M.; de Haan, C. *Livestock's Long Shadow: Environmental Issues and Options*; FAO: Rome, Italy, 2006; ISBN 9251055718.
3. FAO. *Greenhouse Gas Emissions from the Dairy Sector. A Life Cycle Assessment*; Food and Agriculture Organization: Rome, Italy, 2010.
4. Gerber, P.J.; Hristov, A.N.; Henderson, B.; Makkar, H.; Oh, J.; Lee, C.; Meinen, R.; Montes, F.; Ott, T.; Firkins, J.; et al. Technical options for the mitigation of direct methane and nitrous oxide emissions from livestock: A review. *Anim. Int. J. Anim. Biosci.* **2013**, *7*, 220–234. [CrossRef] [PubMed]
5. Herrero, M.; Wirsenius, S.; Henderson, B.; Rigolot, C.; Thornton, P.; Havlík, P.; de Boer, I.; Gerber, P.J. Livestock and the environment: What have we learned in the past decade? *Annu. Rev. Environ. Resour.* **2015**, *40*, 177–202. [CrossRef]
6. Intergovernmental Panel on Climate Change (IPCC). *Guidelines for National Greenhouse Gas Inventories*; Institute for Global Environmental Strategies (IGES) for the IPCC: Kanagawa, Japan, 2006.
7. Intergovernmental Panel on Climate Change (IPCC). *2019 Refinement to the 2006 IPCC Guidelines for National Greenhouse Gas Inventories*; IGES: Kanagawa, Japan, 2019.
8. Storm, I.M.L.D.; Hellwing, A.L.F.; Nielsen, N.I.; Madsen, J. Methods for measuring and estimating methane emission from ruminants. *Animals* **2012**, *2*, 160–183. [CrossRef]
9. Crosson, P.; Shalloo, L.; O'Brien, D.; Lanigan, G.J.; Foley, P.A.; Boland, T.M.; Kenny, D.A. A review of whole farm systems models of greenhouse gas emissions from beef and dairy cattle production systems. *Anim. Feed Sci. Technol.* **2011**, *166*, 29–45. [CrossRef]
10. Rotz, C.A. Modeling greenhouse gas emissions from dairy farms. *J. Dairy Sci.* **2018**, *101*, 6675–6690. [CrossRef]
11. Del Prado, A.; Crosson, P.; Olesen, J.E.; Rotz, C.A. Whole-farm models to quantify greenhouse gas emissions and their potential use for linking climate change mitigation and adaptation in temperate grassland ruminant-based farming systems. *Animal* **2013**, *7*, 373–385. [CrossRef]
12. Chianese, D.S.; Rotz, C.A.; Richard, T.L. Whole-farm greenhouse gas emissions: A review with application to a Pennsylvania dairy farm. *Appl. Eng. Agric.* **2009**, *25*, 431–442. [CrossRef]
13. Rotz, C.A.; Montes, F.; Chianese, D.S. The carbon footprint of dairy production systems through partial life cycle assessment. *J. Dairy Sci.* **2010**, *93*, 1266–1282. [CrossRef]
14. Cordeiro, M.R.C.; Rotz, A.; Kroebel, R.; Beauchemin, K.A.; Hunt, D.; Bittman, S.; Koenig, K.M.; McKenzie, D.B. Prospects of Forage Production in northern regions under climate and land-use changes: A case-study of a dairy farm in Newfoundland, Canada. *Agronomy* **2019**, *9*, 31. [CrossRef]
15. Sejian, V.; Prasadh, R.S.; Lees, A.M.; Lees, J.C.; Al-Hosni, Y.A.S.; Sullivan, M.L.; Gaughan, J.B. Assessment of the carbon footprint of four commercial dairy production systems in Australia using an integrated farm system model. *Carbon Manag.* **2018**, *9*, 57–70. [CrossRef]
16. MARM Anuario de Estadística. Ministerio de Medio Ambiente y Medio Rural y Marino. 2019. Available online: https://www.mapa.gob.es/es/estadistica/temas/publicaciones/anuario-de-estadistica/2019/default.aspx?parte=3 (accessed on 21 September 2020).

17. Del Prado, A.; Mas, K.; Pardo, G.; Gallejones, P. Modelling the interactions between C and N farm balances and GHG emissions from confinement dairy farms in northern Spain. *Sci. Total Environ.* **2013**, *465*, 156–165. [CrossRef] [PubMed]
18. Laca, A.; Gómez, N.; Laca, A.; Díaz, M. Overview on GHG emissions of raw milk production and a comparison of milk and cheese carbon footprints of two different systems from northern Spain. *Environ. Sci. Pollut. Res.* **2020**, *27*, 1650–1666. [CrossRef] [PubMed]
19. Consultants, P.B. V SimaPro. Available online: https://network.simapro.com/esuservices (accessed on 21 September 2020).
20. Brown, L.; Scholefield, D.; Jewkes, E.C.; Lockyer, D.R.; del Prado, A. NGAUGE: A decision support system to optimise N fertilisation of British grassland for economic and environmental goals. *Agric. Ecosyst. Environ.* **2005**, *109*, 20–39. [CrossRef]
21. Del Prado, A.; Misselbrook, T.; Chadwick, D.; Hopkins, A.; Dewhurst, R.J.; Davison, P.; Butler, A.; Schröder, J.; Scholefield, D. SIMS(DAIRY): A modelling framework to identify sustainable dairy farms in the UK. Framework description and test for organic systems and N fertiliser optimisation. *Sci. Total Environ.* **2011**, *409*, 3993–4009. [CrossRef] [PubMed]
22. Álvaro-Fuentes, J.; Del Prado, A.; Yáñez-Ruiz, D.R. Greenhouse gas mitigation in the agricultural sector in Spain. *Mitig. Adapt. Strateg. Glob. Chang.* **2016**, *21*, 969–973. [CrossRef]
23. Hemme, T. IFCN Dairy Report: For a better understanding of the dairy world. *Dairy Res. Netw.* **2019**, *10*, 224.
24. Rotz, C.A.; Corson, M.S.; Chianese, D.S.; Montes, F.; Hafner, S.D.; Jarvis, R.; Coiner, C.U. *The Integrated Farm System Model. Reference Manual, Version 3.4. Pasture Systems and Watershed Management Research Unit*; Agricultural Research Service, USDA: Washington, DC, USA, 2011.
25. Fox, D.G.; Tedeschi, L.O.; Tylutki, T.P.; Russell, J.B.; Van Amburgh, M.E.; Chase, L.E.; Pell, A.N.; Overton, T.R. The cornell net carbohydrate and protein system model for evaluating herd nutrition and nutrient excretion. *Anim. Feed Sci. Technol.* **2004**, *112*, 29–78. [CrossRef]
26. Higgs, R.J.; Chase, L.E.; Ross, D.A.; Van Amburgh, M.E. Updating the cornell net carbohydrate and protein system feed library and analyzing model sensitivity to feed inputs. *J. Dairy Sci.* **2015**, *98*, 6340–6360. [CrossRef]
27. Belflower, J.B.; Bernard, J.K.; Gattie, D.K.; Hancock, D.W.; Risse, L.M.; Rotz, C.A. A case study of the potential environmental impacts of different dairy production systems in Georgia. *Agric. Syst.* **2012**, *108*, 84–93. [CrossRef]
28. Rotz, C.A.; Skinner, R.H.; Stoner, A.M.K.; Hayhoe, K. Farm simulation: A tool for evaluating the mitigation of greenhouse gas emissions and the adaptation of dairy production to climate change. In Proceedings of the ASABE 1st Climate Change Symposium: Adaptation and Mitigation Conference Proceedings American Society of Agricultural and Biological Engineers, Chicago, IL, USA, 3–5 May 2015; pp. 1–3. Available online: https://elibrary.asabe.org/abstract.asp?aid=45511 (accessed on 21 September 2020).
29. Hawkins, J.; Weersink, A.; Wagner-Riddle, C.; Fox, G. Optimizing ration formulation as a strategy for greenhouse gas mitigation in intensive dairy production systems. *Agric. Syst.* **2015**, *137*, 1–11. [CrossRef]
30. Thompson, L.R.; Rowntree, J.E. Invited review: Methane sources, quantification, and mitigation in grazing beef systems. *Appl. Anim. Sci.* **2020**, *36*, 556–573. [CrossRef]
31. Grainger, C.; Beauchemin, K.A. Can enteric methane emissions from ruminants be lowered without lowering their production? *Anim. Feed Sci. Technol.* **2011**, *166*, 308–320. [CrossRef]
32. Boadi, D.A.; Wittenberg, K.M. Methane production from dairy and beef heifers fed forages differing in nutrient density using the sulphur hexafluoride (SF6) tracer gas technique. *Can. J. Anim. Sci.* **2002**, *82*, 201–206. [CrossRef]
33. Beauchemin, K.A.; Kreuzer, M.; O'mara, F.; McAllister, T.A. Nutritional management for enteric methane abatement: A review. *Aust. J. Exp. Agric.* **2008**, *48*, 21–27. [CrossRef]
34. Appuhamy, J.R.N.; Strathe, A.B.; Jayasundara, S.; Wagner-Riddle, C.; Dijkstra, J.; France, J.; Kebreab, E. Anti-methanogenic effects of monensin in dairy and beef cattle: A meta-analysis. *J. Dairy Sci.* **2013**, *96*, 5161–5173. [CrossRef]
35. Capper, J.L.; Cady, R.A.; Bauman, D.E. The environmental impact of dairy production: 1944 compared with 2007. *J. Anim. Sci.* **2009**, *87*, 2160–2167. [CrossRef]

36. Aguirre-Villegas, H.A.; Larson, R.A. Evaluating greenhouse gas emissions from dairy manure management practices using survey data and lifecycle tools. *J. Clean. Prod.* **2017**, *143*, 169–179. [CrossRef]
37. Rotz, C.A.; Stout, R.C.; Holly, M.A.; Kleinman, P.J.A. Regional environmental assessment of dairy farms. *J. Dairy Sci.* **2020**, *103*, 3275–3288. [CrossRef]

3

Comparison of In Vivo and In Vitro Digestibility in Donkeys

Sonia Tassone [1], Riccardo Fortina [1,*], Emanuela Valle [2], Laura Cavallarin [3], Federica Raspa [2], Silvia Boggero [1], Domenico Bergero [2], Mauro Giammarino [4] and Manuela Renna [2]

1 Department of Agriculture, Forest, and Food Sciences, University of Turin, 10095 Grugliasco, TO, Italy; sonia.tassone@unito.it (S.T.); silvia.boggero@gmail.com (S.B.)

2 Department of Veterinary Sciences, University of Turin, 10095 Grugliasco, TO, Italy; emanuela.valle@unito.it (E.V.); federica.raspa@unito.it (F.R.); domenico.bergero@unito.it (D.B.); manuela.renna@unito.it (M.R.)

3 Institute of Sciences of Food Production, National Research Council, 10095 Grugliasco, TO, Italy; laura.cavallarin@ispa.cnr.it

4 Department of Prevention, Asl TO3, Veterinary Service, Area Animal Sanity, Via Torino 62, 10045 Piossasco, TO, Italy; mgiammarino@aslto3.piemonte.it

* Correspondence: riccardo.fortina@unito.it

Simple Summary: Traditional in vivo methods of determining digestibility of feeds are expensive and time-consuming, and very few data are available for donkeys. The aim of this study was to verify if the in vitro method developed for the Ankom Daisy^{II} Incubator could produce accurate estimates of the in vivo dry matter and neutral detergent fiber digestibility of diets fed to donkeys. Four donkeys and four diets were used in the trial, and the experiment was repeated four times. Buffered donkey feces were used as an inoculum source for the assessment of digestibility in vitro, with an incubation time of 60 h. The obtained results showed that the Ankom Daisy^{II} Incubator ranked the diets in the same order as the in vivo method. However, in vitro values were lower than those obtained in vivo. The regression analyses used to predict in vivo estimates from in vitro data gave poor results and low precision. In conclusion, further studies, using different sample size and digestion times in vitro, are needed to verify if accurate prediction of in vivo feed digestibility can be obtained using the Ankom Daisy^{II} Incubator and donkey feces as inoculum.

Abstract: We compared in vivo and in vitro dry matter (DM) and neutral detergent fiber (NDF) digestibility in donkeys using feces as microbial inoculum. Four donkeys were used in a 4 × 4 Latin square design with a 2 × 2 factorial arrangement of treatments. The animals were fed two types of hay, with or without flaked barley. For the in vivo procedure, total feces were collected for 6 days from each donkey; digestibility was calculated as the difference between ingested and excreted DM and NDF. For the in vitro procedure, donkey feces were buffered and used as microbial inoculum in an Ankom Daisy^{II} Incubator; digestibility was estimated after 60 h of incubation. In vivo results showed that the addition of barley to hays did not change the digestibility values. In vivo estimates were higher than in vitro ones. The equations used to predict in vivo estimates from in vitro data were not reliable (R^2 = 0.47 and 0.21; P = 0.003 and 0.078 for NDF and DM digestibility, respectively). Further studies need to evaluate different sample size and digestion times.

Keywords: donkey; in vivo digestibility; in vitro digestibility; Daisy^{II} incubator; fecal inoculum

1. Introduction

To date, in the world there are 43–44 million donkeys used for work; pack transport; pulling carts; farm tillage; drawing water and milling; cosmetics or pharmaceutical industry; milk and meat; and—just in industrialized countries—for recreation, breeding, and companionship [1]. The donkey is a monogastric herbivore, evolved to have a steady flow of dietary fiber moving through the gut at all times [2]. In comparison with horses, donkeys have a higher efficiency in digesting fiber of poor nutritional quality [3]. Their large intestine leads to a higher digestibility of organic matter, gross energy and fiber fractions than in horses or ponies [4–6]. According to Edwards et al. [7,8], donkeys have a fecal microbiota concentration and a community composition that differ from those of other equine types, and this diversity plays an essential role in fiber digestion. Liu et al. [9] studied the microbiota at different sites of the gastrointestinal tract of donkeys, finding a higher microbial concentration in the hindgut than in the foregut.

Despite the increasing interest in donkeys, studies on feed digestibility in this species are very limited. Most of the available data resulted from in vivo feeding trials [4,10–13]. Liu et al. [14] studied the effect of dietary forage on digestibility using the mobile nylon bag technique. Kidane et al. [15] predicted diet digestibility via fecal-NIR, with few successes. Carretero-Roque et al. [11] and Wood [16] studied the in vitro digestibility of feedstuffs for donkeys using the neutral cellulase plus gamanase technique developed by Ankom Technology [17]. Tassone et al. [18] demonstrated that donkey's digestibility can be predicted, with good repeatability and reproducibility, using the Daisy^{II} Incubator (Ankom Technology) and donkey feces as source of microbial inoculum. These authors also showed that the digestibility of different feeds for donkeys requires different incubation times. Leng et al. [19] proposed an in vitro model to study equine gut microbiota that represented an alternative to in vivo research. Despite these studies, a reliable tool for predicting feed digestibility in donkeys is not yet available.

The objective of this study was to compare the in vivo dry matter (DM) and neutral detergent fiber (NDF) digestibility of some diets commonly used in donkey nutrition with in vitro data obtained using the Daisy^{II} Incubator and donkey feces as a source of microbial inoculum. We hypothesized that the in vitro method developed for the Daisy^{II} Incubator produces accurate estimates of the in vivo DM and NDF digestibility of diets fed to donkeys.

2. Materials and Methods

The trial was carried out at the Veterinary Teaching Special Unit of the Department of Veterinary Sciences (DVS) of the University of Turin (Italy). The protocol was designed according to the guidelines of the current European Directive (2010/63/EU) on the care and protection of animals used for scientific purposes.

2.1. Animals and Diets

Four female Ragusano donkeys (A, B, C, D), aged from 12 to 16 months, were selected from a private farm ("Asi Lait"; San Benigno Canavese (TO), Italy) and brought to the experimental barn of the DVS 4 weeks before the beginning of the trial for adaptation. Upon arrival, the donkeys were individually weighed. The initial live weight was equal to 193 ± 24.3 kg. The donkeys were dosed with a broad-spectrum deworming product ("Eqvalan Duo", Boehringer Ingelheim Animal Health Italia S.p.A., Noventana (PD), Italy) at the arrival and regularly checked for fecal parasites. They were housed in a stable in 3 m × 4 m single boxes, on wood shavings. The donkeys had free access to clean and fresh water. During the whole trial, the donkeys had access to a sandy paddock: they were let to exercise and socialize in pairs for 1 hour during the adaptation periods, while they were hand walked by an operator for 10 minutes during the collection periods. The operator took care of avoiding intake of sand by the donkeys during paddock turnout.

The animals were fed four diets (6 kg as fed/head × day) according to a 4 × 4 Latin square experimental design with a 2 × 2 factorial arrangement of treatment. The diets were: (i) 1st cut meadow hay (H1); (ii) 88% 1st cut meadow hay + 12% flaked barley (H1B); (iii) 2nd cut meadow hay (H2); and (iv) 88% 2nd cut meadow hay + 12% flaked barley (H2B). In a previous digestibility trial conducted with horses [20], cereals were given at a rate commonly used in equine rations (i.e., 30% oat and 70% hay). In the current trial, we decided to use a lower concentration of cereals since the starch limit for digestion is not known in donkeys. In addition, we wanted to set up a forage-based diet to avoid using a concentrate diet that could potentially be harmful for the animals and lead to a reduction in feeding consumption time. In H1B and H2B, hay and barley were given together. The diets were divided in three meals a day and the leftovers were collected and weighed daily to calculate the daily dry matter intake (DMI). Each of the four experimental periods (4 × 28 days) included: 14 days of adaptation to the new diet (days 1 to 14), 6 days of total feces collection for the in vivo digestibility assessment (days 15 to 20), 1 day of feces collection from each donkey for the in vitro digestibility assessment (day 21), and 7 days of wash out (days 22 to 28). The week before the beginning of the first collection period as well as during the wash out, each donkey was fed the hay used for the subsequent collection period.

Before every period of feces collection, representative samples of the experimental diets were collected, then ground using a cutting mill (MLI 204; Bühler AG, Uzwil, Switzerland) and analyzed for DM, ash and crude protein (CP) according to AOAC International [21]. Ether extract (EE) was analyzed according to AOAC International [22]. The detailed methods are reported in Fortina et al. [23]. The Ankom^{200} Fiber Analyzer (Ankom Technology, Macedon, NY, USA) was used to determine NDF, acid detergent fibre (ADF) and acid detergent lignin (ADL), following the procedures of Mertens [24] for NDF and Van Soest et al. [25] for ADF and ADL. Non-fibrous carbohydrates (NFC) were calculated with the following formula: NFC = 100-(NDF + ash + CP + EE).

2.2. *In Vivo Digestibility*

For the in vivo digestibility assessment, the donkeys were equipped with harnesses ("Steed Apple Catcher"; La Calèche, Bessemerstraat 370, 3620 Lanaken, Belgium) to allow total feces collection (Figure 1). The harnesses were adapted to the animals, according to their size, and fixed by an elastic rug girth fixed with a polo breastplate. The harnesses avoided the contamination of feces with urine. All the equipment was covered by a sheepskin cover tube as a pad protection.

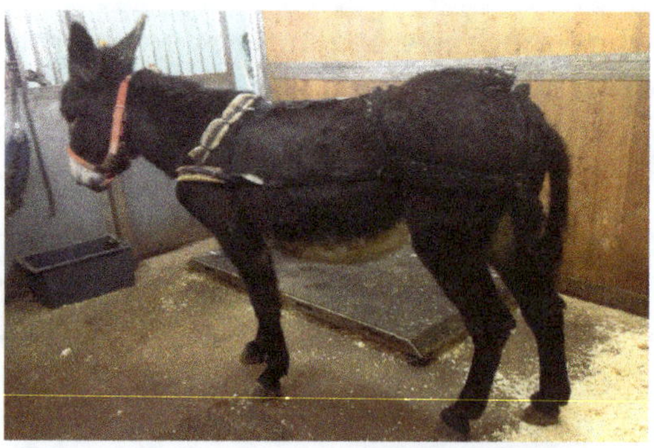

Figure 1. Donkey equipped with an adapted "Steed Apple Catcher".

For 6 consecutive days, the feces were removed from the harness five times a day (at 06:30, 09:00, 13:00, 17:00 and 22:00 h), then weighed and mixed. A daily 10% feces pool from each animal was dried at 50 °C; ground to pass 1 mm screen; and analyzed for DM, ash, CP, EE, NDF, ADF and ADL, following the same analytical procedures implemented for feed analyses [18].

The in vivo apparent DM digestibility ($ADMD_{vv}$) was calculated as follows:

$$ADMD_{vv}\ (\%\ DM) = ((DM_i - DM_e)/DM_i) \times 100 \quad (1)$$

where:

DM_i (%) = ingested DM

DM_e (% DM) = excreted DM.

Similarly, the in vivo NDF digestibility ($NDFD_{vv}$, % NDF) was calculated considering the ingested and excreted NDF, using the same calculation shown for $ADMD_{vv}$.

2.3. In Vitro Digestibility

In vitro digestibility was performed using a Daisy^{II} Incubator (Ankom Technology Corporation, Fairport, USA) and donkey feces as source of microbial inoculum. The Ankom Daisy^{II} Incubator (Ankom Technology Corporation, Fairport, NY, USA) is an alternative to traditional in vitro procedures. It is a thermostatically controlled chamber with four rotating digestion jars, where up to 92 samples (individually weighed in bags) could be analyzed simultaneously with inoculum and buffer solution. With this method, the material that disappears from the bag during the incubation is considered digestible. For the in vitro procedure, feces collection occurred on the 21st day of each experimental period, directly from the rectum of each donkey. While feces collection for the in vivo digestibility assessment occurred throughout the whole day, the feces used for the in vitro procedure were always collected at 09:00 h, with the aim of using the inoculum as soon as possible as to avoid microbial content decreases during time. Immediately after collection, the feces were sealed in individual airtight bags expelling air as much as possible to maintain anaerobic conditions [20,26], purged with CO_2 and immediately transferred to the lab (approximately 3 min travel from the barn to the lab). During transport, the temperature of the feces was maintained at about 39 °C using a warm water-containing cooler.

Each jar of the incubator was assigned to a donkey, using its individual feces, and filled with buffer solution [27]. The buffer solution was obtained mixing two solutions in a 5:1 ratio (1500 mL of solution A: KH_2PO_4 10 g/L, $MgSO_4{\cdot}7H_2O$ 0.5 g/L, NaCl 0.5 g/L, $CaCl_2{\cdot}2H_2O$ 0.1 g/L, CH_4N_2O 0.5 g/L; 300 mL of solution B: Na_2CO_3 15.0 g/L, $Na_2S{\cdot}9H_2O$ 1.0 g/L). The inoculum was prepared mixing 200 g of feces of each donkey and 400 mL of buffer solution. The mixture was blended for 30 s (Osterizer Cyclo-Trol Eight; Oster, Moncalieri (TO), Italy), purged with CO_2 for 15 s, and transferred to the respective digestion jar. The pH was adjusted to 7.0 using additional amounts of the buffer solution as suggested by Earing et al. [20]. Representative samples of the four diets were weighed (0.50 g) in Ankom F57 filter bags (Ankom Technology Corporation, Fairport, USA) and heat sealed. Each jar contained eight samples (two replicates/diet) and two blanks. All the diets were incubated together as no interaction effects were expected between grains and forage [28].

The incubation time was chosen considering previous studies that used the Ankom Daisy^{II} Incubator and equine feces as inoculum source [18,25,26]. In particular, in donkeys, Tassone et al. [18] showed that the in vitro DM and NDF digestibility generally increased from 30 to 72 h of incubation, but with remarkable variations among different feedstuffs. However, when evaluating the relationships existing between in vivo and in vitro (at 24-, 30-, 48- and 72-h incubation) digestibility estimates in horses, other authors obtained higher model fits when predicting in vivo estimates from in vitro data at 48 h rather than at 24 h [26] or at 72 h [20]. Considering that in donkeys the retention of feedstuffs is longer than in horses [4], and considering the lack of available comparisons between in vivo and in vitro estimates at incubation times included between 48 to 72 h in equines, we decided to use an incubation time of 60 h.

At the end of the digestion, the bags were removed from each jar, rinsed thoroughly with cold tap water and placed in a 50 °C forced-air oven to dry for 24 h. The bags were weighed and then analyzed for NDF with the Ankom^{200} Fiber Analyzer (Ankom Technology Corporation, Fairport, USA).

The in vitro apparent DM digestibility ($ADMD_{vt}$) and NDF digestibility ($NDFD_{vt}$) were calculated as follows:

$$ADMD_{vt}\ (\%\ DM) = 100 \times (DM_{0h} - DM_{residue})/DM_{0h} \quad (2)$$

$$NDFD_{vt}\ (\%\ NDF) = 100 \times (NDF_{0h} - NDF_{residue})/NDF_{0h}, \quad (3)$$

where:

DM_{0h} (%) = dry matter ante incubation
$DM_{residue}$ (% DM) = dry matter post incubation
NDF_{0h} (% DM) = neutral detergent fibre ante incubation
$NDF_{residue}$ (% NDF) = neutral detergent fibre post incubation.

2.4. Statistical Analysis

All data were statistically analyzed using SAS [29].

The proximate constituents of the diets were analyzed by one-way ANOVA, using the following model:

$$Y_{ij} = \mu + \alpha_i + \varepsilon_{ij}, \quad (4)$$

where: Y_{ij} = observation; μ = overall mean; α_i = fixed effect of diet (H1, H1B, H2, H2B); ε_{ij} = residual error. The Tukey's post hoc test was used for multiple comparisons.

Digestibility and DMI data were analyzed with the PROC GLM procedure considering diet (α) as a fixed factor, while donkey (β) and period (δ) were considered as random factors. The following model was used:

$$Y_{ijkl} = \mu + \alpha_i + \beta_j + \delta_k + \varepsilon_{ijkl}, \quad (5)$$

where: Y_{ijkl} = observation; μ = overall mean; α_i = effect of diet (H1, H1B, H2, H2B); β_j = effect of donkey (A, B, C, D); δ_k = effect of period (1, 2, 3, 4); ε_{ijkl} = residual error. The *t* test was used for multiple comparisons.

For each diet, linear regression analysis (PROC REG procedure in SAS) was used to assess the relationships existing between in vivo and in vitro ADMD and NDFD.

Significance was set at $p < 0.05$.

3. Results

3.1. Diet Characteristics and Dry Matter Intake

The chemical composition of the experimental diets is reported in Table 1.

Table 1. Chemical composition of the diets (g/100 g DM, unless otherwise stated; n = 4).

Item	H1	H1B	H2	H2B	SEM	*p*
DM (g/100 g as fed)	89.9	89.7	89.9	89.5	0.8	0.899
Ash	6.4 ^{c}	6.1 ^{c}	9.1 ^{a}	8.0 ^{b}	0.2	<0.001
CP	8.5 ^{d}	9.7 ^{c}	18.6 ^{a}	17.5 ^{b}	0.1	<0.001
EE	1.3 ^{b}	1.5 ^{b}	2.3 ^{a}	2.1 ^{a}	<0.1	<0.001
NDF	59.4 ^{a}	54.2 ^{c}	58.2 ^{b}	49.8 ^{d}	0.2	<0.001
ADF	36.6 ^{a}	29.7 ^{c}	33.3 ^{b}	27.3 ^{d}	<0.1	<0.001
ADL	4.4 ^{a}	2.4 ^{b}	4.2 ^{a}	2.1 ^{b}	0.2	<0.001
NFC	24.4 ^{b}	28.5 ^{a}	11.8 ^{d}	22.6 ^{c}	0.4	<0.001

Abbreviations: H1, 1st cut meadow hay; H1B, 88% 1st cut meadow hay + 12% flaked barley; H2, 2nd cut meadow hay; H2B, 88% 2nd cut meadow hay + 12% flaked barley; DM, dry matter; CP, crude protein; EE, ether extract; NDF, neutral detergent fiber; ADF, acid detergent fiber; ADL, acid detergent lignin; NFC, non-fibrous carbohydrates; SEM, standard error of the mean. Different superscripts ($^{a–d}$) within a row indicate significant differences.

Hays (H1 and H2) showed significant differences in their chemical composition. In particular, H1 showed higher levels of NDF and ADF and NFC, and contemporarily lower levels of ash, CP and EE than H2. Comparable levels of DM and ADL were observed between hays. The same differences and similarities were observed when comparing diets H1B and H2B. Compared to hays, the addition of barley resulted in a significant increase of NFC, and a decrease of NDF, ADF and ADL. When barley was added to the 1st cut meadow hay (H1B), CP increased; on the contrary, the amount of CP decreased when barley was added to the 2nd cut meadow hay (H2B). The EE was not significantly affected by the inclusion of barley in the diets.

The characteristics of the hays influenced the average DMI of the donkeys ($p < 0.001$). The DMI values were higher with the 2nd cut meadow hay (H2 and H2B diets; 5.0 ± 0.58 kg/head × day and 5.0 ± 0.37 kg/head × day respectively) than with the 1st cut meadow hay (H1 and H1B diets; 4.5 ± 0.79 kg/head × day and 4.4 ± 0.91 kg/head × day, respectively).

3.2. *Effect of Diet on Dry Matter and Neutral Detergent Fiber Digestibility*

Table 2 shows the variation of $ADMD_{vv}$, $ADMD_{vt}$, $NDFD_{vv}$, and $NDFD_{vt}$ among the four experimental diets. The $ADMD_{vv}$ was significantly higher in H2 and H2B when compared to H1, while H1B showed intermediate values. The $ADMD_{vt}$ ranked the diets in the following order: H2B > H2 > H1B = H1. The H2 and H2B diets showed significantly higher $NDFD_{vv}$ values than H1 and H1B diets. The $NDFD_{vt}$ significantly differed among all the diets, with the highest value found for H2, followed by H2B, then by H1, and finally H1B.

Table 2. Effect of diet on in vivo and in vitro dry matter and neutral detergent fiber digestibility. Comparison between in vivo and in vitro dry matter digestibility ($ADMD_{vv}$ *vs* $ADMD_{vt}$) and neutral detergent fiber digestibility ($NDFD_{vv}$ *vs* $NDFD_{vt}$) for each diet.

Diets	$ADMD_{vv}$ (% DM)	$ADMD_{vt}$ (% DM)	t	p	$NDFD_{vv}$ (% NDF)	$NDFD_{vt}$ (% NDF)	t	p
H1	60.4 ^{b}	53.3 ^{c}	6.34	<0.001	51.7 ^{b}	38.3 ^{c}	8.14	<0.001
H1B	63.1 ^{a,b}	55.4 ^{c}	5.47	<0.001	51.9 ^{b}	34.3 ^{d}	8.91	<0.001
H2	64.6 ^{a}	60.5 ^{b}	2.36	0.022	64.3 ^{a}	51.7 ^{a}	4.75	<0.001
H2B	66.6 ^{a}	63.1 ^{a}	2.19	0.033	63.3 ^{a}	47.1 ^{b}	6.21	<0.001
SEM	5.514	4.527			7.409	7.809		

Abbreviations: $ADMD_{vv}$, apparent in vivo dry matter digestibility; $ADMD_{vt}$, apparent in vitro dry matter digestibility; $NDFD_{vv}$, in vivo neutral detergent fiber digestibility; $NDFD_{vt}$, in vitro neutral detergent fiber digestibility; DM, dry matter; NDF, neutral detergent fiber; H1, 1st cut meadow hay; H1B, 88% 1st cut meadow hay + 12% flaked barley; H2, 2nd cut meadow hay; H2B, 88% 2nd cut meadow hay + 12% flaked barley; SEM, standard error of the mean. Different superscripts ($^{a–d}$) within a column indicate significant differences among the diets ($p < 0.05$). Within each row and parameter (ADMD or NDFD), means with a p value lower than 0.05 are significantly different.

3.3. *Comparison between In Vivo and In Vitro Digestibility*

Table 2 shows the comparison between in vivo and in vitro ADMD and NDFD for each experimental diet. In all cases, in vitro values were significantly lower than in vivo values.

The prediction equations obtained for the estimation of in vivo digestibility from in vitro values are the following ones (Figure 2):

$$ADMD_{vv}\ (\%\ DM) = 37.155 + 0.4569 \times ADMD_{vt}\ (\%\ DM)\ (R^2 = 0.21;\ SE = 0.240;\ P = 0.078) \quad (6)$$

$$NDFD_{vv}\ (\%\ NDF) = 21.787 + 0.8405 \times NDFD_{vt}\ (\%\ NDF)\ (R^2 = 0.47;\ SE = 0.238;\ P = 0.003). \quad (7)$$

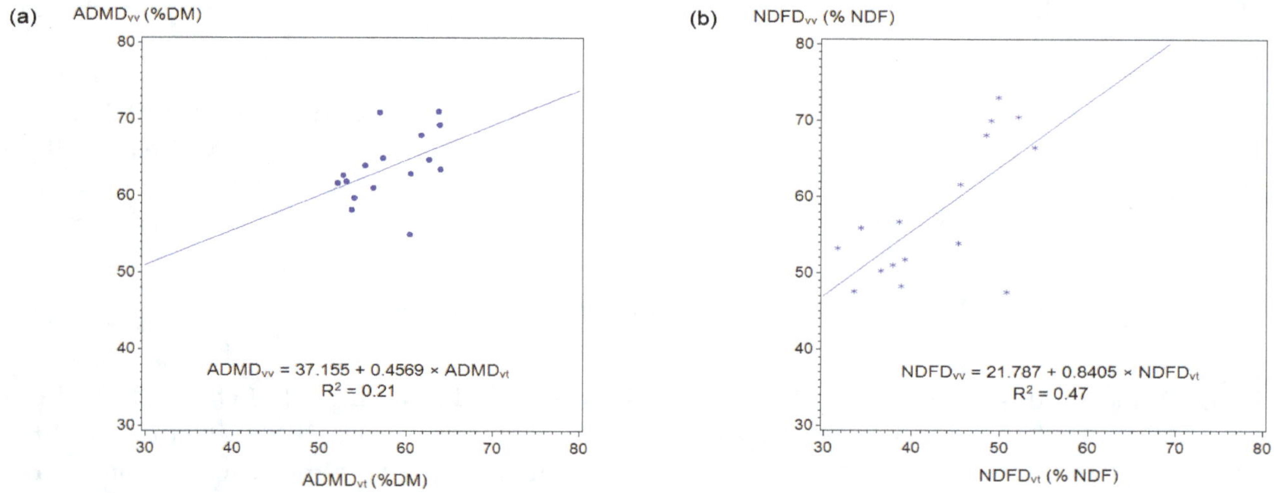

Figure 2. Relationship between 60-h in vitro digestibility using the Daisy^{II} Incubator and in vivo digestibility. (**a**) Dry matter digestibility. (**b**) Neutral detergent fiber digestibility. ADMDvv, apparent in vivo dry matter digestibility; $ADMD_{vt}$, apparent in vitro dry matter digestibility; $NDFD_{vv}$, in vivo neutral detergent fiber digestibility; $NDFD_{vt}$, in vitro neutral detergent fiber digestibility. DM, dry matter; NDF, neutral detergent fiber.

The equation for NDFD was statistically significant and showed higher fitting ($R^2 = 0.47$) than the equation for ADMD ($R^2 = 0.21$). However, in both cases, the regression analyses gave poor results and the model fits were considered not reliable.

4. Discussions

4.1. Dry Matter Intake

Regardless of the type of diet, the DMI of the donkeys was higher than 2% body weight (BW). According to Raspa et al. [30], typical DMI in donkeys ranges from 1.3% BW in summer to 1.7% BW in winter. The DMI values observed in this trial were higher than those reported by Smith and Burden [3] (3 kg/head × day) and by Burden and Thiemann [31] (1.3 to 1.8% BW), but similar to those observed by Smith and Pearson [32], who reported values ranging between 1.75% and 2.25% BW. For lactating Amiata donkeys (BW of 309 kg), Gatta et al. [13] reported an average DMI of 8 kg/day for diets characterized by average NDF and ADF values equal to 49% DM and 31% DM, respectively. Agbagla-Dohnani et al. [33] measured an average DMI of rice and barley straws equal to 16.2 g/kg BW. Pearson et al. [12] observed that the DMI decreased with the lowering of feed quality (lucerne > temperate grass hay > straw), with values ranging between 21.4 and 10.4 g/kg BW. Differences in DMI are common in donkeys and are related to body weight, diet composition, and physiological and environmental conditions. In our study, the observed differences in DMI are related to the characteristics of the hays that influenced the average DMI of the donkeys. In fact, the DMI values were higher with the 2nd cut meadow hay than with the 1st cut meadow hay.

4.2. Effect of Diet on Dry Matter and Neutral Detergent Fiber Digestibility

The digestibility of feeds is generally reported to be higher in donkeys than other equines thanks to their capacity to digest low-quality feeds, to the high retention time of feeds in the gut [10] and, probably, to differences in the composition of the hindgut microbiota. Cudderford et al. [4] carried out a trial to determine the differences between Thoroughbreds, Highland ponies, Shetland ponies, and donkeys in their ability to digest diets containing different levels of fiber and protein (molassed alfalfa and oat straw at different proportions). The results showed that all animals digested the components of the high-fiber diets less well than those of the low-fiber diets; donkeys were more efficient in digesting the fiber and retained feed residues longer in the gastro-intestinal tract than the other equids. Pearson et al. [10] provided molassed dehydrated alfalfa or oat straw ad libitum or given at 70% ad libitum intake to ponies and donkeys. The results confirmed that ponies ingested more forages than donkeys per unit BW at both levels of feeding, but the apparent digestibility of dietary DM, energy and fiber fractions were always higher in donkeys. Izraely et al. [34] fed five donkeys with alfalfa and hay, and the $NDFD_{VV}$ values ranged from 50.9 to 54.2% NDF, with no significant differences between the fiber digestibility of the two fodders. These authors concluded that donkeys are able to compensate for the adverse nutritional conditions like wild ruminants rather than the other domestic equines. Edwards et al. [7] observed that both fecal microbial concentrations and community composition of donkeys were generally most distinct from those of other equine types (horse, donkey, horse × donkey and zebra), even if a core fecal microbiota exists across all the equines. Most of the bacterial taxa of this core lack cultured representation; their availability is essential to gain fundamental knowledge of the microbial functions that underpin the equine hindgut ecosystem and that of the donkey in particular. In another trial, Edwards et al. [8] found that donkey fecal microbiota differed from that of both pony and pony × donkey, and these differences were related to a higher abundance and diversity of taxa with known (or speculated) roles in fiber degradation. The findings of these trials suggest that the specific composition of hindgut microbiota of donkeys plays an essential role in fiber digestion.

In our study, $NDFD_{VV}$ values of H1 and H1B diets (average: 51.8% NDF) were comparable to those obtained by Izraely et al. [34], while higher values were found for diets containing 2nd cut meadow hay (H2 and H2B; average: 63.8% NDF). In Mexican donkeys, Carretero-Roque et al. [11] found higher DM digestibility in vitro than in vivo for diets based on oat straw (average 53.4% vs 45.0%) and maize stover (average 58.3% vs 43.3%), both supplemented with 15% alfalfa hay. Pearson et al. [12] compared in vivo digestibility of forages of different quality (low: barley straw; medium: temperate meadow hay; high: lucerne) in different species, including donkeys, fed ad libitum or restricted to about 75% ad libitum intake. They found that DM digestibility decreased with the decrease of forage quality. Our results confirm such findings, as ADMD was significantly lower in the diets containing 1st cut meadow hay, which showed lower quality when compared to the 2nd cut meadow hay (Table 1). In donkeys, the apparent digestibility coefficients for DM ranged from 0.67 to 0.48. For NDF, the lowest and highest digestibility coefficients were found in lucerne (0.47) and temperate meadow hay (0.59), respectively. In the present trial, the average $ADMD_{VV}$ value (63.7% DM) was similar to that obtained by Pearson et al. [12] for lucerne.

In our trial, the addition of 12% flaked barley (720 g/d) to hays did not significantly modify the $ADMD_{VV}$ (Table 2). Similarly, Liu et al. [14] did not find any difference in the digestibility of diets characterized by different forage-to-concentrate ratios fed to donkeys. In contrast, other authors showed that the addition of a concentrate increased the digestibility of diets fed to horses up to approximately 10%. Palmgren Karlsson et al. [35] observed that the digestibility of DM, organic matter and energy increased in a curvilinear pattern with increasing inclusion levels of oat to grass-based diets fed to adult Standardbred geldings. However, the digestibility of fibrous components lowered at increasing inclusion levels of oats because of negative associative effects between grass hay and oats. Peiretti et al. [36] compared the in vivo digestibility of DM, organic matter, CP, NDF, and ADF of rations fed to horses based on hay and hay supplemented with flaked or crushed barley. They concluded that the addition of technological treated barley, and in particular flaked barley, increased total digestibility.

De Marco et al. [37] compared the in vivo and in vitro (pepsin-cellulase technique) apparent digestibility of broken rice added to a hay-based diet fed to adult geldings. The cereal showed high organic matter digestibility coefficients (92.6% in vivo and 87.1% in vitro) and its supplementation increased the digestibility of the diet.

High amounts of grains or starch may lead to intestinal disorders and have negative impacts on digestibility of DM and dietary fiber in equids [38,39]. We did not find studies related to this topic on donkeys. However, the impact of high amounts and different barley forms on total tract apparent digestibility of dietary fiber in horses was assessed by Philippeau et al. [40]. In this trial, in the morning meal the animals were fed a concentrate (1.7% BW) made of 80% barley, distributed either as whole grain or as ground, pelleted or steam-flaked. Results showed that diets including treated barley forms led to a higher total tract apparent digestibility of NDF. Providing horses with high amounts of pelleted or flaked barley limited the negative impact of starch on fiber digestibility when animals were fed a high level of starch in the morning meal.

4.3. Comparison between In Vivo and In Vitro Digestibility

This is the first trial comparing in vivo and in vitro digestibility in donkeys using the Ankom Daisy^{II} Incubator and feces as a source of microbial inoculum. In 2005, Carretero-Roque et al. [11] also compared in vivo digestibility in donkeys with an in vitro method, but in that case the neutral cellulase plus gamanase technique [17] was used. Only two published studies estimated feed digestibility in equines using this closed system fermentation apparatus developed by Ankom Technology, and both of them were conducted on horses [20,26].

In our trial, in vivo values were higher than in vitro estimates. The in vitro method ranked the diets in the same order as the in vivo method for both ADMD and NDFD. Similarly, Carretero-Roque et al. [11] found the same trend between in vivo and in vitro digestibility estimates. Earing et al. [20] compared the in vivo and in vitro DMD and NDFD of four diets for horses at three incubation periods (30, 48 and 72 h) using the Daisy^{II} Incubator and horse feces as inoculum source. For three diets, the 72-h in vitro DMD and in vivo DMD values were not significantly different. The 30- and 48-h in vitro estimates were consistently lower than the in vivo estimates, but they ranked diets in the same order as the in vivo method. For NDFD, only the 72-h in vitro NDFD estimate of one diet was not significantly different when compared to the in vivo estimate. Earing et al. [20] concluded that the horse feces can be used in the Daisy^{II} Incubator to estimate in vivo DMD of horse feeds, and the 72-h incubation period provided digestibility estimates most similar to the in vivo data.

Comparing in vivo and in vitro digestibility estimates in donkeys, Carretero-Roque et al. [11] showed a significant ($p < 0.05$) regression equation with reliable model fit ($R^2 = 0.66$). These authors concluded that in vitro analysis using the neutral cellulase plus gamanase technique could predict in vivo digestibility of donkeys fed low-quality forages. Conversely, in the current trial, the regression analysis showed that 60-h incubation length in a Daisy^{II} Incubator did not provide accurate estimation of in vivo DM and NDF digestibility ($R^2 = 0.21$ and 0.47 for AMDM and NDFD, respectively). When compared to our results on donkeys, Earing et al. [20] showed in horses higher model fit for ADMD at all incubation times ($R^2 = 0.71$, 0.70 and 0.63 at 30-, 48- and 72-h of incubation, respectively). On the contrary, for NDFD, the prediction equation obtained for donkeys in our trial showed less bad fit than those obtained by Earing et al. [20] for horses ($R^2 = 0.18$, 0.06 and 0.06 at 30-, 48- and 72-h of incubation, respectively).

Digestibility measured in vitro using the Ankom Daisy^{II} Incubator is influenced by different sources of variation, such as type of inoculum, time and temperature of its storage, type of bag, quantity of sample, time of digestion, physical conditions, and diet of the donor [41]. In our study, most of these conditions were fixed and under control. We can hypothesize that the limitations could be attributed to the weight of the sample and/or the time of digestion. Lattimer et al. [26] showed that in vivo digestibility on horses was predicted more accurately with an in vitro incubation of 48 h than 24 h ($p < 0.05$) and using 0.25 g instead of 0.50 g of sample. According to Cuddeford et al. [11], the estimated retention time (h) in donkeys fed with diets based on alfalfa or oat straw is equal to 76.7 and 53.8, respectively. An in vitro incubation time longer than 60 h may be required to obtain reliable prediction of in vivo digestibility data. In particular, further studies should be designed to determine the retention time of the feeds in donkeys, then using those times as incubation periods for in vitro digestibility trials. Another possibility could be that soluble substances were escaped from the F57 filter bags, influencing microbial population and consequently digestion, as found by Wilman and Adesogan [42].

5. Conclusions

In vivo DM and NDF digestibility values obtained through total feces collection were higher than the corresponding in vitro values obtained incubating donkey feces as inoculum source for 60 h in a Daisy^{II} Incubator. The in vitro method ranked the experimental diets in the same order as the in vivo method. However, the regression analysis did not show reliable prediction of in vivo estimates from in vitro data for both digestibility parameters. Further studies are needed to assess if the use of different sample size and/or digestion times with the Daisy^{II} Incubator would produce accurate estimates of in vivo DM and NDF digestibility when using donkey feces as inoculum source.

Author Contributions: Conceptualization, S.T., R.F., E.V., F.R., S.B., and M.R.; Investigation, S.T., R.F., E.V., L.C., F.R., S.B. and M.R.; Resources, R.F., E.V., L.C. and D.B.; Data curation, F.R., S.B. and M.G.; Formal analysis, M.G. and M.R.; Writing-original draft preparation, S.T., R.F. and M.R.; Writing-review and editing, S.T., R.F., E.V., L.C., D.B., M.G. and M.R.; Supervision, S.T., R.F., E.V. and M.R. All authors have read and agreed to the published version of the manuscript.

Acknowledgments: The authors would like to thank Andrea Gamarra for providing the donkeys. Giovanni Perona, Stefano Nurisso, Francesca Spada, Barbara Negri, and Vanda Maria Malfatto are also gratefully acknowledged for animal care and analytical help.

References

1. Martin-Rosset, W. Donkey nutrition and feeding: Nutrient requirements and recommended allowances—A review and prospect. *J. Equine Vet. Sci.* **2018**, *65*, 75–85. [CrossRef]
2. Polidori, P.; Vincenzetti, S. Farm management and feeding strategies for donkey milk production. In *Agricultural Research Updates*; Gorawala, P., Mandhatri, S., Eds.; Nova Science Publishers, Inc.: Hauppauge, NY, USA, 2017; Volume 14, pp. 93–109.
3. Smith, D.G.; Burden, F.A. Practical donkey and mule nutrition. In *Equine Applied and Clinical Nutrition;* Geor, R., Coenen, M., Harris, P., Eds.; Sauders: Philadelphia, PA, USA, 2013; pp. 304–316.
4. Cuddeford, D.; Pearson, R.A.; Archibald, R.F.; Muirhead, R.H. Digestibility and gastrointestinal transit time of diets containing different proportions of alfalfa and oat straw given to Thoroughbreds, Shetland ponies, Highland ponies and donkeys. *Anim. Sci.* **1995**, *61*, 407–417. [CrossRef]
5. Pearson, R.; Merrit, J.B. Intake, digestion and gastrointestinal transit time in resting donkeys and ponies and exercised donkeys given *ad libitum* hay and straw diets. *Equine Vet. J.* **1991**, *23*, 339–343. [CrossRef] [PubMed]
6. Wilson, R.T. The donkey. In *An Introduction to Animal Husbandry in the Tropics*; Payne, W.J.A., Ed.; Longman: Harlow, UK, 1990; pp. 581–603.
7. Edwards, J.E.; Shetty, S.A.; van den Berg, P.; Burden, F.; van Doorn, D.A.; Pellikaan, W.F.; Dijkstra, J.; Smidt, H. Multi-kingdom characterization of the core equine fecal microbiota based on multiple equine (sub)species. *Anim. Microbiome* **2020**, *2*, 6. [CrossRef]
8. Edwards, J.E.; Schennink, A.; Burden, F.; Long, S.; van Doorn, D.A.; Pellikaan, W.F.; Dijkstra, J.; Saccenti, E.; Smidt, H. Domesticated equine species and their derived hybrids differ in their fecal microbiota. *Anim. Microbiome* **2020**, *2*, 8. [CrossRef]
9. Liu, G.; Bou, G.; Su, S.; Xing, J.; Qu, H.; Zhang, X.; Wang, X.; Zhao, Y.; Dugarjaviin, M. Microbial diversity within the digestive tract contents of Dezhou donkeys. *PLoS ONE* **2019**, *14*, e0226186. [CrossRef]
10. Pearson, R.A.; Archibald, R.F.; Muirhead, R.H. The effect of forage quality and level of feeding on digestibility and gastrointestinal transit time of oat straw and alfalfa given to ponies and donkeys. *Br. J. Nutr.* **2001**, *8*, 599–606. [CrossRef]
11. Carretero-Roque, L.; Colunga, B.; Smith, D.G.; Gonzales-Ronquillo, M.; Solis-Mendez, A.; Castelan-Ortega, O. Digestible energy requirements of mexican donkeys fed oat straw and maize stover. *Trop. Anim. Health Prod.* **2005**, *37*, 123–142. [CrossRef]
12. Pearson, R.A.; Archibald, R.F.; Muirhead, R.H. A comparison of the effect of forage type and level of feeding on the digestibility and gastrointestinal mean retention time of dry forages given to cattle, sheep, ponies and donkeys. *Brit. J. Nutr.* **2006**, *95*, 88–98. [CrossRef]

13. Gatta, D.; Casini, L.; Magni, L.; Liponi, G.B. Apparent digestibility of three diets in the Amiata breed donkey during lactation. *Ital. J. Anim. Sci.* **2009**, *8*, 706–708. [CrossRef]
14. Liu, L.L.; Zhou, X.L.; Yang, H.J. Effect of dietary forage: Concentrate ration on pre-caecal and total digestive tract digestibility of diverse feedstuffs in donkeys as measured by the mobile nylon bag technique. *Animals* **2020**, *10*, 1070. [CrossRef] [PubMed]
15. Kidane, N.F.; Stuth, J.W.; Tolleson, D.R. Predicting diet quality of donkeys via fecal-NIRS calibrations. *Rangel. Ecol. Manag.* **2008**, *61*, 232–239. [CrossRef]
16. Wood, S.J. Some Factors Affecting the Digestible Energy, Requirements and Dry Matter Intake of Mature Donkeys and Comparison with Normal Husbandry Practices. Ph.D. Thesis, University of Edinburgh, Edinburgh, Scotland, 2010.
17. Ankom Technology. Method for Determining Neutral Cellulase plus Gamanase (NCGD). Available online: www.ankom.com/sites/default/files/document-files/NCGD%20Procedure.pdf (accessed on 16 July 2018).
18. Tassone, S.; Renna, M.; Barbera, S.; Valle, E.; Fortina, R. In vitro digestibility measurement of feedstuffs in donkeys using the Daisy^{II} incubator. *J. Equine Vet. Sci.* **2019**, *75*, 122–126. [CrossRef]
19. Leng, J.; Walton, G.; Swann, J.; Dardby, A.; La Ragione, R.; Proudmann, C. "Bowel on the Bench": Proof of concept of a three-stage, In Vitro fermentation model of the equine large intestine. *Appl. Environ. Microbiol.* **2019**, *86*, e02093. [CrossRef] [PubMed]
20. Earing, J.E.; Cassill, B.D.; Hayes, S.H.; Vanzant, E.S.; Lawrence, L.M. Comparison of in vitro digestibility estimates using the Daisy^{II} incubator with in vivo digestibility estimates in horses. *J. Anim. Sci.* **2010**, *88*, 3954–3963. [CrossRef] [PubMed]
21. AOAC International. *Official Methods of Analysis of AOAC International*, 17th ed.; AOAC International: Gaithersburg, MD, USA, 2000.
22. AOAC International. *Official Methods of Analysis of AOAC International*, 17th ed.; 2nd revision; AOAC International: Gaithersburg, MD, USA, 2003.
23. Fortina, R.; Gasmi-Boubaker, A.; Lussiana, C.; Malfatto, V.; Tassone, S.; Renna, M. Nutritive value and energy content of the straw of selected *Vicia* L. taxa from Tunisia. *Ital. J. Anim. Sci.* **2015**, *14*, 280–284. [CrossRef]
24. Mertens, D.R. Gravimetric determination of amylase-treated neutral detergent fiber in feeds using refluxing in beakers or crucibles: Collaborative study. *J. AOAC Int.* **2002**, *85*, 1217–1240.
25. Van Soest, P.J.; Robertson, J.B.; Lewis, B.A. Methods for dietary fiber, neutral detergent fiber, and nonstarch polysaccharides in relation to animal nutrition. *J. Dairy Sci.* **1991**, *74*, 3583–3597. [CrossRef]
26. Lattimer, J.M.; Cooper, S.R.; Freeman, D.W.; Lalman, D.L. Effect of yeast culture on in vitro fermentation of a high-concentrate or high-fiber diet using equine fecal inoculum in a Daisy II incubator. *J. Anim. Sci.* **2007**, *85*, 2484–2491. [CrossRef]
27. Ankom Technology. In Vitro True Digestibility Using the DaisyII Incubator. Available online: www.ankom.com/sites/default/files/document-files/Method_3_Invitro_D200_D200I.pdf (accessed on 21 December 2018).
28. Holden, L.A. Comparison of methods of in vitro dry matter digestibility for ten feeds. *J. Dairy Sci.* **1999**, *82*, 1791–1794. [CrossRef]
29. *The SAS System for Windows*; Release 9.4; SAS Institute Inc.: Cary, NC, USA, 2020.
30. Raspa, F.; Cavallarin, L.; McLean, A.K.; Bergero, D.; Valle, E. A review of the appropriate nutrition welfare criteria of dairy donkeys: Nutritional requirements, farm management requirements and animal-based indicators. *Animals* **2019**, *9*, 315. [CrossRef] [PubMed]
31. Burden, F.; Thiemann, A. Donkeys are different. *J. Equine Vet. Sci.* **2015**, *35*, 376–382. [CrossRef]
32. Smith, D.G.; Pearson, R.A. A review of the factors affecting the survival of donkeys in semi-arid regions of sub-Saharian Africa. *Trop. Anim. Health Prod.* **2005**, *37*, 1–19. [CrossRef] [PubMed]
33. Agbagla-Dohnani, A.; Cornu, A.; Nozière, P.; Besle, J.-M.; Dulphy, J.P.; Doreau, M.; Grenet, E. Microbial degradation of rice and barley straws in the sheep rumen and the donkey caecum. *J. Sci. Food Agric.* **2003**, *83*, 383–394. [CrossRef]
34. Izraely, H.; Choshniak, I.; Stevens, C.E.; Demment, M.W.; Shkolnik, A. Factors determining the digestive efficiency of the domesticated donkey (*Equus asinus asinus*). *Q. J. Exp. Physiol.* **1989**, *74*, 1–6. [CrossRef] [PubMed]
35. Palmgren Karlsson, C.; Lindberg, J.E.; Rundgren, M. Associative effects on total tract digestibility in horses fed different rations of grass hay and whole oats. *Livest. Prod. Sci.* **2000**, *65*, 143–153. [CrossRef]
36. Peiretti, P.G.; Miraglia, N.; Meineri, G.; Costantini, M.; Bergero, D. Digeribilità in vivo di orzi schiacciati e fioccati in razioni per cavalli. *Ippologia* **2009**, *3*, 17–21.

37. De Marco, M.; Peiretti, P.G.; Miraglia, N.; Bergero, D. Apparent digestibility of broken rice in horses using in vivo and in vitro methods. *Animal* **2014**, *8*, 245–249. [CrossRef]
38. Clarke, L.L.; Roberts, M.C.; Argenzio, R.A. Feeding and digestive problems in horses. Physiologic responses to a concentrated meal. *Vet. Clin. North Am. Equine Pract.* **1990**, *6*, 433–450. [CrossRef]
39. de Fombelle, A.; Julliand, V.; Drogoul, C.; Jacotot, E. Feeding and microbial disorders in horses: 1-effects of an abrupt incorporation of two levels of barley in a hay diet on microbial profile and activities. *J. Equine Vet. Sci.* **2001**, *21*, 439–445. [CrossRef]
40. Philippeau, C.; Sadet-Bourgeteau, S.; Varloud, M.; Julliand, V. Impact of barley form on equine total tract fibre digestibility and colonic microbiota. *Animal* **2015**, *9*, 1943–1948. [CrossRef] [PubMed]
41. Tassone, S.; Fortina, R.; Peiretti, P.G. In Vitro Techniques Using the Daisy^{II} Incubator for the Assessment of Digestibility: A Review. *Animals* **2020**, *10*, 775. [CrossRef] [PubMed]
42. Wilman, D.; Adesogan, A. A comparison of filter bag methods with conventional tube methods of determining the in vitro digestibility of forages. *Anim. Feed Sci. Technol.* **2000**, *84*, 33–47. [CrossRef]

4

In Vitro Digestibility, In Situ Degradability, Rumen Fermentation and N Metabolism of Camelina Co-Products for Beef Cattle Studied with a Dual Flow Continuous Culture System

Hèctor Salas, Lorena Castillejos *, Montserrat López-Suárez and Alfred Ferret

Animal Nutrition and Welfare Service (SNIBA), Department of Animal and Food Science, Universitat Autonòma de Barcelona, 08193 Bellaterra, Spain; hectorsalasolive@gmail.com (H.S.); Montserrat.Lopez.Suarez@uab.cat (M.L.-S.); Alfred.Ferret@uab.cat (A.F.)

* Correspondence: Lorena.Castillejos@uab.cat;

Simple Summary: Currently, vegetable protein sources such as soybean meal and rapeseed meal are expensive and with volatile prices. These economic circumstances are driving the research of potential new protein resources for beef cattle diets that can reduce the ration cost without compromising animal productive yields. As possible candidates, camelina meal and camelina expeller have been studied; they are co-products with a high protein percentage, obtained after oil extraction from the oil seeds of *Camelina sativa*. The objectives of this study were to characterize these camelina co-products and ascertain if they could be useful ingredients for beef cattle diets. The results indicate that the diets formulated with camelina meal and camelina expeller do not show differences in the efficiency of microbial protein synthesis compared to the current reference proteins, camelina meal diet being the most similar to soybean meal and rapeseed meal diets, and camelina expeller the diet with the highest fermentation potential. The results of soybean meal as an individual ingredient reveal more differences with camelina co-products. In vivo studies are necessary to draw conclusions, but in vitro results obtained suggest that camelina meal and camelina expeller are potential substitutes for rapeseed meal in beef cattle diets.

Abstract: Camelina meal (CM) and camelina expeller (CE) were compared with soybean meal (SM) and rapeseed meal (RM). Trial 1 consisted of a modified Tilley and Terry in vitro technique. Trial 2 was an in situ technique performed by incubating nylon bags within cannulated cows. Trial 3 consisted in dual-flow continuous culture fermenters. In Trial 1, CM, CE and RM showed similar DM digestibility and OM digestibility, and SM was the most digestible ingredient ($p < 0.05$). Trial 2 showed that CE had the numerically highest DM degradability, but CP degradability was similar to RM. Camelina meal had a DM degradability similar to SM and RM and had an intermediate coefficient of CP degradability. In Trial 3, CE diet tended to present a higher true OM digestibility than SM diet ($p = 0.06$). Total volatile fatty acids (VFA) was higher in CE and CM diets than in SM diet ($p = 0.009$). Crude protein degradation tended to be higher ($p = 0.07$), and dietary nitrogen flow tended to be lower ($p = 0.06$) in CE diet than in CM diet. The efficiency of microbial protein synthesis was not affected by treatment ($p > 0.05$). In conclusion, CE and CM as protein sources differ in CP coefficient of degradability but their results were similar to RM. More differences were detected with regard to SM.

Keywords: beef cattle; protein sources; camelina co-products; rumen microbial fermentation

1. Introduction

Meeting protein requirements in ruminant nutrition can be costly. The main reasons are the high and unstable prices of protein sources, such as soybean meal, and their availability, which is affected by global trade [1]. This situation makes it necessary to search for new alternatives to replace totally or partially the protein sources currently used in ruminant diets. In this line of research, *Camelina sativa* co-products could be one such alternative. *Camelina sativa* or false flax is an oilseed crop of the Brassica family, which originates from the Mediterranean and Central Asia. It is an annual or overwintering herb with low agronomic requirements [2] and is more tolerant to frost, heat, and drought than other plants of the same family [3], such as rapeseed meal. The biofuel industry's growing interest in its cultivation is attributable to the 40% oil content of the seed, which is used to produce biodiesel [4–8]. When oil is extracted from the seed, camelina expeller (CE) and camelina meal (CM) are produced, the former being obtained after mechanical oil extraction of the seed and the latter after mechanical and subsequent chemical oil extraction. According to Zubr [9], the resulting meal after oil extraction contains 30–40% CP and 12% fiber. However, the main limitation of using CM and CE is related to the presence of anti-nutritional components. Camelina co-products contain glucosinolates and erucic acid [10], which, according to Tripathi and Mishra [11] affect the thyroid and the cardiovascular system. That said, CM has been used in beef steers without effect on growth performance or thyroid function [12,13]. Our hypothesis was that given the high protein content of both co-products [14], CE and CM could be alternative protein sources in ruminant nutrition. The aims of this study were to characterize these camelina co-products and to compare in vitro beef cattle diets made either with them or with more commonly used protein sources like soybean meal and 00-rapeseed meal.

2. Materials and Methods

Animal procedures were approved by the Institutional Animal Care and Use Committee (reference CEEAH 1585) of the Universitat Autònoma de Barcelona (Spain), in accordance with the European directive 2010/63/EU.

Three different trials were conducted to achieve the objectives. Trial 1 was performed to determine the in vitro digestibility of four protein ingredients commonly used in cattle diets (Table 1). Two of them were sources supplied by the feed industry and frequently used for beef cattle: Soybean meal 44% CP (SM) and 00-rapeseed meal (RM), while the other two were CE and CM, supplied by Camelina Company, S.L. (Madrid, Spain). The CE and CM samples were selected according to their chemical composition similarity to others studies in the case of CE [8,15] and to RM in the case of CM. To select the SBM and RM samples, we performed chemical analyses of different SBM and RM varieties and selected those most similar to the most commonly used varieties in beef diets in our commercial conditions. In Trial 2, the in situ technique was used to estimate the dry matter (DM) and the crude protein (CP) effective degradability of these protein ingredients. Finally, in Trial 3 a dual-flow continuous culture system was used to study the true digestibility, rumen fermentation and nitrogen metabolism of four isoenergetic and isonitrogenous diets (Table 2). Each one was formulated with one of these protein sources and for a targeted gain of 1.4 kg/day according to Fundación Española para el Desarrollo Animal (FEDNA) [16] recommendations for beef cattle. All diets were designed with a 90:10 concentrate to barley straw ratio.

Table 1. Chemical composition of protein sources [1] (DM basis).

Item	SM	CM	CE	RM
Chemical composition (g/kg)				
DM	881	915	928	883
OM	824	858	880	814
CP	467	395	351	398
EE	26.1	12.9	135.7	15.5
NDF	107	375	327	358
ADF	120	174	144	214
Lignin	5.3	40.3	25.8	88.2
NDICP	29.3	118.8	107.1	86.4
ADICP	13.5	62.7	63.0	58.2
Gross energy (Mcal/Kg)	4.38	4.35	4.96	4.29
Antinutritional factors				
Allyl isothiocyanate (mg/g)	-	0.09	0.09	0.13
Erucic acid (g/100g fat)	-	0.05	0.04	<0.01

[1] SM, Soybean meal 44% CP; CM, Camelina meal; CE, Camelina expeller; RM, Rapeseed meal 00.

Table 2. Ingredients and chemical composition of treatment diets [1] tested with the dual-flow continuous culture system.

Item	SMD [1]	CMD	CED	RMD
Ingredients (g/kg DM)				
Corn grain	376	379	343	364
Barley grain	318	272	308	286
Soybean meal 44%	114	-	-	-
Camelina meal	-	152	-	-
Camelina expeller	-	-	163	-
Rapeseed 00 meal	-	-	-	149
Barley straw	100	100	100	100
Palm oil	10	27	10	28
Soybean hulls	69	56	62	60
Calcium carbonate	9	10	10	9
Vitamin and mineral premix [2]	4	4	4	4
Chemical composition (g/kg DM)				
DM	889	894	893	889
OM	852	860	858	854
CP	125	132	132	130
EE	35.8	43.1	43.3	45.9
NDF	217	224	242	221
ADF	112	117	124	124
Lignin	12.2	104	20.4	21.7
Gross energy (Mcal/Kg)	3.97	4.07	4.08	4.07

[1] SMD, diet with soybean meal 44% CP; CMD, diet with camelina meal; CED, diet with camelina expeller; RMD, diet with rapeseed meal 00. [2] Vitamin and mineral premix: Vitamin A 8400 IU/Kg; Vitamin D3 1680 IU/Kg; Zinc oxide 85 mg/Kg; Iron carbonate 39 mg/Kg; Manganese oxide 30 mg/Kg.

2.1. Chemical Composition

Chemical composition of protein sources used in Trial 1 and Trial 2, and composition of ingredients and chemical composition of diets in Trial 3 are described in Tables 1 and 2, respectively. Samples, of either ingredients or diets, were ground to pass through a 1-mm sieve and analyzed in duplicate. The dry matter (DM) content was determined by drying samples for 24 h in a 103 °C forced air oven, and organic matter (OM) was determined after ignition of a sample in a muffle furnace at 550 °C overnight according to AOAC ID 950.05 [17]. Nitrogen content was determined by the Kjeldahl procedure AOAC ID 976.05 [17]. Ether extract (EE) was performed according to AOAC ID 920.30 [17].

The neutral detergent fiber (NDF) and acid detergent fiber (ADF) contents were determined sequentially by using an Ankom Fiber Analyzer (Ankom Technology, Fairport, NY, USA) in accordance with the methodology provided by the company. This is based on the procedure of Van Soest et al. [18] using a thermostable α–amylase and sodium sulfite. Ash determination was performed at the end of the sequential process to express fiber results on an ash-free basis. In the case of both camelina co-products, the concentration of the α–amylase and sodium sulfite were doubled compared to the original procedure to achieve acceptable repeatability of the determination. The lignin content was determined after fiber procedures using sulfuric acid 72%. Neutral detergent insoluble crude protein (NDICP) and acid detergent insoluble crude protein (ADICP) were determined by the aforementioned Kjeldahl procedure in residues obtained after fiber determinations. Gross energy content was measured by completely burning a sample of the ingredients to measure the heat produced using a bomb calorimeter (Parr 6300, Parr Instrument Company, Moline, IL, USA). The allyl isothiocyanate level was determined by a destilation-volumetry procedure according to the European Directive 71/250/EEC. The erucic acid content was analyzed by chromatography (Model 6890, Hewlett Packard, Palo Alto, CA, USA), according to American Oil Chemists' society (AOCS) method CE 2-66 [19].

2.2. Trial 1: In Vitro Digestibility

In vitro DM and OM digestibility of samples were obtained following the Tilley and Terry [20] procedure modified by Stern and Endres [21]. The experiment was performed in two periods with three replicates per period. The rumen liquid obtained from two different cows fed a 60:40 forage to concentrate ratio diet was diluted with the buffer solution proposed by McDougall [22] at a ratio of 1:4 (rumen fluid:buffer solution). A sample of 0.5 g was incubated in each tube with 50 mL of mixed solution in continuous agitation and anaerobiosis, which was achieved with the introduction of CO_2 at 39 °C. After 48 h of incubation, 0.2 g of pepsin and 2 mL of HCl were added in each tube. Tubes were maintained in continuous agitation at 39 °C for 24 h. The content of each tube was filtered using filtration crucibles, and the DM and OM of the residue obtained were determined as mentioned previously.

2.3. Trial 2: In Situ Degradability

Two cannulated cows fed ad libitum with a 60:40 forage to concentrate ratio diet were used. The DM and CP in situ degradability analysis was carried out by incubating nylon bags (Ankom Technology Corporation, Fairport, NY, USA) in the rumen, which were 5 cm × 10 cm with a pore size of 50 μm containing 1 g of sample that was ground to pass through a 1.5-mm sieve. In three repeated periods, one bag per sample, animal and hour was inserted into the rumen at 8.30 am directly before feeding. Incubation periods were 2, 4, 6, 8, 12, 24, 48 and 72 h. After rumen incubation, bags were subjected to a washing procedure consisting of three washing cycles of 5 min, and immediately frozen at −18 °C for further analyses. The 0-h nylon bags were treated with the same methodology but without passing through the rumen. Degradation of DM and CP parameters was calculated using the equation of Ørskov and McDonald [23]:

$$D = a + b\,(1 - e^{-ct}) \tag{1}$$

where D is the disappearance of either DM or CP at time t; a is an intercept representing the DM or CP soluble fraction, b is the fraction of insoluble but degradable DM or CP, c is the rate of disappearance of fraction b, and t is the time of incubation. The non-linear parameters a, b and c were estimated by an iterative least-squares procedure of SAS (v. 9.1; SAS Inst. Inc., Cary, NC, USA). The effective degradability (ED) of DM and CP were calculated using the equation:

$$ED = a + [bc/(c + k)] \tag{2}$$

where *a*, *b* and *c* are the same parameters as described earlier and k is the estimated rate of passage of 0.06/h according to Institut National de la Recherche Agronomique (INRA) [24].

2.4. Trial 3: Dual-Flow Continuous Culture System

2.4.1. Fermenters

Eight 1320 mL dual-flow continuous culture fermenters developed by Hoover et al. [25] were used in two replicated periods. Each experimental period consisted of 5 days for adaptation and 3 days for sampling. Fermenters were inoculated with ruminal liquid from two dairy cows fed a 60:40 forage to concentrate ratio diet. Fermentation conditions were maintained constant with a temperature of 39 °C, and pH at 6.1 ± 0.05 by infusion of 3 N HCl or 5 N NaOH, monitored and controlled by a computer. Liquid and solid constant dilution rates (0.1 and 0.05 h-1, respectively) were obtained by a continuous infusion of artificial saliva [26]. Finally, anaerobic conditions were maintained by infusion of N2 gas at a rate of 40 mL/min. Fermenters were daily fed 90 g of diet (DM basis) in three equidistant doses of 30 g.

2.4.2. Sample Collection

On the sampling days, collection vessels were maintained at 4 °C to prevent microbial activity. Solid and liquid effluents were mixed and homogenized for 1 min, and a 600 mL sample was removed by aspiration. Upon completion of each period, effluents from each fermenter collected over the three sampling days were composited and mixed and homogenized for 1 min. Subsamples were taken for total N, ammonia N, and VFA analyses. The remainder of the sample was lyophilized. Dry samples were analyzed for DM, ash, NDF, ADF, and purine contents.

Bacteria were isolated from the fermenter contents on the last day of each experimental period. Solid and liquid associated bacteria were isolated using a combination of several detachment procedures, which were selected to obtain the maximum detachment without affecting cell integrity [27]. To remove attached bacteria, 100 mL of a 2 g/L methylcellulose solution and small marbles (thirty measuring 2 mm in diameter, and fifteen of 4 mm) were added to each fermenter, incubated in the same fermenter contents at 39 °C, and mixed for 1 h. After incubation, fermenter contents were refrigerated for 24 h at 4 °C, and subsequently agitated for 1 h to dislodge loosely attached bacteria. Finally, the fermenter contents were filtered through cheesecloth and washed with saline solution (8.5 g/L NaCl). Bacterial cells were isolated within 4 h by differential centrifugation at 1000× *g* for 10 min, to obtain a supernatant without feed particles, which was then centrifuged at 20,000× *g* for 20 min to isolate bacterial cells. Pellets were rinsed twice with saline solution and recentrifuged at 20,000× *g* for 20 min. The final pellet was recovered with distilled water to prevent contamination of bacteria with ash. Bacterial cells were lyophilized and analyzed for DM, ash, N, and purine contents. Digestion of DM, OM, NDF, ADF and CP, and flows of total, non-ammonia, microbial, and dietary N were calculated as described by Stern and Hoover [28].

2.4.3. Chemical Analyses

Effluent DM was determined by lyophilizing 200 mL aliquots in triplicate. The DM, OM, total N, NDF, ADF, lignin and EE content of the lyophilized effluents, bacterial samples and diets were determined as described in Section 2.1. Effluent ammonia N was analyzed by colorimetry as described by Chaney and Marbach [29], where 4 mL of a 0.2 N HCl solution were added to 4 mL of filtered rumen fluid and frozen at −20 °C until later analysis. Samples were centrifuged at 15,000× *g* for 15 min, and the supernatant was used to determine ammonia N. Effluent samples for VFA determination were prepared as described by Jouany [30] and analyzed by gas chromatography: 1 mL of a solution made up of a 2 g/L solution of mercuric chloride, 0.002 mg/L of 4-methylvaleric acid as an internal standard, and 2 g/L orthophosphoric acid was added to 4 mL of filtered rumen fluid and frozen at −20 °C until later analysis. Samples were centrifuged at 3000× *g* for 30 min, and the supernatant analyzed by gas chromatography (Model 6890, Hewlett Packard, Palo Alto, CA, USA) using a polyethylene glycol TPA treated capillary column (BP21, SGE, Europe Ltd., Buckinghamshire, UK). The dimensions of the column were 30 m × 0.25 mm ID, with 0.25 μm film thickness. The injector was set at 275 °C with a split ratio of 4:1. Helium was used as the carrier gas. The initial temperature was 85 °C for 1 min, and increased by 3C/min until a final temperature of 220 °C was reached, and held for a further 2 min. The detector temperature was 275 °C. Samples of lyophilized effluent and bacterial cells were analyzed for adenine and guanine content by HPLC as described by Balcells et al. [31], using allopurinol as internal standard.

2.5. Statistical Analysis

Data from the in vitro digestibility experiment were analyzed using the MIXED procedure of SAS (v. 9.1; SAS Inst. Inc., Cary, NC, USA). The model contained the fixed effect of treatment, and the random effect of period. Statistical analysis of the dual-flow continuous culture system data was performed using the MIXED procedure, and multiple comparisons were performed by LSMEANS adjusted with the Tukey test using SAS (v. 9.1; SAS Inst. Inc., Cary, NC, USA). The model contained treatment as the fixed effect and period as the random effect. Statistical significance was declared at $p < 0.05$ and differences among means with $0.05 < p < 0.10$ were considered tendencies.

3. Results

3.1. In Vitro Digestibility with the Two-Stage Tilley and Terry Technique

In vitro DM digestibility (DMD) and OM digestibility (OMD) values are presented in Table 3. Soybean meal was the protein source with the highest digestibility ($p < 0.05$), both DMD and OMD. There were no significant differences ($p > 0.05$) among CM, CE and RM in DMD, and OMD values.

Table 3. In vitro digestibility of protein sources [1].

Item	SM	CM	CE	RM	SEM [2]	*p*-Value
Coefficients of digestibility						
Dry matter	0.74 ^{a3}	0.65 ^{b}	0.64 ^{b}	0.62 ^{b}	0.160	<0.001
Organic matter	0.87 ^{a}	0.72 ^{b}	0.70 ^{b}	0.72 ^{b}	0.131	<0.001

[1] SM, soybean meal 44% CP; CM, camelina meal; CE, camelina expeller; RM, rapeseed meal 00. [2] SEM, standard error of mean. [3] Means with different superscript differ statistically ($p < 0.05$).

3.2. In Situ Rumen Degradation

Nonlinear parameter estimates and ED values are presented in Table 4, and ruminal degradation kinetics of DM and CP are shown in Figures 1 and 2, respectively. The greatest DM soluble fraction was recorded with CE, while the remaining protein sources showed similar values. In contrast, the greatest fraction b was obtained in SM, being intermediate in CM and RM, and lowest in CE. The DM rate of the disappearance of fraction b was similar in CM and CE, and in an intermediate position between SM, with the lowest value, and RM, with the greatest. The ED of DM was similar in SM, CM and RM, and the greatest value was recorded in CE. In descending order, the CP a fraction values came out as follows: CE, RM, CM and SM. Soybean meal showed the highest b value, followed by CM, RM and CE. The lowest rate of CP degradation was found in SM, followed by CM and RM, and the greatest was recorded in CE. The lowest ED of CP was obtained in SM, the greatest in CE and RM, and the value was intermediate for CM.

Table 4. In situ nonlinear estimates [1] and effective degradability values of dry matter (ED) and crude protein (EDCP) of protein sources [2].

Item	SM	CM	CE	RM
Dry matter				
a	0.32 ± 0.012 [3]	0.35 ± 0.022	0.50 ± 0.013	0.37 ± 0.025
b	0.57 ± 0.015	0.44 ± 0.025	0.33 ± 0.015	0.40 ± 0.028
c (/h)	0.077 ± 0.001	0.135 ± 0.017	0.132 ± 0.014	0.180 ± 0.027
ED	0.64	0.66	0.72	0.67
Crude Protein				
a	0.24 ± 0.017	0.29 ± 0.027	0.47 ± 0.030	0.38 ± 0.019
b	0.79 ± 0.022	0.66 ± 0.031	0.48 ± 0.034	0.57 ± 0.017
c (/h)	0.064 ± 0.001	0.160 ± 0.017	0.249 ± 0.038	0.199 ± 0.013
ED	0.64	0.77	0.85	0.82

[1] *a*: Soluble fraction; *b*: Insoluble but degradable fraction; *c*: The rate (/h) of disappearance of *b* fraction; [2] SM, soybean meal 44% CP; CM, camelina meal; CE, camelina expeller; RM, rapeseed meal 00; [3] Mean ± standard error.

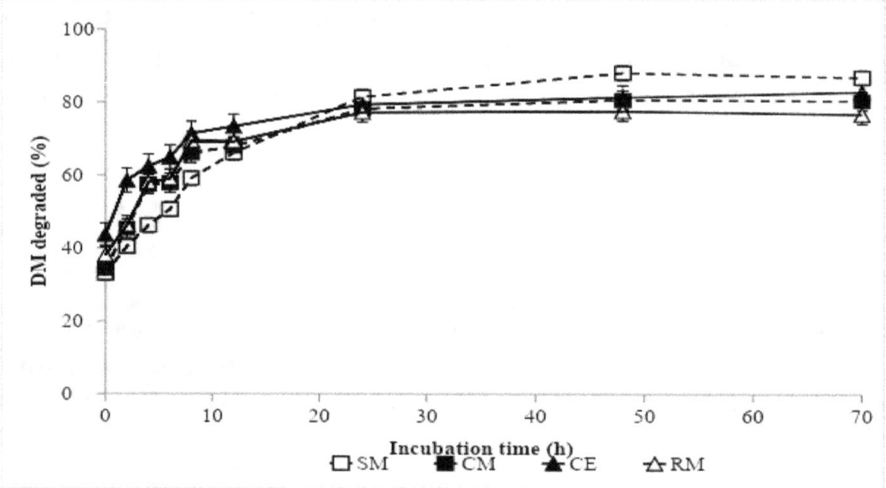

Figure 1. In situ dry matter degradation curve of protein sources: SM, Soybean meal 44% CP; CM, camelina meal; CE, camelina expeller; RM, rapeseed meal 00.

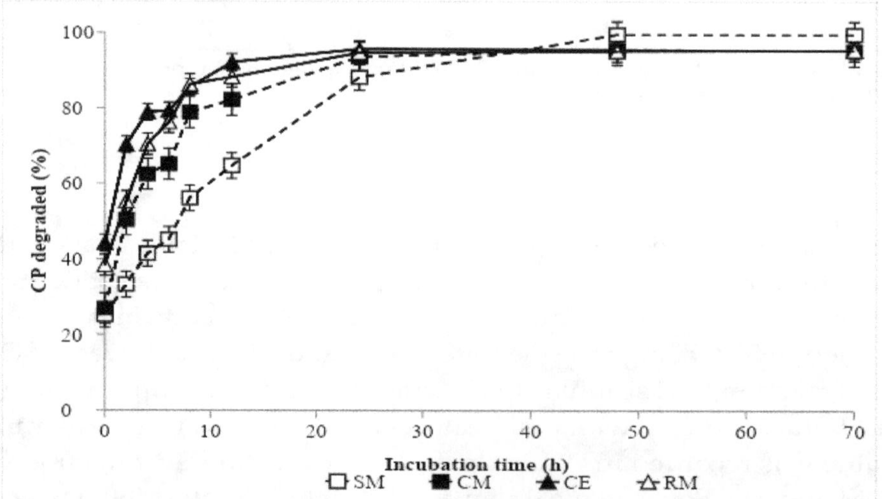

Figure 2. In situ crude protein degradation curve of protein sources: SM, Soybean meal 44% CP; CM, camelina meal; CE, camelina expeller; RM, rapeseed meal 00.

3.3. *Dual-Flow Continuous Culture System*

Results of dual-flow continuous culture system are presented in Tables 5–7.

Table 5. Effect of treatment diet on apparent digestibility and true digestibility in the dual-flow continuous culture system.

Item	Diet [1]				SEM [2]	*p*-Value
	SMD	CMD	CED	RMD		
Apparent digestibility (g/kg)						
DM	442	431	413	412	34.6	0.73
OM	346	352	372	360	22.2	0.58
NDF	380	357	408	359	67.2	0.83
ADF	370	385	421	345	77.8	0.76
True digestibility (g/kg)						
DM	548	531	563	521	37.8	0.66
OM	480 ^{b3}	486 ^{ab}	574 ^{a}	494 ^{ab}	37.1	0.06

[1] SMD, diet with soybean meal; CMD, diet with camelina meal; CED diet with camelina expeller; RMD, diet with rapeseed meal. [2] SEM, standard error of the mean. [3] Means with different superscript tended to differ ($p < 0.10$).

Table 6. Effect of treatment diet on volatile fatty acids (VFA) concentration and profile in the dual-flow continuous culture system.

Item	Diet [1]				SEM [2]	*p*-Value
	SMD	CMD	CED	RMD		
Total VFA (mM)	119.6 ^{b3}	135.6 ^{a}	138.8 ^{a}	121.0 ^{ab}	7.06	0.009
VFA (mol/100 mol)						
Acetate	45.2	46.0	46.2	44.0	2.44	0.777
Propionate	32.1	35.9	34.6	33.6	2.01	0.236
Butyrate	16.4 ^{a}	12.1 ^{b}	13.1 ^{ab}	15.0 ^{ab}	1.46	0.016
Iso-butyrate	0.39	0.41	0.45	0.43	0.04	0.357
Valerate	5.23	4.42	4.39	5.89	1.18	0.494
Iso-valerate	0.74	1.14	1.25	1.11	0.36	0.412
BCVFA [4] (mM)	1.14	1.55	1.70	1.54	0.40	0.399
Acetate:propionate ratio	1.42	1.31	1.35	1.32	0.12	0.773

[1] SMD, diet with soybean meal 44% CP; CMD, diet with camelina meal; CED, diet with camelina expeller; RMD, diet with rapeseed meal 00. [2] SEM, standard error of mean. [3] Means with different superscripts differ statistically ($p < 0.05$). [4] BCVFA, Branch-chained VFA.

Table 7. The effect of treatment diet on N metabolism in the dual-flow continuous culture system.

Item	Diet [1]				SEM [2]	*p*-Value
	SMD	CMD	CED	RMD		
NH3-N (mg/100 mL)	1.60	1.37	1.80	2.23	0.51	0.42
N flow (g/d)						
Total	2.37	2.49	2.46	2.46	0.07	0.34
Ammonia	0.05	0.04	0.06	0.07	0.02	0.41
Non-ammonia	2.32	2.44	2.36	2.39	0.09	0.51
Bacterial	0.86	0.83	1.23	0.87	0.19	0.13
Dietary	1.46 ^{ab3}	1.61 ^{a}	1.13 ^{b}	1.52 ^{ab}	0.18	0.06
Crude protein degradation (g/kg)	229 ^{ab}	196 ^{b}	423 ^{a}	236 ^{ab}	90.62	0.07
EMPS [4]	35.27	35.61	37.06	34.12	4.19	0.89

[1] SMD, diet with soybean meal 44% CP; CMD, diet with camelina meal; CED, diet with camelina expeller; RMD, diet with rapeseed meal 00. [2] SEM, standard error of mean. [3] Means with different superscript tended to differ statistically ($p < 0.10$). [4] The efficiency of microbial protein synthesis (g bacterial N/Kg organic matter truly digested).

Apparent digestibility of DM, NDF and ADF, and true DM digestibility were not different among diets (Table 5). In contrast, true OM digestibility tended to be affected by diet, with the highest value in CED and the lowest in SMD. Total VFA was higher in CED and CMD than in SMD ($p = 0.009$), but there was no difference between these diets and RMD (Table 6). Butyrate proportion was lower in CMD than in SMD ($p = 0.016$), but the proportion detected in CED and RMD was not different from that detected in CMD and SMD. There were no differences among diets in acetate, propionate, iso-butyrate, valerate, and iso-valerate proportions, the BCVFA, and the acetate to propionate ratio. Table 7 shows the results on nitrogen metabolism. There were no diet effects on total NH3-N, and flows of total, ammonia, non-ammonia and bacterial N, and EMPS. However, dietary N flow tended to be lower in CED than in CMD ($p = 0.06$), and CP degradation tended to be higher in CED than in CMD ($p = 0.07$).

4. Discussion

Chemical composition of feedstuffs commonly used in beef cattle diets, like SM and RM, was in accordance with published values [24,32]. Comparing both *Camelina sativa* co-products, CM contained higher values of CP, NDF, ADF and lignin compared to CE, but lower EE content. The CP content of the CM used in the present study was similar to the values referenced in the literature [8,33]. On the contrary, the EE content of these references was closer to the value found in CE. In reference to the anti-nutritional factors, the content of allyl isothiocyanate, as a major metabolite of glucosinolates of CM and CE, was below that of RM, an ingredient that is considered to have a negligible amount of glucosinolates. Tripathi and Mishra [11] obtained values between 0.3 mg/g and 2.1 mg/g of allyl isothiocyanate in different varieties of RM obtained in diverse oil extraction processes. Values of erucic acid presented by CM and CE were below 1% of the fat fraction that is considered to be the threshold of a rapeseed meal zero erucic acid variety [16]. Therefore, the comparison between camelina co-products and RM suggests that their use would not represent a nutritional problem for beef cattle nutrition.

In vitro DM and OM digestibility were not different in both camelina co-products. These digestibility values were similar to those recorded with RM but lower than in SM. The higher CP content and the lower fiber content of SM would explain these higher digestibility coefficients. Yong-Gang Liu et al. [33] and Moss and Givens [34], also using in vitro studies, reported equivalent OMD coefficients for SM and RM (0.89 and 0.74, respectively) to those reported in the present study.

Effective DM degradation was higher in CE than in CM. This difference would be related with the *a* and *b* fractions, which were higher and lower, respectively, in CE than in CM, but without differences in the rate of DM disappearance. Dry matter ED of CM was close to the values recorded in RM and SM. However, this result was obtained with different kinetic parameters because although the *a* fraction was similar among these protein sources, the b fraction was higher in SM and lower in CE, while rate of DM disappearance was higher in RM and lower in SM. In the comparison between SM and RM, our results agree with Heendeniya et al. [35] and Wulf and Südekum [36], who presented similar DM kinetic parameters between SM and RM, with a higher insoluble but degradable fraction and slower degradation rate in SM compared with RM. In contrast, Prestløkken [37] (1999) did not report differences between SM and RM, and Maxin et al. [38] found a greater DM degradation rate and ED in SM than in RM. The highest *b* fraction recorded in SM could be explained by its chemical composition: High CP content, low fiber content, and low content of NDICP and ADICP, resulting in the highest DMD and OMD in comparison with the remaining protein sources.

Effective CP degradation of CE was higher than that of CM, with a higher *a* fraction and rate of CP disappearance and a lower *b* fraction. Lawrence and Anderson [39] studied the kinetic parameters of a CM and recorded CP degradability very similar to the result obtained in the present study. However, the chemical composition of their CM was closer to our CE, because their EE content was 143 g/kg DM, similar to the 135 g/kg DM of our CE and different from the 13 g/kg DM of our CM. Camelina expeller and RM showed the same ED of CP, but the *a* fraction and rate of disappearance were lower and the *b* fraction higher in RM than in CE. The lowest ED of CP was found in SM, in comparison with the remaining protein sources. The values obtained with SM were in accordance with the literature [36,37,40]. Conversely, RM soluble fraction was very high in comparison with Prestløkken [37], Wulf and Südekum [36] and Heendeniya et al. [35], and close to the result obtained by Maxin et al. [38].

Chemical composition of the diets tested in the dual-flow continuous culture system confirmed that they were formulated to be isonergetic and isonitrogenous, with only a slight decrease in the CP content in the SMD diet. No differences among diets in apparent digestibilities and true DMD were observed. In contrast, Brandao et al. [41], in a dual-flow continuous culture system trial, observed lower NDF digestibility in diets with either 50% or 100% of CE instead of RM, meaning that RM was included at 10.3% and 0% (on DM basis), respectively, compared to the control diet with RM, in which RM was included at 20.6% as the main protein source. The lack of differences in our Trial between SMD and RMD digestibilities is in accordance with Paula et al. [42] who also compared, in a dual-flow continuous culture system, a SM diet with RM diets with different rumen-undegradable protein (RUP) content. However, true OMD tended to be affected by diet. The trend detected in the highest true OMD value in CED compared to SMD, although not statistically different from CMD and RMD, could be related with the highest fermentation activity observed in this diet, where high values of total VFA were recorded in comparison with SMD. With similar true OMD to SMD but also CED, CMD and RMD did not differ from CED in total VFA concentration. While there were no differences among diets in acetate and propionate proportion, butyrate proportion in CMD was lower than in SMD. During the synthesis of acetate and butyrate, there is also a methane generation that reduces its energy efficiency compared with the synthesis of propionate, in the formation of which there is no loss of any carbon during the reaction [43]. Therefore, it seems that CMD could be more efficient in energy usage. This could be related with the higher total VFA concentration detected in CMD and CED than in SMD. Although there was no different BCVFA content among diets, the numerically higher value found in CED diet could be related with the high CP degradability recorded in this diet, because BCVFA are produced when rumen microbes degrade protein [44]. The highest CP degradation detected in the CED would coincide with the highest CP degradation recorded in the in situ trial. In contrast, the same did not occur in the case of SMD and RMD when comparing in situ and fermenter results. In addition we obtained a lower CP degradation in CMD than in CED, and the same result between SMD and RMD. In the in vitro experiment of Brandao et al. [41], there were also no differences between RMD and CED treatments. In the in situ trial, in accordance with Huhtanen et al. [45] and most feed tables [24,46], a greater RUP content for SM compared with RM was recorded. However, Brito et al. [47] and Brito and Broderick [48] reported no differences in omasal RUP flows when isonitrogenous RM and SM diets were compared, in accordance with our fermenter results. Higher RUP of SMD, RMD and CMD could explain the effects of diet on dietary N flow, in which a tendency was found for the lowest flow in CED, highest in CMD, and intermediate in SMD and RMD diets. This difference in the RUP content between CED and CMD could be due to the second oil-extraction chemical treatment performed in CM, which is not applied in CE. This second extraction could modify the rumen digestibility of protein by decreasing the accessibility of rumen microorganisms to protein. Chemical and physical extractions are strategies commonly used to reduce ruminal CP degradability and increase RUP [45,49]. Moreover, this could also explain the numerical results observed in bacterial N flows, which were numerically higher in CED than in the remaining diets. When comparing a high RUP diet with a basal diet, Ipharraguerre and Clark [50] reported that a significant decrease in the flow of microbial N to the small intestine occurred with RUP supplements. Hoover and Stokes [51] concluded that when rumen degradable protein is replaced by RUP, the microbial growth in the rumen can decrease. The high OMD, the high CP degradation and the low RUP of CED could have promoted the microbial growth. This would be in agreement with Santos et al. [49], who concluded that high RUP diets resulted in decreased microbial protein synthesis. However, there was no significant difference between treatments in EMPS. It is important to remark that CED and CMD presented equivalent results in EMPS than diets formulated with SM and RM, the most common protein sources used in beef cattle. The lack of differences between treatments could be related with the low NH3-N concentration recorded in all effluents, which did not attain the 5 mg/dL, concentration usually recommended to ensure maximum microbial growth [52]. However, Russell et al. [53] reported no difference in microbial growth when NH3-N concentration was below 5 mg/dL or greater than 16 mg/dL. Owens and Bergen [54] concluded that the minimum amount of NH3-N to maximize bacterial growth was 2.5 mg/dL. Other studies have also reported NH3-N concentrations below 5 mg/ dL without reporting differences in microbial yield [55]. Finally, considering that barley protein is more degradable than corn protein [56,57], the differences in CP degradation detected in the present study could also be explained by the different barley to corn ratio of the diets, this ratio being 0.90, 0.85, 0.79, and 0.72 for CED, SMD, RMD, and CMD, respectively.

The results for digestibility, rumen fermentation and nitrogen flow recorded in CMD were similar to those found in diets formulated with standard proteins such as SMD and RMD, but with an increase in total VFA concentration. These similar results could be related to the close chemical composition of the main protein sources of these diets, and especially the very similar CP digestibility detected for these diets compared with CED. As has been argued previously in the specific case of the comparison between CMD and CED, the chemical oil extraction process could decrease the protein availability for the rumen microorganisms.

5. Conclusions

In conclusion, although both camelina co-products showed some differences between them in CP degradability and N flows, when used as a main protein source in isocaloric and isonitrogenous diets, neither of them differed from rapeseed meal in coefficients of digestibility, rumen fermentation and nitrogen metabolism. Thus, camelina co-products could be an alternative ingredient for beef cattle diets. Further in vivo research is necessary to confirm these results and to ascertain the suitable level of inclusion of these co-products.

Author Contributions: Conceptualization, L.C., A.F.; methodology, H.S, M.L.-S., L.C., A.F.; formal analysis, H.S, M.L.-S.; investigation, H.S, L.C., A.F.; data curation, H.S, L.C., A.F.; writing—original draft preparation, H.S.; L.C. and A.F. critically reviewed the manuscript. All authors revised and approved the final manuscript.

Acknowledgments: This manuscript had been proofread by Phil Grayston, a native English-speaking instructor.

References

1. Zagorakis, K.; Liamadis, D.; Milis, C.; Dotas, V.; Dotas, D. Nutrient digestibility and in situ degradability of alternatives to soybean meal protein sources for sheep. *Small Rumin. Res.* **2015**, *124*, 38–44. [CrossRef]
2. Putnam, D.H.; Budin, J.T.; Filed, L.A.; Breene, W.M. A promising low input oilseed. *New Crops. J.* **1993**, 314–322.
3. Wittkop, B.; Snowdon, R.J.; Friedt, W. Status and perspectives of breeding for enhanced yield and quality of oilseed crops for Europe. *Euphytica* **2009**, *170*, 131–140. [CrossRef]
4. Fröhlich, A.; Rice, B. Evaluation of *Camelina sativa* oil as a feedstock for biodiesel production. *Ind. Crop. Prod.* **2005**, *21*, 25–31. [CrossRef]
5. Moser, B.R. Camelina (*Camelina sativa* L.) oil as a biofuels feedstock: Golden opportunity or false hope? *Lipid Technol.* **2010**, *22*, 270–273. [CrossRef]
6. Shonnard, D.R.; Williams, L.; Kalnes, T.N. Camelina-derived jet fuel and diesel: Sustainable advanced biofuels. *Environ. Prog. Sustain. Energy* **2010**, *29*, 382–392. [CrossRef]
7. Waraich, E.A.; Ahmed, Z.; Ahmad, R.; Yasin Ashraf, M.; Saifullah Naeem, M.S.; Rengel, Z. *Camelina sativa*, a climate proof crop, has high nutritive value and multiple-uses: A review. *Aust. J. Crop Sci.* **2013**, *7*, 1551–1559.
8. Hurtaud, C.; Peyraud, J.L. Effects of Feeding Camelina (Seeds or Meal) on Milk Fatty Acid Composition and Butter Spreadability. *J. Dairy Sci.* **2007**, *90*, 5134–5145. [CrossRef]
9. Zubr, J. Oil-seed crop: *Camelina sativa*. *Ind. Crop. Prod.* **1997**, *6*, 113–119. [CrossRef]

10. Betz, J.M.; Fox, W.D. High-performance liquid chromatographic determinations of glucosinolates in *Brassica* vegetables. Food Phytochemicals I: Fruits and vegetables. In *Food Phytochemicals for Cancer Prevention I*; American Chemical Society: Washington, DC, USA, 1994; pp. 181–196.
11. Tripathi, M.K.; Mishra, A.S. Glucosinolates in animal nutrition: A review. *Anim. Feed Sci. Technol.* **2007**, *132*, 1–27. [CrossRef]
12. Moriel, P.; Nayigihugu, V.; Cappellozza, B.I.; Gonçalves, E.P.; Krall, J.M.; Foulke, T.; Hess, B.W. Camelina meal and crude glycerin as feed supplements for developing replacement beef heifers. *J. Anim. Sci.* **2011**, *89*, 4314–4324. [CrossRef] [PubMed]
13. Cappellozza, B.I.; Cooke, R.F.; Bohnert, D.W.; Cherian, G.; Carroll, J.A. Effects of camelina meal supplementation on ruminal forage degradability, performance, and physiological responses of beef cattle. *J. Anim. Sci.* **2012**, *90*, 4042–4054. [CrossRef] [PubMed]
14. Bonjean, A.; Le Goffic, F. La caméline—*Camelina sativa* (L.) Crantz: Une opportunité pour l'agriculture et l'industrie Européennes. *Oléagineux Corps Gras Lipides* **1999**, *6*, 28–34.
15. Lawrence, R.D.; Anderson, J.L.; Clapper, J.A. Evaluation of camelina meal as a feedstuff for growing dairy heifers. *J. Dairy Sci.* **2016**, *99*, 6215–6228. [CrossRef] [PubMed]
16. Ferret, A.; Calsamiglia, S.; Bach, A.; Devant, M.; Fernández, C.; García-Rebollar, P. *Necesidades Nutricionales Para Rumiantes de Cebo. Normas FEDNA*; Ed Fundación Española para el Desarrollo de la Alimentación Animal: Madrid, Spain, 2008.
17. AOAC. *Official Methods of Analysis*, 15th ed.; Association of Official Analytical Chemists: Arlington, VA, USA, 1990.
18. Van Soest, J.P.; Robertson, J.B.; Lewis, B.A. Methods of dietary fiber, neutral detergent fiber, and nonstarch polysaccharides in relation to animal nutrition. *J. Dairy Sci.* **1991**, *71*, 1587–1597. [CrossRef]
19. American Oil Chemists' Society. *Official Methods and Recommended Practices of the AOCS*; American Oil Chemists' Society: Champaign, IL, USA, 1998.
20. Tilley, J.M.A.; Terry, R.A. A two-stage technique for the in vitro digestion of forage crops. *J. Br. Grassl. Soc.* **1963**, *18*, 104–111. [CrossRef]
21. Stern, M.D.; Endres, M.I. *Laboratory Manual*; Department of Animal Science, University of Minnesota: Heights, MN, USA, 1991; pp. 90–92.
22. McDougall, E.I. Studies on ruminant saliva. 1. The composition and output of sheep's saliva. *Biochem. J.* **1948**, *43*, 99–109. [CrossRef]
23. Orskov, E.R.; Mcdonald, I. The estimation of protein degradability in the rumen from incubation measurements weighted according to rate of passage. *J. Agric. Sci.* **1979**, *92*, 499–503. [CrossRef]
24. INRA. *Nutrition of Cattle, Sheep and Goats: Animal Needs-Values of Feeds*; Quae Editions: Paris, France, 2002.
25. Hoover, W.H.; Crooker, B.A.; Sniffen, C.J. Effects of Differential Solid-Liquid Removal Rates on Protozoa Numbers in Continous Cultures of Rumen Contents. *J. Anim. Sci.* **1976**, *43*, 528–534. [CrossRef]
26. Weller, R.A.; Pilgrim, A.F. Passage of Portozoa and volatile fatty acids from the rumen of the sheep and from a continuos in vitro fermentation system. *Br. J. Nutr.* **1974**, *32*, 341. [CrossRef]
27. Whitehouse, N.L.; Olson, V.M.; Schwab, C.G.; Chesbro, W.R.; Cunningham, K.D.; Lykos, T. Improved techniques for dissociating particle-associated mixed ruminal microorganisms from ruminal digesta solids. *J. Anim. Sci.* **1994**, *72*, 1335–1343. [CrossRef] [PubMed]
28. Stern, M.D.; Hoover, W.H. The dual flow continuous culture system. In Proceedings of the Continuous Culture Fermenters: Frustation or fermentation, Northeast ADSA-ASAS Regional meeting, Chazy, NY, USA, 8 July 1990; pp. 17–32.
29. Chaney, A.L.; Marbach, E.P. Modified reagents for determination of urea and ammonia. *Clin. Chem.* **1962**, *8*, 130–132. [PubMed]
30. Jouany, J.P. Volatile fatty acids and alcohol determination in digestive contents, silage juice, bacterial cultures and anaerobic fermentor contents. *Sci. Aliment.* **1982**, *2*, 131–144.
31. Balcells, J.; Guada, J.; Peiró, J.M.; Parker, D.S. Simultaneous determination of allantoin and oxypurines in biological fluids by high performance liquid chromatography. *J. Chromatogr.* **1992**, *575*, 153–157. [CrossRef]
32. NRC. *Nutrient Requirements of Dairy Cattle*, 8th ed.; National Academy of Sciences: Washington, DC, USA, 2016.
33. Liu, Y.G.; Steg, A.; Hindle, V.A. Rumen degradation and intestinal digestion of crambe and other oilseed by-products in dairy cows. *Anim. Feed. Sci. Technol.* **1994**, *45*, 397–409. [CrossRef]

34. Moss, A.R.; Givens, D.I. The Chemical-Composition, Digestibility, Metabolizable Energy Content and Nitrogen Degradability of Some Protein-Concentrates. *Anim. Feed. Sci. Technol.* **1994**, *47*, 335–351. [CrossRef]
35. Heendeniya, R.G.; Christensen, D.A.; Maenz, D.D.; McKinnon, J.J.; Yu, P. Protein fractionation byproduct from canola meal for dairy cattle. *J. Dairy Sci.* **2012**, *95*, 4488–4500. [CrossRef]
36. Wulf, M.; Südekum, K.H. Effects of chemically treated soybeans and expeller rapeseed meal on in vivo and in situ crude fat and crude protein disappearance from the rumen. *Anim. Feed. Sci. Technol.* **2005**, *118*, 215–227. [CrossRef]
37. Prestløkken, E. In situ ruminal degradation and intestinal digestibility of dry matter and protein in expanded feedstuffs. *Anim. Feed. Sci. Technol.* **1999**, *77*, 1–23. [CrossRef]
38. Maxin, G.; Ouellet, D.R.; Lapierre, H. Ruminal degradability of dry matter, crude protein, and amino acids in soybean meal, canola meal, corn, and wheat dried distillers grains. *J. Dairy Sci.* **2013**, *96*, 5151–5160. [CrossRef]
39. Lawrence, R.D.; Anderson, J.L. Ruminal degradation and intestinal digestibility of camelina meal and carinata meal compared with other protein sources. *Prof. Anim. Sci.* **2018**, *34*, 10–18. [CrossRef]
40. Batajoo, K.K.; Shaver, R.D. In situ dry matter, crude protein, and starch degradabilities of selected grains and by-product feeds. *Anim. Feed Sci. Technol.* **1998**, *71*, 165–176. [CrossRef]
41. Brandao, V.L.N.; Silva, L.G.; Paula, E.M.; Monteiro, H.F.; Dai, X.; Lelis, A.L.J.; Faccenda, A.; Poulson, S.R.; Faciola, A.P. Effect of replacing canola meal with solvent-extracted camelina meal on microbial fermentation in a dual-flow continuous culture system. *J. Dairy Sci.* **2018**, *101*, 9028–9040. [CrossRef] [PubMed]
42. Paula, E.M.; Monteiro, H.F.; Silva, L.G.; Benedeti, P.D.B.; Daniel, J.L.P.; Shenkoru, T.; Faciola, A.P. Effects of replacing soybean meal with canola meal differing in rumen-undegradable protein content on ruminal fermentation and gas production kinetics using 2 in vitro systems. *J. Dairy Sci.* **2017**, *100*, 5281–5292. [CrossRef] [PubMed]
43. McDonald, P. *Animal Nutrition*, 6th ed.; Pearson education: Edinburg, UK, 2002; pp. 84–215.
44. Bryant, M.P. Nutritional Requirements of the Predominant Rumen Cellulolytic Bacteria. *Fed. Proc.* **1973**, *32*, 1809–1813. [PubMed]
45. Huhtanen, P.; Hetta, M.; Swensson, C. Evaluation of canola meal as a protein supplement for dairy cows: A review and a meta-analysis. *Can. J. Anim. Sci.* **2011**, *91*, 529–543. [CrossRef]
46. NRC. *Nutrient Requirements of Dairy Cattle*, 7th ed.; National Academy of Sciences: Washington, DC, USA, 2001.
47. Brito, A.F.; Broderick, G.A. Effects of Different Protein Supplements on Milk Production and Nutrient Utilization in Lactating Dairy Cows. *J. Dairy Sci.* **2007**, *90*, 1816–1827. [CrossRef]
48. Brito, A.F.; Broderick, G.A.; Reynal, S.M. Effects of Different Protein Supplements on Omasal Nutrient Flow and Microbial Protein Synthesis in Lactating Dairy Cows. *J. Dairy Sci.* **2007**, *90*, 1828–1841. [CrossRef]
49. Santos, F.A.P.; Santos, J.E.P.; Theurer, C.B.; Huber, J.T. Effects of rumen-undegradable protein on dairy cow performance: A12-year literature review. *J. Dairy Sci.* **1998**, *81*, 3182–3213. [CrossRef]
50. Ipharraguerre, I.R.; Clark, J.H. Impacts of the Source and Amount of Crude Protein on the Intestinal Supply of Nitrogen Fractions and Performance of Dairy Cows. *J. Dairy Sci.* **2005**, *88*, E22–E37. [CrossRef]
51. Hoover, W.H.; Stokes, S.R. Balancing carbohydrates and proteins for optimum rumen microbial yield. *J. Dairy Sci.* **1991**, *74*, 3630–3644. [CrossRef]
52. Satter, L.D.; Slyter, L.L. Effect of ammonia concentration on rumen microbial protein production in vitro. *Br. J. Nutr.* **1974**, *32*, 199. [CrossRef] [PubMed]
53. Russell, J.B.; Sniffen, C.J.; Van Soest, J. Effect of carbohydrate limitation on degradation and utilization of casein by mixed rumen bacteria. *J. Dairy Sci.* **1983**, *66*, 763–775. [CrossRef]
54. Owens, F.N.; Bergen, W.G. Nitrogen metabolsim of ruminant animals: Historical Perspective, Current Understanding and Future Implications. *J. Anim. Sci.* **1983**, *57*, 498–518. [PubMed]
55. Bach, A.; Stern, M.D. Effects of different levels of methionine and ruminally undegradable protein on the amino acid profile of effluet from continuous culture feremeters. *J. Anim. Sci.* **1999**, *77*, 3377–3384. [CrossRef]
56. Spicer, L.A.; Theurer, C.B.; Sowe, J.; Noon, T.H. Ruminal and post-ruminal utilization of nitrogen and starch from sorghum grain -corn- and barley-based diets by beef steers. *J. Anim. Sci.* **1986**, *62*, 521. [CrossRef]
57. Herrera-Saldana, R.E.; Huber, J.T.; Poore, M.H. Dry Matter, Crude Protein, and Starch Degradability of Five Cereal Grains. *J. Dairy Sci.* **1990**, *73*, 2386–2393. [CrossRef]

5

The Role of Condensed Tannins in the In Vitro Rumen Fermentation Kinetics in Ruminant Species

Ives C. S. Bueno [1], Roberta A. Brandi [1], Gisele M. Fagundes [2,*], Gabriela Benetel [1] and James Pierre Muir [3]

1. Department of Animal Science, University of São Paulo, Pirassununga, São Paulo 1365-900, Brazil; ivesbueno@usp.br (I.C.S.B.); robertabrandi@usp.br (R.A.B.); gabriela.benetel@usp.br (G.B.)
2. Department of Animal Science, Federal University of Roraima—UFRR, BR 174, km 12, Boa Vista, Roraima 69300-000, Brazil
3. Texas A&M AgriLife Research, Texas A&M University, Stephenville, TX 76401, USA; Jim.Muir@ag.tamu.edu

* Correspondence: gisele.fagundes@ufrr.br;

Simple Summary: Inoculum from different feeding types of the ruminant species host has unequal tolerance and effects to condensed tannin (CT) due to their respective feeding strategies behavior producing different ruminal microbiota profiles. This paper describes that in long term incubation, CT plant extract addition affects in vitro fermentation kinetics more severely in grazing ruminant than browsing ruminants.

Abstract: Animal feeding behavior and diet composition determine rumen fermentation responses and its microbial characteristics. This study aimed to evaluate the rumen fermentation kinetics of domestic ruminants feeding diets with or without condensed tannins (CT). Holstein dairy cows, Nelore beef cattle, Mediterranean water buffalo, Santa Inês sheep and Saanen goats were used as inoculum donors (three animals of each species). The substrates were maize silage (*Zea mays*), fresh elephant grass (*Pennisetum purpureum*), Tifton-85 hay (*Cynodon* spp.) and fresh alfalfa (*Medicago sativa*). Acacia (*Acacia molissima*) extract was used as the external CT source. The in vitro semi-automated gas production technique was used to assess the fermentation kinetics. The experimental design was completely randomized with five inoculum sources (animal species), four substrates (feeds) and two treatments (with or without extract). The inclusion of CT caused more severe effects in grazing ruminants than selector ruminants.

Keywords: condensed tannins; fermentability; gas production; grazing ecology; ruminant; microbial responses

---

1. Introduction

Tropical shrubs and trees are important feed for livestock because they are sources of protein [1], minerals and vitamins as well as playing important roles in ruminant feeding systems. However, many of these plant species have secondary compounds capable of changing the utilization of nutrients by mammalian herbivores.

Because plants developed defense mechanisms against herbivores and pathogens, animals have developed mechanisms to nullify or restrict the toxic and negative effects of ingested plant secondary compounds such as condensed tannins (CT) [2,3]. Ruminant herbivores and plant CT coexist and adapt natural evolutionary processes. Some ruminant feeders, especially goats, developed physiological adaptations, and even dependence on CT-rich legumes, selectively including such plants in their selector habits [4]. In the case of some browsers such as goats, this adaptation takes the form of salivary glands that produce large amounts of mucus-containing enzymes that can bind to CT to increase palatability and leaving plant proteins more available for digestion [5].

The evolution of different feeding strategies among domestic ruminant species implies differing microbial interactions with CT and, consequently, the diversity of rumen microorganisms and digestive capacity. Thus, we hypothesize that the role of CT on ruminal fermentability may vary depending on the species of the ruminant and, in the case of in vitro fermentation, the species of the rumen fluid donor. Comparative studies with different rumen-fluid donor species with previous pre-adapted to tannin feeding have been published [6–9]. However, the ability of the rumen microbes of different non-CT-adapted ruminant species to adapt to CT has not been fully investigated. Therefore, the objective of our study was to compare the effects of plant extracts containing CT on ruminal fermentation kinetics of taurine dairy cattle (*Bos taurus taurus*), zebu beef cattle (*Bos taurus indicus*), water buffaloes (*Bubalus bubalis*), sheep (*Ovis aries*) and goats (*Capra hircus*) without previous CT feeding exposure.

2. Material and Methods

2.1. Animals

Rumen fluid donors included three Holstein dairy cows (*B. taurus taurus*), three Nelore beef cows (*B. taurus indicus*), three Mediterranean water buffalo cows, three Santa Inês ewes and three Saanen goats. Diets were formulated to meet the nutritional requirements of each animal species/breed and consisted of 60% to 80% forage and 20% to 40% concentrate with ground maize and soybean meal. All animals were allowed free access to water and mineral mixture. All methods and animal care were performed in accordance with the relevant guidelines and regulations of the Ethic Committee on Animal Use of the School of Animal Science and Food Engineering, (São Paulo University).

2.2. Substrates

Four forages were evaluated as substrates: maize silage (*Zea mays*), fresh elephant grass (*Pennisetum purpureum*), Tifton-85 hay (*Cynodon* spp.) and fresh alfalfa (*Medicago sativa*). Forage samples were dried at 55 °C and ground through a 1-mm sieve. All were analyzed (Table 1) for dry matter (DM), mineral matter (MM), crude protein (CP) and acid-detergent fiber (ADF) with residual ash according to AOAC [10]. Neutral-detergent fiber (NDF) was estimated according to the methodology described by Mertens [11]. Total phenol (TP) concentrations were determined by the Folin-Ciocalteau reagent method [12] and total tannins (TT) were estimated as the difference in TP concentration before and after the treatment with insoluble polyvinylpolypirrolidone [13], using tannic acid as standard. Condensed tannin concentrations were determined by the butanol-HCl method [12], using leucocyanidin as a standard.

Table 1. Chemical composition of substrates used during the assay of in vitro fermentation kinetics.

Composition	Substrates (1)				
	ALF	**ELE**	**TIF**	**SIL**	**ACA**
organic matter (2)	916.82	897.75	936.45	964.64	978.86
ether extract (2)	84.04	46.72	57.92	62.98	n.d. (5)
crude protein (2)	278.97	60.28	158.02	82.02	n.d. (5)
neutral-detergent fiber (2)	735.62	770.03	795.29	563.28	n.d. (5)
acid-detergent fiber (2)	510.25	519.52	428.92	332.30	n.d. (5)
acid-detergent lignin (2)	126.69	121.63	133.08	71.35	n.d. (5)
total phenols (3)	13.60	5.47	5.32	10.18	558.63
total tannins (3)	8.14	3.05	2.82	6.58	519.58
condensed tannins (4)	0.25	0.10	0.10	0.15	235.87

(1) ALF: fresh alfalfa; ELE: fresh elephant grass; TIF: Tifton-85 hay; SIL: maize silage; ACA: Acacia tannin extract. (2) expressed as g/kg DM. (3) expressed as eq-g tannic acid/kg DM. (4) expressed as eq-g leucocyanidin/kg DM. (5) n.d.: not determined.

Acacia (*Acacia molissima*) extract (Seta® Estância Velha, Brazil) was added to the diets to raise the CT concentration to 50 eq-g of leucocyanidin per kg of feed DM. This CT concentration has been appointed as the minimum to cause harmful effects to ruminants [14]. The chemical characterization of substrates and *Acacia* extract were performed at the Laboratory of Animal Nutrition, Center for Nuclear Energy in Agriculture, University of São Paulo.

2.3. In Vitro Gas Production Assay

Ruminal contents were collected through permanent ruminal cannulas from each animal. Equal volumes of liquid and solid phases sample were homogenized in a blender for 10 s. The resulting material was filtered through three layers of cotton (cheese cloth) tissue [15]. Filtered fractions were kept in a water bath at 39 °C and CO_2 saturation until introduced into the in vitro system.

The in vitro gas production described by Theodorou et al. [16] and Maurício et al. [17] was used to compare disappearance rates. Each 160-mL fermentation flask received 0.5 g of the substrate with or without 0.3 g of extract CT. The inoculation was performed by injecting 10 mL of inoculum in 90 mL of buffered mineral solution [16] into each fermentation flask. The flasks were sealed with rubber stoppers, then shaken and incubated in an oven with forced-air circulation at 39 °C, thus allowing gas accumulation within each flask.

The pressure of the generated gases was measured at 4, 8, 12, 16, 24, 30, 46, 59, 72 and 96 h after inoculation, using a "transducer" (PressData 800®). After each measuring, the accumulated gas was released from all bottles. These values were used to calculate the volume of gas produced. At the end of this bioassay (96 h), 2 mL of liquid phase were sampled with a syringe and frozen until analysis of short-chain fatty acids (SCFA). After 96 h, residual material was filtered through sintered crucibles to determine in vitro dry matter degradability (IVDMD) and in vitro organic matter degradability (IVOMD). The partitioning factor (PF), calculated by relating DM degradation and OM degradation to total gas production, was used to compare microbial efficiency [18].

Gas production data were used to determine fermentation kinetics based on the model of Ørskov and McDonald [19] modified by McDonald [20] as:

$$p = 0 \text{ to } t < t_0 \quad (1)$$

$$p = a + b\,(1 - exp^{-ct}) \text{ to } t \geq t_0 \quad (2)$$

where p is gas production (ml) in time t; a and b are constants of the model; c is the gas production rate (h^{-1}); $a + b$ is the potential gas production (mL); t_0 is the lag time (h).

2.4. Short-Chain Fatty Acids Determination

Short-chain fatty acid analysis was measured by gas chromatography (GC-2014, Shimadzu®, Tokyo, Japan), split-injector, flame ionization detector and capillary column (Stabilwax®, Restek, Bellefonte, PA, USA) at 145 °C (isothermal) according to Erwin et al. (1961) [21], with adaptations by Getachew et al. [22]. Acetic, propionic, isobutyric, butyric, isovaleric and valeric acid (99.5% purity, Chem service, USA) were used as a quantitative external standard. The operational conditions were: injector and detector temperatures were 250 °C; helium was the carrier gas at 8.01 mL/min; hydrogen flow to the flame jet at 60 kPa and synthetic air at 40 kPa. The samples were thawed at room temperature and centrifuged at 14,500× g for 10 min. The supernatant (800 μL) was transferred to a dry and clean flask with 200 μL formic acid (98–100%) and 100 μL of the internal standard (100 mM 2-ethyl butyric acid, Chem service, USA).

2.5. Experimental Design and Statistical Analysis

The experiment tested three factors and their interactions. These included four forage substrates, five sources of inoculum (animal species) and two levels of CT extract. Individuals ($n = 3$) of each species constituted the experimental units for three replications.

The statistical design was completely randomized, according to the following model:

$$Y_{ijk} = \mu + F_i + S_j + T_k + (F_i \times S_j) + (S_j \times T_k) + (F_i \times T_k) + e_{ijk} \quad (3)$$

where Y_{ijk} is the dependent variable; μ the overall mean; F the effect of feeds (substrates) (i = 1 to 4); S the effect of animal species (j = 1 to 5); T the effect of CT (k = 1 to 2); F × S the interaction of feeds and animal species; S × T the interaction of animal species and CT; F×T the interaction of feeds and CT; e_{ijk} the residual error of the model. Results were compared by a Tukey test, using the software SAS for Windows® [23]. Results were considered different at $p \leq 0.05$.

3. Results

3.1. Tannin Extract Effects

There was an effect ($p \leq 0.001$) of CT inclusion on IVDMD and IVOMD regardless of the rumen fluid source. The inclusion of CT inhibited fermentation as indicated by the model parameters (Table 2), visualized in Figure 1. The partitioning factor is a measure created to bring together two variables: degradability and gas production. It was greater ($p \leq 0.001$) when CTs were included in the diet (Table 2).

Table 2. Effect of condensed tannins (CT) on degradability, fermentability and microbial efficiency during in vitro rumen fermentation in four ruminant species.

Variables	no CT	CT	SEM (1)	*p*-Value (2)
in vitro DM degradability	0.646 ^{a}	0.466 ^{b}	0.0090	***
in vitro OM degradability	0.644 ^{a}	0.458 ^{b}	0.0095	***
partitioning factor (mg DMD/mL)	3.40 ^{b}	7.68 ^{a}	0.234	***
partitioning factor (mg OMD/mL)	3.14 ^{b}	6.94 ^{a}	0.206	***
Model parameters (3)				
a	−27.39 ^{b}	−9.06 ^{a}	0.816	***
b	281.89 ^{a}	133.47 ^{b}	7.023	***
c	0.0281	0.0258	0.00130	ns
a + b	254.50 ^{a}	124.40 ^{b}	7.097	***
t_0	4.39 ^{a}	3.06 ^{b}	0.149	***

(1) SEM: standard error of means. (2) ns: not significant ($p > 0.05$); *: $p \leq 0.05$; **: $p \leq 0.01$; ***: $p \leq 0.001$. (3) Model of Ørskov and McDonald (1979), modified by McDonald (1981): $p = a + b\,(1 - exp^{-ct})$, where p = gas production (mL), in time t; a and b = constants of model; c = production gas rate (h^{-1}); $a + b$ = potential gas production (mL); t_0 = *lag time* (h) ^{a,b} means followed by distinct superscripts, within rows, are different (Tukey test at $p \leq 0.05$).

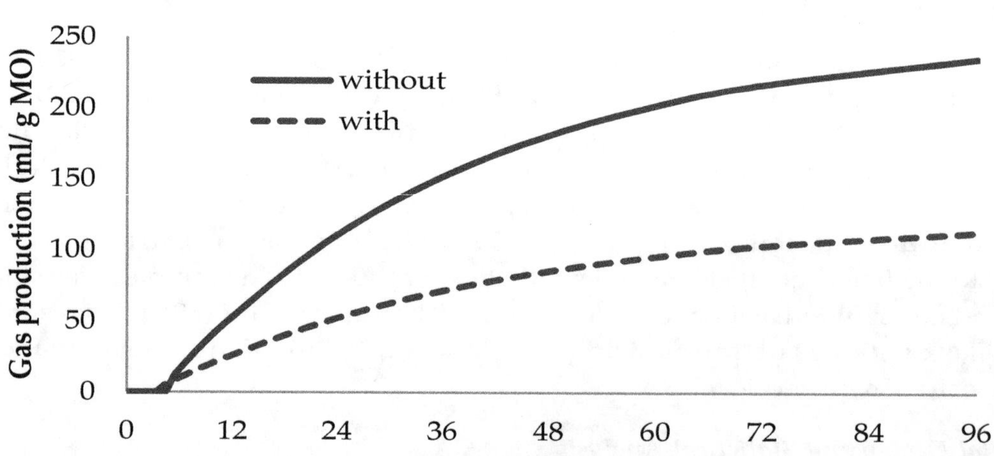

Figure 1. Effects of condensed tannin in diet (with and without) on gas production profiles during in vitro rumen fermentation (pooled over ruminant species).

Short-chain fatty acids production was also affected by the inclusion of CT extracts (Table 3). The addition of CT promoted a greater ($p \leq 0.001$) proportion of propionic acid and less ($p \leq 0.001$) acetic acid.

Table 3. Effect of tannins on short-chain fatty acids (SCFA) production, evaluated during in vitro organic matter rumen fermentation with and without condensed tannins (CT).

Variables	no CT	CT	SEM (1)	*p*-Value (2)
SCFA production (mmol/g OMD)				
acetic acid	6.57 ^{a}	2.83 ^{b}	0.26	***
propionic acid	2.46 ^{a}	1.30 ^{b}	0.09	***
iso-butyric acid	0.11 ^{a}	0.02 ^{b}	0.01	***
butyric acid	1.07 ^{a}	0.42 ^{b}	0.04	***
iso-valeric acid	0.15 ^{a}	0.05 ^{b}	0.01	***
valeric acid	0.19 ^{a}	0.10 ^{b}	0.01	***
total SCFA	10.51 ^{a}	4.69 ^{b}	0.39	***

(1) SEM: standard error of means. (2) ns: not significant ($p > 0.05$); *: $p \leq 0.05$; **: $p \leq 0.01$; ***: $p \leq 0.001$. ^{a,b} means followed by distinct superscripts, within rows, are different (Tukey test at $p \leq 0.05$).

3.2. *Animal Species Effects*

At the end of the 96-h in vitro fermentation, no differences ($p > 0.05$) in degradability and partitioning factor were observed among animals (Table 4). All species degraded similar proportions of feeds (IVDMD and IVOMD from 0.54 to 0.56) with similar microbial efficiency (PF). However, differences ($p \leq 0.01$) among model parameters were observed (Figure 2). The production of acetic, propionic and butyric acids by each unit of the degraded substrate was greater ($p \leq 0.001$) in large than in small ruminants and this was reflected in the total amount of SCFA produced (Table 5). Acetate and butyrate to total acid production were not different.

Table 4. Effect of animal species on degradability, fermentability and microbial efficiency assayed in four forages during in vitro rumen fermentation.

Variable (1)	Animal Species (2)					SEM (3)	*p* Value (4)
	Goats	Sheep	Buffalo	Taurine Cattle	Zebu Cattle		
IVDMD	0.556	0.565	0.543	0.564	0.554	0.014	Ns
IVOMD	0.551	0.560	0.538	0.559	0.547	0.015	Ns
PF (mg DMD/mL)	5.15	4.94	6.19	5.70	5.69	0.37	Ns
PF (mg OMD/mL)	4.65	4.48	5.70	5.31	5.07	0.33	Ns
Model parameters (5)							
A	−15.15 ^{a}	−22.01 ^{c}	−15.92 ^{ab}	−20.92 ^{bc}	−17.13 ^{abc}	1.29	***
B	185.82 ^{b}	197.70 ^{ab}	238.02 ^{a}	194.83 ^{ab}	222.00 ^{ab}	11.10	**
C	0.0348 ^{a}	0.0314 ^{a}	0.0210 ^{b}	0.0267 ^{ab}	0.0209 ^{b}	0.0021	***
a + b	170.67 ^{b}	175.69 ^{b}	222.10 ^{a}	173.91 ^{b}	204.88 ^{ab}	11.22	**
t_0	2.67 ^{b}	3.42 ^{ab}	4.24 ^{a}	4.27 ^{a}	4.03 ^{a}	0.24	***

(1) IVDMD: in vitro dry matter degradability; IVOMD: in vitro organic matter degradability; GP: gas production; PF: partitioning factor. (2) breeds—goats: Saanen; sheep: Santa Inês; buffalo: Mediterranea; taurine cattle: Holstein; zebu cattle: Nelore. (3) SEM: standard error of means. (4) ns: not significant ($p > 0.05$); *: $p \leq 0.05$; **: $p \leq 0.01$; ***: $p \leq 0.001$. (5) Model of Ørskov and McDonald (1979), modified by McDonald (1981): $p = a + b\,(1 - exp^{-ct})$, where p = gas production (mL), in time t; a and b = constants of model; c = production gas rate (h^{-1}); $a + b$ = potential gas production (mL); t_0 = *lag time* (h). ^{a,b,c} means followed by distinct superscripts, within rows, are different (Tukey test at 5%).

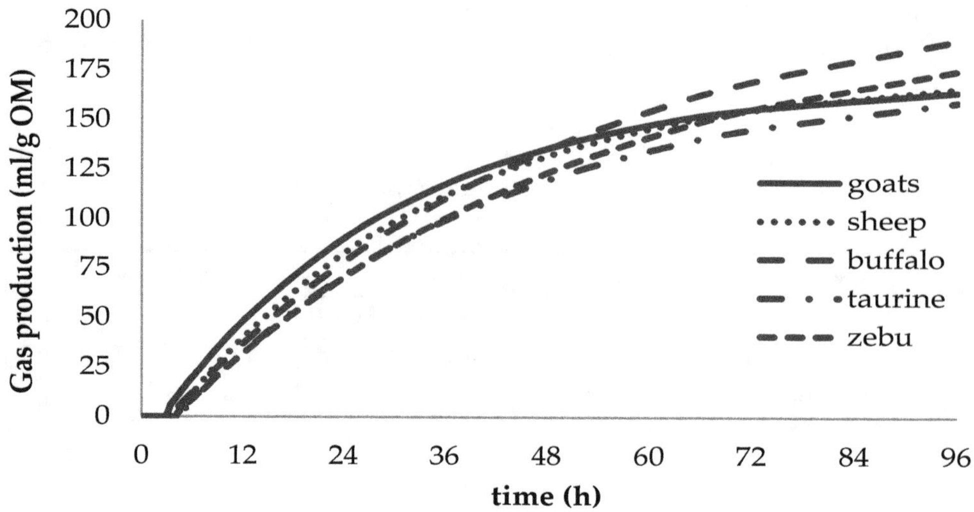

Figure 2. Effects of animal species on gas production profiles during in vitro rumen fermentation (pooled over condensed tannin treatments in four forages).

Table 5. Effect of animal species on short-chain fatty acids (SCFA) production evaluated during in vitro rumen fermentation.

Variables	Animal Species (1)					SEM (2)	p-Value (3)
	Goats	Sheep	Buffalo	Taurine Cattle	Zebu Cattle		
SCFA production (mmol/g OMD)							
acetic acid	3.37 c	3.92 bc	5.44 ab	5.70 a	5.06 ab	0.41	***
propionic acid	1.30 b	1.45 b	2.25 a	2.28 a	2.11 a	0.14	***
iso-butyric acid	0.09 a	0.06 abc	0.07 ab	0.02 c	0.04 bc	0.01	***
butyric acid	0.53 c	0.61 bc	0.90 a	0.88 a	0.82 ab	0.07	***
iso-valeric acid	0.10	0.10	0.10	0.12	0.10	0.01	ns
valeric acid	0.12 bc	0.10c	0.15 abc	0.19 a	0.17 ab	0.02	***
total SCFA	5.41 c	6.21 bc	8.89 a	9.17 a	8.30 ab	0.61	***

(1) breeds—goats: Saanen; sheep: Santa Inês; buffalo: Mediterranean; taurine cattle: Holstein; zebu cattle: Nelore. (2) SEM: standard error of means. (3) ns: not significant ($p > 0.05$); *: $p \leq 0.05$; **: $p \leq 0.01$; ***: $p \leq 0.001$. a,b,c means followed by distinct superscripts, within rows, are different (Tukey test at $p \leq 0.05$).

3.3. Interactions

CT extract was effective ($p \leq 0.001$) in reducing IVDMD and IVOMD in all four ruminant species evaluated (Figure 3). However, there were no differences in degradability among animal species between the paired treatment groups receiving CT and those which did not ($p > 0.05$). Among the treatment combinations including CT, buffalo and cattle showed greater microbial efficiency ($p \leq 0.01$) than sheep or goats (Figure 3). The same was not observed for the partitioning factor by species not receiving CT because all species showed similar values. Condensed tannin addition to the diet increased the partitioning factor in large and small ruminant species.

The non-CT diet showed greater ($p \leq 0.001$) levels of SCFA in cattle and buffalo than in sheep and goats. However, for animals receiving CT, there were no differences ($p > 0.05$) in SCFA between large and small ruminants (Figure 4).

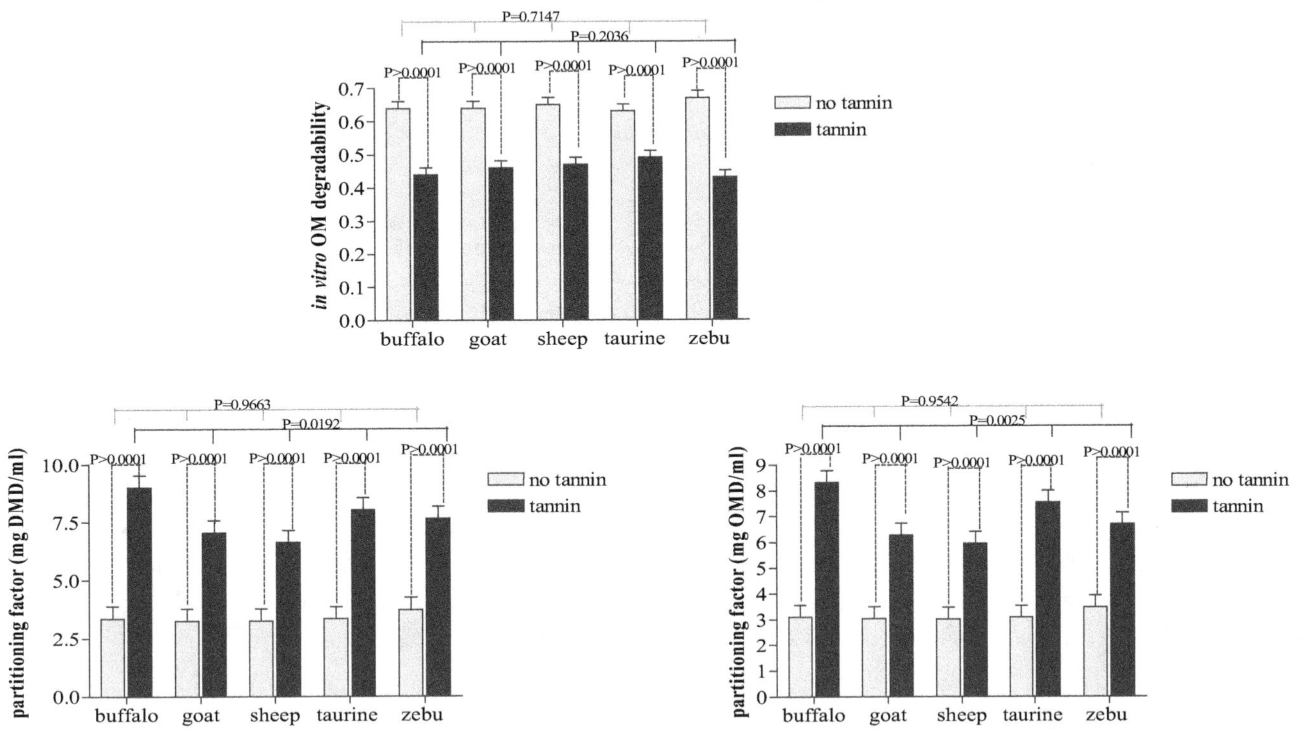

Figure 3. Effects of condensed tannin and animal species interactions on the organic matter degradability and partitioning factor during in vitro rumen fermentation.

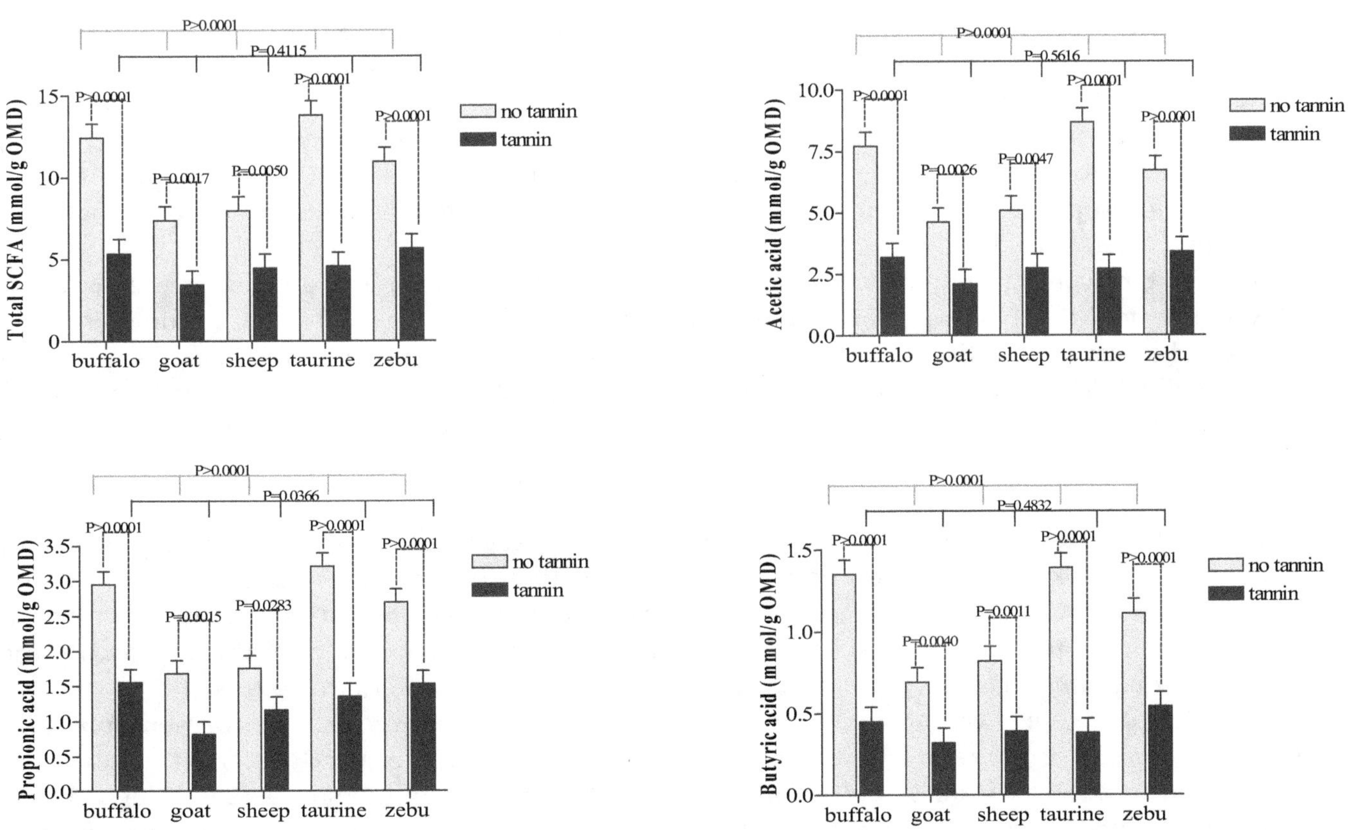

Figure 4. Effects of condensed tannin and animal species interaction on short-chain fatty acids (SCFA) production during in vitro rumen fermentation.

4. Discussion

4.1. Condensedtannin Extract Effects

Flavonoids from *Acacia* spp. have potential biological effects on rumen fermentation [24]; thus, gas production reduction from diets containing CT extract was expected due to its effect on rumen microbiota. The regression equation for *Acacia* CT extract had a slow initial slope (12 h) followed by a slight increase, which allowed for a constant long-term assay up to 96 h. The slow but steady gas production observed throughout the total incubation period reflects the lower ruminal digestion of organic matter in diets containing CT as reported elsewhere in the literature [25,26]. The inhibitory effects of CT on rumen gas production can be explained by a direct effect of CT-protein complexes on the rumen fibrolytic microbe and consequently lower cell wall degradation. The greater SCFA production and organic matter degradability for tannin-free diets also highlights the interference of CT on cell wall degradability. Previous in vitro studies support our findings that CT may reduce or slow organic matter degradability in the rumen, as reflected in less total SCFA production [27–29].

The addition of CT resulted in greater partitioning factors, expressed as mg OM degraded per mL of produced gases. The greater partitioning factor observed in this trial could be due to the initial washing losses by a substrate that might provide greater values of organic matter degradability in poorly substrate degradation samples with CT and therefore with less gas production. However, this assumption needs to be confirmed by future in vitro trials.

4.2. Animal Species Effects

Despite their different feeding behavior and digestive capacity, there were similar nutrient degradability and microbial efficiencies among ruminant species studied, which can be explained by the prolonged incubation period of 96 h. The incubation duration used in our study may have allowed the adaptation of the rumen bacterial community to diets, eventually resulting in equivalent substrate fermentation. Contrary to our observations, Calabrò et al. [30] found differences in rumen fermentation between buffaloes and cattle; however, we believe that differences in microbial ecology and their donors probably are strongly reflected in short term incubation [31]. This emphasizes the importance of appropriate incubation duration in studies comparing substrates exposed to inoculants from donors fed different diets.

Due to better fiber digestion capacity, buffaloes, taurine and zebu cattle produced greater amounts of SCFA (per mmol MOD) than small ruminants, regardless of donor diet. Within the ruminant species, goats showed lower production of the major SCFA, including acetic, propionic and butyric acids as well as total SCFA. These results confirm the lower ruminal and omasal capacity and its less efficient degradability of poor quality forages by small-bodied browsers [2,3].

4.3. Interactions

The evolution of different feeding behaviors in ruminant species has required anatomic, physiologic and microbial gastrointestinal adaptations to their respective dietary niches [2,32]. In accordance with this hypothesis, Clauss et al. [33] reported clear effects of the ecology of wild ruminants on the dynamics of their protozoal fauna. In this context, the interaction between tanniniferous plants and browsers likely resulted in greater tolerance to flavonoid compounds than in grazers.

The major objective of our study was to assess possible CT-tolerant microbes present in rumen fluid from browsers or grazers. As a result, our 96-h incubation produced different results than those reported for 24 h by Bueno et al. [31], with CT addition resulting in similar total degradation of substrates in vitro during rumen fermentation among grazing and browsing ruminants. In other words, CT was effective in reducing cell wall degradability rates and SCFA production in all species evaluated. This finding suggests that an incubation time of 96 h was insufficient to alter rumen microbial dynamics associated with changing diets, resulting in an incomplete adaptation of ruminal microbial community to the new diet.

Comparisons across CT treatment groups revealed greater propionic acid production and microbial efficiency in buffalo compared to cattle/taurine. Both results suggest that microbial populations in buffalo appear to be less sensitive to dietary CT. Although this contradicts most published research on the topic, it agrees with results reported by Salem [34] that buffalos were more tolerant of CT than other large grazers.

Despite the few differences observed as a result of our longer assay duration, results reported from in vivo animal trials must consider factors other than simply diet. These include anatomical adaptions of the mouth, teeth, salivary glands, body mass and digestive system altering selectivity, forage intake, passage rate and rumination time among the different feeding categories. As a result, degradability trials should account for differentiated rumen fermentation responses in browsers and grazers and selective versus bulk feeders when assaying CT-rich substrates.

5. Conclusions

The initial differences in the microbial communities resulting from feeding of donor species provide different responses between large and small ruminants in vitro tannin-rich diet fermentation. Inoculum from sheep and goats is less affected by the addition of CT than buffalo, zebu and taurine.

Author Contributions: Designed the study I.C.S.B.; performed the experiment and collected the samples, R.A.B. and I.C.S.B.; performed the laboratory analyses, G.M.F., G.B., and J.P.M.; analyzed and interpreted the data, G.M.F., R.A.B. and I.C.S.B.; wrote the manuscript: G.M.F. All authors have read and agreed to the published version of the manuscript.

Acknowledgments: The authors acknowledge the financial support of Fundação de Amparo à Pesquisa do Estado de São Paulo (FAPESP), Cristina Teresa Alves (Embrapa, São Carlos, Brazil), Francisco Palma Rennó (FMVZ-Pirassununga, Brazil) and Paulo Henrique Mazza Rodrigues (FMVZ, Pirassununga, Brazil), who kindly collaborated with this study, and Seta® (Estância Velha, Brazil) for the CT extract.

References

1. Fagundes, G.M.; Modesto, E.C.; Fonseca, C.E.M.; Lima, H.R.P.; Muir, J.P. Intake, digestibility and milk yield in goats fed *Flemingia macrophylla* with or without polyethylene glycol. *Small Rumin. Res.* **2014**, *116*, 88–93. [CrossRef]
2. Van Soest, P.J. *Nutritional Ecology of the Ruminant*, 2nd ed.; Cornell University Press: Ithaca, NY, USA, 1994; 476p.
3. Lamy, E.; Rawel, H.; Schweigert, F.J.; Capela e Silva, F.; Ferreira, A.; Costa, A.R.; Antunes, C.; Almeida, A.M.; Coelho, A.V.; Sales-Baptista, E. The effect of tannins on Mediterranean ruminant ingestive behavior: The role of the oral cavity. *Molecules* **2011**, *16*, 2766–2784. [CrossRef] [PubMed]
4. Muir, J.P. The multi-faceted role of condensed tannins in the goat ecosystem. *Small Rumin. Res.* **2011**, *98*, 115–120. [CrossRef]
5. Hofmann, R.R. Evolutionary steps of ecophysiological adaptation and diversification of ruminants: A comparative view of their digestive system. *Oecologia* **1989**, *78*, 443–457. [CrossRef] [PubMed]
6. Osawa, R.O.; Sly, L.I. Occurence of tannin-protein complex degrading Streptococcus sp. in feces of various animals. *Syst. Appl. Microbiol.* **1992**, *15*, 144–147. [CrossRef]
7. Odenyo, A.A.; McSweeney, C.S.; Palmer, B.; Negassa, D.; Osuji, P.O. In vitro screening of rumen fluid samples from indigenous African ruminants provides evidence for rumen fluid with superior capacities to digest tannin-rich fodders. *Aust. J. Agric. Res.* **1999**, *50*, 1147–1157. [CrossRef]
8. Jones, R.J.; Meyer, J.H.F.; Bechaz, F.M.; Stolzt, M.A.; Palmer, B.; van der Merwe, G. Comparison of rumen fluid from South African game species and from sheep to digest tanniniferous browse. *Aust. J. Agric. Res.* **2001**, *52*, 453–460. [CrossRef]

9. Odenyo, A.A.; Bishop, R.; Asefa, G.; Jamnadass, R.; Odongo, D.; Osuji, P.O. Characterization of tannin-tolerant bacterial isolates from East African ruminants. *Anaerobe* **2001**, *7*, 5–156. [CrossRef]

10. Association of Official Analytical Chemists (AOAC). *Official Methods of Analysis*, 16th ed.; AOAC: Arlington, VA, USA, 1995.

11. Mertens, D.R. Gravimeric determination of amylase-treated neutral detergent fibre in feeds with refluxing beakers or crucibles: Collaborative study. *J. Assoc. Off. Anal. Chem.* **2002**, *85*, 1217–1240.

12. Makkar, H.P. *Quantification of Tannins in Tree and Shrub Foliage: A Laboratory Manual*; Springer: Berlin/Heidelberg, Germany, 2003.

13. Makkar, H.P.S.; Blümmel, M.; Borowy, N.K.; Becker, K. Gravimetric determination of tannins and their correlations with chemical and protein precipitation methods. *J. Sci. Food Technol.* **1993**, *61*, 161–165. [CrossRef]

14. Minho, A.P.; Bueno, I.C.S.; Louvandini, H.; Jackson, F.; Gennari, S.M.; Abdalla, A.L. Effect of *Acacia molissima* tannin extract on the control of gastrointestinal parasites in sheep. *Anim. Feed Sci. Technol.* **2008**, *147*, 172–181. [CrossRef]

15. Bueno, I.C.S.; Cabral Filho, S.L.S.; Gobbo, S.P.; Louvandini, H.; Vitti, D.M.S.S.; Aballa, A.L. Influence of inoculum source in a gas production method. *Anim. Feed Sci. Technol.* **2005**, *123–124*, 96–105. [CrossRef]

16. Theodorou, M.K.; Williams, B.A.; Dhanoa, M.S.; McAllan, A.B.; France, J. A simple gas production method using a pressure transducer to determine the fermentation kinetics of ruminant feeds. *Anim. Feed Sci. Technol.* **1994**, *48*, 185–197. [CrossRef]

17. Mauricio, R.M.; Mould, F.L.; Dhanoa, M.S.; Owen, E.; Channa, K.S.; Theodorou, M.K. A semi-automated in vitro gas production technnique for ruminant feedstuff evaluation. *Anim. Feed Sci. Technol.* **1999**, *79*, 321–330. [CrossRef]

18. Blümmel, M.; Makkar, H.P.S.; Becker, K. In vitro gas production: A technique revisited. *J. Anim. Physiol. Anim. Nutr.* **1997**, *77*, 24–34. [CrossRef]

19. Ørskov, E.R.; Mcdonald, I. The estimation of protein degradability in the rumen from incubation measurements weighted according to rate of passage. *J. Am. Sci.* **1979**, *92*, 449–453. [CrossRef]

20. Mcdonald, I. A revised model for estimation of protein degradability in the rumen. *J. Agric. Sci.* **1981**, *96*, 251–252. [CrossRef]

21. Erwin, E.S.; Marco, G.J.; Emery, E.M. Volatile fatty acid analyses of blood and rumen fluid by gas chromatography. *J. Dairy Sci.* **1961**, *44*, 1768–1771. [CrossRef]

22. Getachew, G.; Makkar, H.P.S.; Becker, K. Tropical browses: Contents of phenolic compounds, in vitro gas production and stoichiometric relationship between short chain fatty acid and in vitro gas production. *J. Agric. Sci.* **2002**, *139*, 341–352. [CrossRef]

23. SAS INSTITUTE. *The SAS System for Windows*; Release 8.01; Sas Institute: Cary, NC, USA, 2000.

24. Bueno, I.C.S.; Vitti, D.M.; Louvandini, H.; Abdalla, A.L. A new approach for in vitro bioassay to measure tannin biological effects based on a gas production technique. *Anim. Feed Sci. Technol.* **2008**, *141*, 153–170. [CrossRef]

25. Patra, A.K.; Saxena, J. Exploitation of dietary tannins to improve rumen metabolism and ruminant nutrition. *J. Sci. Food Agric.* **2011**, *9*, 24–37. [CrossRef]

26. Tedeschi, L.O.; Ramírez-Restrepo, C.A.; Muir, J.P. Developing a conceptual model of possible benefits of condensed tannins for ruminant production. *Animal* **2014**, *8*, 1095–1105. [CrossRef] [PubMed]

27. Hervás, G.; Frutos, P.; Serrano, E.; Mantecón, A.R.; Giráldez, F.J. Effect of tannic acid on rumen degradation and intestinal digestion of treated soya bean meals in sheep. *J. Agric. Sci.* **2000**, *135*, 305–310.

28. Tiemann, T.T.; Lascano, C.E.; Wettstein, H.R.; Mayer, A.C.; Kreuzer, M.; Hess, H.D. Effect of the tropical tannin-rich shrub legumes *Calliandra calothyrsus* and *Flemingia macrophylla* on methane emission and nitrogen and energy balance in growing lambs. *Animal* **2008**, *2*, 790–799. [CrossRef] [PubMed]

29. Hariadi, B.T.; Santoso, B. Evaluation of tropical plants containing tannin on in vitro methanogenesis and fermentation parameters using rumen fluid. *J. Sci. Food Agric.* **2010**, *90*, 456–461. [PubMed]

30. Calabrò, S.; Moniello, G.; Piccolo, V.; Bovera, F.; Infascelli, F.; Tudisco, R.; Cutrignelli, M.I. Rumen fermentation and degradability in buffalo and cattle using the in vitro gas production technique. *J. Anim. Physiol. Anim. Nutr.* **2008**, *92*, 356–362.

31. Bueno, I.C.; Brandi, R.A.; Franzolin, R.; Benetel, G.; Fagundes, G.M.; Abdalla, A.L.; Louvandini, H.; Muir, J.P. In vitro methane production and tolerance to condensed tannins in five ruminant species. *Anim. Feed Sci. Technol.* **2015**, *205*, 1–9. [CrossRef]
32. Gordon, I.J. Browsing and grazing ruminants: Are they different beasts? *Forest. Ecol. Manag.* **2003**, *181*, 13–21. [CrossRef]
33. Clauss, M.; Müller, K.; Fickel, J.; Streich, W.J.; Hatt, J.-M.; Südekum, K.-H. Macroecology of the host determines microecology of endobionts: Protozoal faunas vary with wild ruminant feeding type and body mass. *J. Zool.* **2011**, *283*, 169–185. [CrossRef]
34. Salem, A.F.Z. Impact of season of harvest on in vitro gas production and dry matter degradability of *Acacia saligna* leaves with inoculum from three ruminant species. *Anim. Feed Sci. Technol.* **2005**, *123*, 67–79. [CrossRef]

6

Comparison of Faecal vs. Rumen Inocula for the Estimation of NDF Digestibility

Maria Chiaravalli, Luca Rapetti, Andrea Rota Graziosi, Gianluca Galassi, Gianni Matteo Crovetto and Stefania Colombini *

Dipartimento di Scienze Agrarie e Ambientali, Università degli Studi di Milano, 20133 Milano, Italy; maria.chiaravalli@unimi.it (M.C.); luca.rapetti@unimi.it (L.R.); andrea.rota@unimi.it (A.R.G.); gianluca.galassi@unimi.it (G.G.); matteo.crovetto@unimi.it (G.M.C.)

* Correspondence: stefania.colombini@unimi.it

Simple Summary: The evaluation of fibre digestibility is very important for the formulation of ruminant diets. Fibre digestibility is usually determined in lab with rumen inoculum obtained from cannulated cows. The research of alternative and less invasive inoculum sources is a critical issue that should be addressed. The present study evaluated the potential of faecal inocula, obtained from cows fed different diets, to assess fibre digestibility of different substrates at different incubation times (48, 240 and 360 h). At short incubation time, fibre digestibility obtained with rumen fluid was always higher than those obtained with faecal inocula, confirming a lower activity of the faecal inocula compared with rumen fluid. However, the type of diets fed to the donor animals had a significant effect on fibre digestibility, with a more active faecal inoculum for cows fed a diet based on maize silage. Despite the differences obtained at the short incubation time, the digestibility values at longer intervals showed that faecal inoculum could replace rumen inoculum. As a consequence, faeces may replace rumen fluid as inoculum for end-point measures, avoiding the use of cannulated animals and decreasing the analytical costs.

Abstract: Cow faeces have been investigated as alternative inoculum to replace rumen fluid to determine neutral detergent fibre (NDF) digestibility (NDFD). Aims of this study were to estimate: (1) the NDFD (48 h) of feed ingredients using a rumen inoculum in comparison with faecal inocula from cows fed diets with different forage basis; (2) the undigestible NDF (uNDF) at 240 and 360 h with ruminal fluid and faecal inocula from lactating cows fed two different diets. At 48 h incubation, the NDFD was affected both by feed and type of inoculum ($p < 0.01$) and by their interaction ($p = 0.03$). Overall, the mean NDFD was higher for rumen inoculum than for faecal inocula (585 vs. 389 g/kg NDF, $p < 0.05$), and faecal inoculum obtained from cows fed hay-based diets gave lower NDFD than those from cows fed maize silage (367 vs. 440 g/kg, $p < 0.05$). At long incubation times, the average uNDF was affected by substrate, inoculum and incubation time ($p < 0.01$), but not by their interactions. For each inoculum, significantly lower values were obtained at 360 than at 240 h. Regressions between uNDF with rumen and with the tested faecal inocula resulted in $r^2 \geq 0.98$. Despite the differences at 48 h, the uNDF showed that faecal inoculum could replace rumen fluid at longer incubation times.

Keywords: NDF digestibility; faecal inoculum; diet composition; in vitro

---

1. Introduction

Neutral detergent fibre (NDF) concentration of forages varies from 30% to 80% of DM (Dry Matter) with a wide variation for its digestibility [1]: from less than 40% for highly lignified mature legumes to greater than 90% for unlignified immature grass [2]. The accurate estimation of the NDF digestibility (NDFD) is important because ruminant nutritionists and forage plant breeders use in vitro measures of NDFD to assess forage quality, predict diet digestibility, and select plant genotypes for breeding [2]. For these purposes, NDFD is generally determined at short incubation times (for example: 24, 30, 48 h). However, NDFD can also be measured at longer incubation time to estimate the undegradable NDF (uNDF) of feedstuff and, consequently, to predict the potentially digestible NDF (pdNDF) and to estimate the extent of NDF digestion. Moreover, the uNDF content of faeces and total mixed ration (TMR) is used as a marker to estimate in vivo NDF digestibility with a field application at farm scale.

The in vitro techniques developed to determine NDFD (for example: [3,4]) involve the use of ruminal inoculum. However, in recent time fresh faeces from ruminants have been investigated as alternative inoculum to replace rumen fluid [5–7]. Using a fresh faecal inoculum would have some advantages: it is easier to obtain faeces than rumen liquor and rumen cannulated animals are not needed. Faecal inocula from cows have been used to evaluate the in vitro organic matter digestibility (OMD) of forages [6,8] and gas production (GP) of different feeds [5,7]. For example, Akhter et al. [8] obtained significant regression equations between in vitro OMD of eight forages determined with sheep rumen liquor and cow faeces. Another study [6] showed that the OMD of different forages estimated with faecal inoculum was comparable to the OMD determined with cow rumen fluid after 48 h of incubation. Regarding GP, a study [7] showed that in vitro total GP was greater for feeds incubated with cow rumen inoculum as compared to cow faecal inoculum. Mauricio et al. [5] confirmed these results: they observed that cow faecal inoculum gave lower GP volumes and longer lag time than rumen inoculum; however, potential GP was highly correlated between rumen liquor and faecal inoculum. Overall, all the studies indicate that faeces have the potential to be used as inoculum, but some limitations have been identified; one of the most important is the lower enzymatic activity of faecal inocula compared to rumen liquor [5,8]. In this regard, the type of diets fed to the donor animals can change the microbial population within the digestive tract, and therefore in the faeces of animals. For example, Kim et al. [9] observed that the community structure of cow faecal microbiota is greatly affected by diet and it is particularly associated to the dietary forage and concentrate ratio. These changes should be considered in using faeces as microbial inoculum and should be better evaluated. To the best of our knowledge, the effects of diets fed to cows on faecal inoculum activity were never evaluated. Similarly, as far as we know, there are no studies in literature that determined in vitro NDFD with cow faecal inoculum. Overall further research seems needed to fully understand and develop in vitro digestibility techniques using faecal inoculum.

The aims of this study were: (1) to estimate the NDFD (48 h) of several feed ingredients using a rumen inoculum in comparison with faecal inocula obtained both from dry and lactating cows fed diets characterised by a different forage basis; (2) to estimate the uNDF concentrations of different substrates (faeces and feeds) at two incubation times (240 and 360 h) with ruminal fluid and faecal inocula obtained from lactating cows fed two different diets.

2. Materials and Methods

The study was conducted at Cascina Baciocca, the experimental farm of the University of Milan, with the authorisation of the Ministry of Health, authorisation n. 904/2016-PR.

2.1. *In Vitro Incubations with Faecal and Rumen Inocula at 48 h*

The NDFD analyses were conducted on four feeds and using six inocula (five different faecal inocula and one rumen inoculum). The feeds were selected in order to have different NDF contents, as follows: grass hay (533 aNDF g/kg DM), wheat bran (497 aNDF g/kg DM), maize distiller (387 aNDF g/kg DM), and maize silage (373 aNDF g/kg DM). The feeds were dried at 60 °C for 48 h in a forced-air oven and ground to pass a 1-mm Fritsch mill (Fritsch Pulverisette, Idar-Oberstein, Germany). Each sample was weighed (0.250 g) in duplicate in Ankom F57 bags (Ankom Technology, Macedon, NY, USA); two blank Ankom bags (i.e., bags without sample) were also incubated for each inoculum source. Each bag was placed into a pre-warmed 100 mL Erlenmeyer flask closed by a rubber stopper with a Bunsen valve for gas release and maintained at 39 °C in a water bath with constant agitation.

Different faeces were tested as inoculum for a total of five: faeces collected from two cannulated Holstein dry cows fed a diet composed by (g/kg DM) grass hay (700), lucerne hay (130), maize meal (135), soybean meal (30) and vitamin mineral supplement (5), faeces collected from two Holstein lactating cows fed different diets based on the following forages: maize silage (FL-MS), ryegrass and lucerne silages (FL-GLS), wheat and lucerne silages (FL-WLS), and ryegrass and lucerne hays (FL-GLH). Two different donor cows fed the same diet were used for each incubation run. The composition of the lactating cow diets is in Table 1.

Table 1. Composition of the faecal inoculum donor lactating cow diets (on a dry matter basis)[1].

Ingredient	MS	GLS	WLS	GLH
(g DM/kg total DM)				
maize silage	493			
maize high moisture ear		286	291	
ryegrass hay	172			253
lucerne silage		268	104	
ryegrass silage		191		
wheat silage			200	
lucerne hay			106	253
grass hay				
soybean meal 44% CP			127	
soybean meal 48% CP	157	81.3		89.8
maize grain ground fine	121	114	127	228
maize grain flaked				86.0
maize gluten feed dry				43.9
wheat bran				11.5
molasses cane	30.3	34.6	19.1	21.5
limestone ground	9.59	8.97	9.11	4.74
sodium bicarbonate	6.15	5.75	5.84	3.16
sodium chloride	3.64	3.40	3.46	2.39
magnesium oxide	3.13	2.93	2.97	1.04
dicalcium phosphate	2.10	1.97	2.00	0.27
smartamine M	0.36	0.30	0.34	0.44
minvit suppl.[2]	1.22	1.14	1.16	1.29

[1] MS: maize silage diet; GLS: ryegrass and lucerne silage diet; WLS: wheat and lucerne silage diet; GLH: ryegrass and lucerne hay diet; [2] minvit composition: vitamin A 400,000 UI/kg, vitamin D3 60,000 UI/kg, vitamin E 1000 mg/kg, vitamin B1 60 mg/kg, vitamin PP 6000 mg/kg, biotin 40 mg/kg, copper sulfate pentahydrate 1900 mg/kg, cupric chelate of amino acids hydrate 4900 mg/kg, calcium iodate anhydrous 40 mg/kg, ferrous carbonate 1800 mg/kg, manganese chelate of amino acids hydrate 9000 mg/kg, sodium selenite 31 mg/kg, zinc oxide 1900 mg/kg, zinc chelate of amino acids hydrate 600 mg/kg, propyl gallate 12 mg/kg, butyl hydroxytoluene 34 mg/kg.

All faecal samples were collected immediately after defecation from the floor of the pen and transported in pre-warmed sealed containers to the laboratory for processing. The time lap in faecal collection and processing was within the range of 20 min. Rumen inoculum was collected from the two rumen cannulated donor cows. The rumen fluid was collected before the morning meal and was immediately strained through four layers of cheesecloth into a pre-warmed (39 °C) flask with CO_2 and mixed with the buffer solution [10] in a 1:2 ratio. Faecal samples were first mixed with the same buffer solution (at the same 1:2 ratio) and maintained for thirty minutes at 39 °C with CO_2, and then strained through four layers of cheesecloth [5,6,11].

Following the method described by Spanghero et al. [12], 90 mL of each inoculum was dispensed into the Erlenmeyer flasks under anaerobic conditions, flushing the flask with CO_2. Two runs of incubations were made in a shaking water bath at 39 °C for 48 h. At the end of incubation, the F57 bags were rinsed with cold water until the water ran clear and then placed in a 60 °C forced-air oven to dry. Subsequently, NDF concentration was determined for each bag using the fibre analyser (Ankom Technology, Macedon, NY, USA) and in vitro NDFD was calculated.

2.2. *In Vitro Incubations with Faecal and Rumen Inocula at 240 and 360 h*

The uNDF content was determined on seven substrates using three different inocula (one rumen and two faeces) and two incubation times at 240 and 360 h.

The substrates were: barley meal (293 aNDF g/kg DM), maize silage (505 aNDF g/kg DM), lucerne hay (538 aNDF g/kg DM), lactating cow TMR (321 aNDF g/kg DM) and the faeces (595 aNDF g/kg DM) produced by a lactating Holstein cow fed the same TMR, grass hay (662 aNDF g/kg DM) and the faeces (753 aNDF g/kg DM) produced by a dry Holstein cow fed the same grass hay. Each substrate was dried at 60 °C for 48 h in a forced-air oven and then ground to pass a 1-mm Fritsch mill (Fritsch Pulverisette, Idar-Oberstein, Germany) Incubations were conducted weighting (0.250 g) duplicate samples in F57 bags and using the DaisyII incubator jars (Ankom Technology, Macedon, NY, USA). Each jar contained four F57 bags for each substrate (two bags for each incubation time) and four F57 blank bags (i.e., bags without sample; two blanks for each incubation time).

The inocula RD-GH, FL-MS and FL-GLH previously described were tested. Faeces and rumen liquor were collected and treated as described above. The buffer was composed by two solutions as reported by Ankom protocol [13]. Faeces were mixed with the buffer in a ratio of 450 g/L and processed as decribed by Hughes et al. [6], while rumen fluid was added at a dose of 400 mL/jar, using a 1:3 ratio with the buffer [13].

Each inoculum was poured into one pre-warmed jar and two fermentation runs were carried out. Two F57 bags for each substrate and blank were removed at two times: 240 and 360 h. The inocula were renewed at 120 and 240 h replacing the content of jars with a new inoculum. Upon completion of the incubation, jars were emptied and the F57 bags were treated as described above.

2.3. *Statistical Analysis*

All data were analysed by the mixed procedure of SAS (version 9.4) considering the main effects of inoculum, run, substrate and time and their interactions. Data are reported as LS-MEANS.

Linear regressions between the rumen and the faecal inocula at each incubation time were performed by PROC REG procedure of SAS (version 9.4, SAS Institute Inc., Cary, NC, USA).

3. Results

3.1. *NDFD at 48 h of Incubation Using Rumen and Faecal Inocula*

The results of NDFD for rumen and faecal inocula at 48 h are presented in Table 2.

Table 2. Effect of inocula[1] on neutral detergent fibre (NDF) digestibility (NDFD, g/kg NDF) of different feeds at 48 h of incubation.

Sample	RD-GH	FL-MS	FL-GLS	FL-WLS	FL-GLH	FD-GH	*P*	SE
grass hay	578^{a}	379^{b}	405^{b}	411^{b}	361^{b}	446^{b}	0.03	32.4
wheat bran	521^{a}	431^{b}	425^{b}	415^{b}	408^{b}	407^{b}	0.01	18.0
maize distiller	741^{a}	534^{b}	$467^{b,c}$	$441^{b,c}$	$428^{b,c}$	383^{c}	<0.01	34.9
maize silage	498^{a}	$418^{a,b}$	$288^{b,c}$	$315^{b,c}$	214^{c}	$286^{b,c}$	0.04	50.4
all substrates	585^{a}	440^{b}	$396^{b,c}$	$396^{b,c}$	353^{c}	381^{c}	<0.01	14.9

Mean values in a row with different superscripts (a, b, c) differ significantly ($p < 0.05$); [1]RD-GH: rumen, dry cow—grass hay diet; FL-MS: faeces, lactating cow—maize silage diet; FL-GLS: faeces, lactating cow—ryegrass and lucerne silage diet; FL-WLS: faeces, lactating cow—wheat and lucerne silage diet; FL-GLH: faeces, lactating cow—ryegrass and lucerne hay diet; FD-GH: faeces, dry cow—grass hay diet.

The values of NDFD were affected by inoculum ($p < 0.01$) and by the interaction between inoculum and feed ($p = 0.03$).

Overall, the mean NDFD value was higher for RD-GH (585 g/kg NDF) than for the faecal inocula (389 g/kg NDF, on average) ($p < 0.05$) but there were differences among faecal inocula. Particularly, faecal inocula of cows fed diets based on hays (FD-GH and FL-GLH) resulted, on average, in lower NDFD (367 g/kg NDF) than FL-MS (440 g/kg NDF) ($p < 0.05$). The values obtained using FL-WLS and FL-GLS were intermediate.

Considering each individual feed sample, all feed samples had higher NDFD values using RD-GH treatment ($p < 0.05$), except for maize silage which was not significantly different between RD-GH (498 g/kg) and FL-MS (418 g/kg). The NDFD of maize distillers obtained with FD-GH inoculum was also significantly lower ($p < 0.05$) than the value obtained using FL-MS (383 vs. 534 for FD-GH and FL-MS, respectively).

3.2. Determination of uNDF Using Rumen and Faecal Inocula

The results of uNDF with faecal and rumen inocula at both incubation times (240 and 360 h) are presented in Table 3.

Table 3. Effect of inocula[1] and incubation time on undigestible NDF (uNDF, g/kg DM) of different substrates.

Sample	240 h			360 h			P		SE
	RD-GH	FL-MS	FL-GLH	RD-GH	FL-MS	FL-GLH	inoculum	time	
grass hay	223	287	237	198	233	225	0.16	0.15	22.2
lucerne hay	294	309	298	293	295	303	0.62	0.69	9.16
barley meal	79.7	92.9	81.3	58.7	69.8	67.1	0.43	0.03	8.72
maize silage	155	217	175	136	168	160	0.11	0.12	18.3
lactating cow TMR	78.6	104	84.5	74.4	74.9	83.6	0.33	0.13	8.09
faeces dry cow	400	454	434	372	385	403	0.16	0.02	17.0
faeces lactating cow	322	367	334	282	319	289	0.20	0.04	20.6
all substrates	222[b]	262[a]	235[b]	202[c]	221[b]	219[b,c]	<0.01	<0.01	6.06

Mean values in a row with different superscripts (a, b, c) differ significantly ($p < 0.05$); TMR: total mixed ration; [1]RD-GH: rumen, dry cow—grass hay; FL-MS: faeces, lactating cow—maize silage diet; FL-GLH: faeces, lactating cow—ryegrass and lucerne hay diet.

The average uNDF value was affected by inoculum and incubation time ($p < 0.01$), but not by the interactions between substrate, inoculum and time. Considering the effect of inoculum, the uNDF was not affected by the type of inoculum for RD-GH and FL-GLH; on the contrary, FL-MS had a higher value than RD-GH both at 240 and 360 h, and a higher value at 240 h in comparison with FL-GLH (Table 3). For each inoculum, the time effect was significant with lower values obtained at 360 h (on average, 240 vs. 214 g uNDF/kg DM for 240 and 360 h of incubation, respectively, $p < 0.01$). This time effect has to be ascribed particularly to the faeces samples. Moreover, uNDF decreased more for FL-MS treatment (−15.6%) than for RD-GH (−9.0%) and FL-GLH (−6.8%) from 240 to 360 h of incubation.

Considering the values for each substrate (Table 3), the dry cow faeces had the highest ($p < 0.05$) uNDF concentration (408 g/kg DM), while the lowest concentrations were obtained for barley meal and lactating cow TMR (74.9 g/kg DM and 83.4 g/kg DM, respectively). Intermediate values ($p < 0.05$) were registered for maize silage (169 g/kg DM), grass hay (234 g/kg DM), lucerne hay (298 g/kg DM) and lactating cows faeces (319 g/kg DM).

All linear regressions (Figures 1 and 2) between uNDF determined with rumen and faecal inocula at 240 and 360 h resulted in high r^2 values ($r^2 = 0.98$ for RD-GH vs. FL-MS at 240 h; $r^2 = 0.996$ for RD-GH vs. FL-GLH at 240 h; $r^2 = 0.980$ for RD-GH vs. FL-MS at 360 h; $r^2 = 0.994$ for RD-GH vs. FL-GLH at 360 h) and slopes within the range 0.92–0.97 showing a high relationship between data.

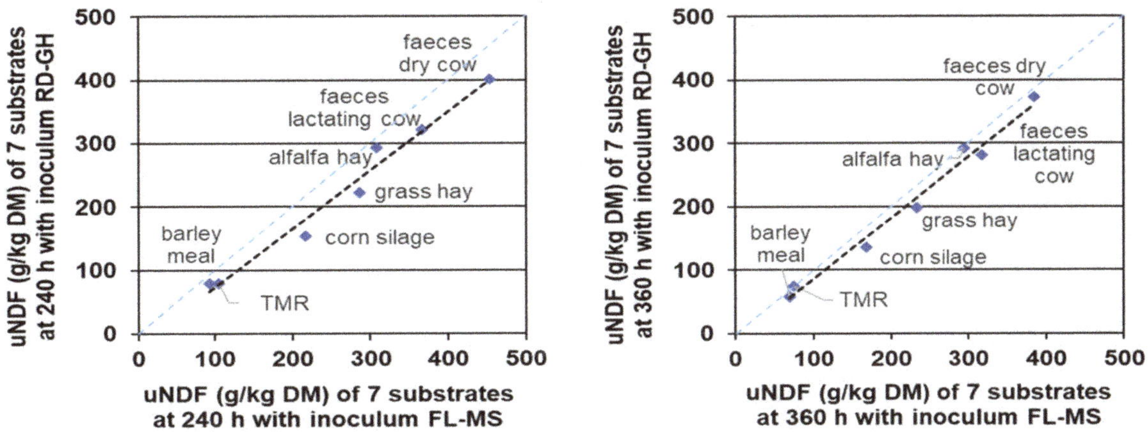

(a) (b)

Figure 1. Linear regression between uNDF (g/kg DM) of the tested substrates with the inocula RD-GH (rumen, dry cow—grass hay diet) and FL-MS (faeces, lactating cow—maize silage diet). (**a**) At 240 h, Y = 0.92X ($p < 0.01$) – 19.5 ($p = 0.35$); $r^2 = 0.98$, MSE = 443; (**b**) At 360 h, Y = 0.97X ($p < 0.01$) – 11.4 ($p = 0.44$); $r^2 = 0.98$, MSE = 276.

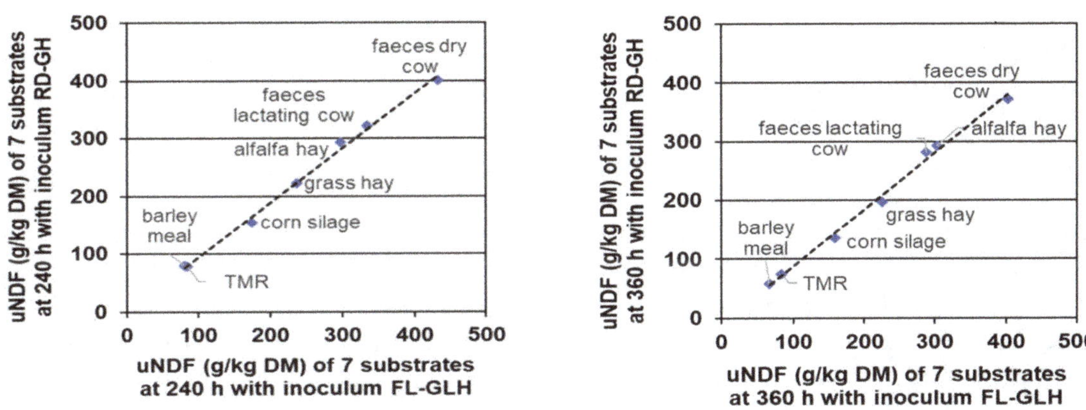

(a) (b)

Figure 2. Linear regression between uNDF (g/kg DM) of the tested substrates with the inocula RD-GH (rumen, dry cow—grass hay) and FL-GLH (faeces, lactating cow—ryegrass and lucerne hay diet). (**a**) At 240 h, Y = 0.94X ($p < 0.01$) + 0.03 ($p = 1.00$); $r^2 = 0.995$, MSE = 823; (**b**) At 360 h, Y = 0.96X ($p < 0.01$) – 8.68 ($p = 0.33$); $r^2 = 0.99$, MSE = 97.0.

4. Discussion

The objective of this study was to evaluate different faecal inocula (obtained from donor cows fed different diets) to determine NDFD at different incubation times (48, 240 and 360 h). It is well known that in ruminants the intestinal microbes can ferment fibre not digested in the rumen, hence there is the interest to evaluate the potential of faecal inoculum to digest NDF. The evaluation was done on a different set of samples including both forages and fibrous by-products, characterised by a wide variation in NDF digestibility and with an NDF content in the range between 373 and 533 g/kg DM in the first experiment.

Cow faeces were used in other studies as inoculum source, however, other parameters such as GP [5,7] or in vitro OM digestibility [6,8] were evaluated. The results of these studies underlined that faeces can be used as source of inoculum although the digestibility and the GP determined with faecal inocula were always lower than those found with rumen liquor. Several studies [7,14] showed a consistently longer lag phase with faecal inocula rather than rumen inocula, and this determined lower GP values with faecal inoculum. Similarly, Jiao et al. [15] evaluated fibre digestion in different segments of the post-ruminal tract demonstrating that detectable changes in hydrolysis of NDF occurred after 24 h of in vitro incubation.

In agreement with the cited studies, the results of the present study showed lower NDFD values at 48 h for cow faecal inocula in comparison with rumen inoculum. The lower NDFD values for faecal inocula were expected since the populations of cellulolytic bacteria are more numerous in rumen inoculum as compared to faecal inoculum [16]. In the rumen, the existence of a greater number of microbial populations is due to the medium stability and various diet compositions, whilst, on the other hand, in the gut nutrients are poor and the nutrient absorption rate is quick [17]. However, Van Vliet et al. [18] showed that diet composition can affect faecal bacterial biomass concentration with highest values associated with high energy diets due to a higher microbial growth in the rumen. Similarly, Lettat et al. [19] showed that increasing the maize silage proportion in the diet reduced the ruminal richness and diversity of the bacterial community but increased the number of total bacteria, which in turn should be reflected in the faecal microbial biomass and hence in the enzymatic activity of inoculum. As far as we know, no studies were conducted to evaluate NDFD using faecal inocula collected from cows fed different diets, although it is known that diet greatly influences the faecal microbiota of cattle [9,20,21]. In the present study FL-MS inoculum significantly affected maize silage NDFD in comparison with the other faecal inocula; it has to be underlined that maize silage was the main ingredient of the diet fed to donor cows of FL-MS inoculum. Although microbial communities were not determined in the present study, it can be speculated that there were some differences in rumen microbial community depending on diet, which in turn affects the faecal microbial community. Lengowski et al. [22] found higher abundance of the cellulolytic species *Fibrobacter succinogenes* during in vitro incubation of maize silage compared to grass silage. Similarly, another study [23] found a larger population of *F. succinogenes* in maize stems compared to other cellulolytic bacteria in in vitro conditions, while Lettat et al. [19] observed higher numbers of *F. succinogenes* in the rumen of dairy cows fed diets high in maize silage compared to those fed diets high in lucerne silage. Maize silage is a C_4 plant; C_4 plants usually have a lower nutritive value than C_3 plants due to a higher proportion of thick-walled cells associated with indigestible fibre in their leaves [24]; therefore, as suggested by Lengowski et al. [25], it may be that *F. succinogenes* has an advantage in degrading maize silage cell walls. The enzymatic system of *F. succinogenes* is more effective at degrading cellulose than the mechanisms used by the other cellulolytic organisms that occupy the same environment [26]. Moreover, *Fibrobacter spp.* have been detected in herbivorous species including the bovine rumen and caecum and in the faeces of different animal species [26]. Overall, it can be speculated that the higher NDFD for FL-MS inoculum can be due both to a higher bacterial concentration and to a possible higher presence of *F. succinogenes* due to the high concentration of maize silage in the diet [19].

Despite the difference between rumen and faecal inocula in the NDFD values at 48 h, the incubation conducted at 240 and 360 h showed promising results for faecal inocula to determine uNDF, especially for the inoculum FL-GLH. For all inocula and considering all the substrates, the average uNDF was affected by incubation time with lower values at 360 than 240 h. The uNDF fraction can be estimated by long-term (240 h) in vitro fermentations [27] or by incubating the samples in bags placed in the rumen for 288 h [28]. In the present study, an unconventional incubation time of 360 h (15 days) was also tested in order to compensate for the longer expected lag time of faecal inocula [5]. Unexpectedly, with the in vitro method used in the present study, the incubation time was significant also for the rumen inoculum; with significantly lower uNDF values at 360 than 240 h. Despite the wide application of uNDF, few documented recommendations of the method exist [28]. Recently, Raffrenato et al. [29] indicated that 288 h (12 days) of fermentation were necessary to reach the maximum extent of NDF digestion for the Daisy incubator; hence the incubation time of 240 h used in the present study seems not adequate to evaluate uNDF for some of the incubated substrates (faeces), confirming the results of Raffrenato et al. [29].

The effect of the diet fed to donor cows was also significant at long incubation times; however, differently from 48 h incubation, in the long incubation runs the FL-GLH inoculum gave results similar to RD-GH. As suggested by Aiple et al. [30], to obtain an active faecal inoculum, microbial populations in the hind gut should be supplied with fermentable substrate from a varied diet. Although knowledge of the nutrient requirements and supply of the hindgut microbes is limited, it is likely that their requirements are similar to those of the ruminal bacteria [31]. In the present study, cows fed GLH diet (inoculum FL-GLH) had a higher DMI and lower total tract organic matter digestibility (data not presented) than cows fed MS diet. As a consequence, a higher quantity of undigested organic matter was present in FL-GLH inoculum which could have better supported the microbial growth in the long time incubations and positively affecting NDFD. However, it has to be underlined that overall both faecal inocula gave results close to the rumen inocula, as confirmed by the slopes obtained with the linear regression analysis. The long time incubation results are promising since in feed evaluation systems the uNDF is normally computed from the ADL content multiplied by 2.4. However, Raffrenato et al. [32] showed that the ratio between ADL and uNDF is not constant, and, as reported also by Colombini et al. [33], a wrong evaluation of uNDF results in a bias in the prediction of fibre digestibility. Therefore, the use of an appropriate faecal inoculum seems a valuable alternative method to estimate the uNDF avoiding the use of cannulated cows.

5. Conclusions

The NDFD results at 48 h obtained with rumen fluid were always higher than those obtained with faecal inocula confirming a lower activity of the faecal inocula compared with rumen fluid. However, the type of diets fed to the donor animals had a significant effect on NDFD values determined by faecal inocula. This should be considered in using faeces as microbial inoculum source. As a prospective view to improve the results at 48 h, the use of fibrolytic enzymes added to the faecal inoculum should be evaluated.

Despite the differences obtained at 48 h, the uNDF results showed that faecal inoculum could replace rumen fluid at longer incubation times. As a consequence, in agreement with the review of Mould et al. [34], faeces may replace rumen fluid as an inoculum for end-point measures. Within the in vitro method applied in the present study, the incubation time of 240 h seems to be not enough to measure uNDF for both faecal and rumen inocula. In the present study, an unconventional time (360 h) was used; hence, further studies seem to be necessary to better define for different substrates (faeces, feeds and TMR) and inoculum the optimum incubation length (in the interval between 240 and 360 h) to evaluate uNDF.

Author Contributions: Conceptualization, S.C. and L.R.; methodology, M.C., A.R.G. and G.G.; software, M.C. and L.R.; validation, M.C. and A.R.G.; formal analysis, M.C.; and A.R.G.; investigation, S.C.; resources, S.C.; data curation, M.C. and S.C.; writing—original draft preparation, M.C.; writing—review and editing, S.C., G.M.C., L.R. and G.G.; visualization, S.C.; supervision, S.C.; project administration, S.C.; funding acquisition, S.C.

References

1. Bender, R.W.; Cook, D.E.; Combs, D.K. Comparison of in situ versus in vitro methods of fiber digestion at 120 and 288 hours to quantify the indigestible neutral detergent fiber fraction of corn silage samples. *J. Dairy Sci.* **2016**, *99*, 5394–5400. [CrossRef] [PubMed]
2. Goeser, J.P.; Combs, D.K. An alternative method to assess 24-h ruminal in vitro neutral detergent fiber digestibility. *J. Dairy Sci.* **2009**, *92*, 3833–3841. [CrossRef] [PubMed]
3. Hall, M.B.; Mertens, D.R. In vitro fermentation vessel type and method alter fiber digestibility estimates. *J. Dairy Sci.* **2008**, *91*, 301–307. [CrossRef] [PubMed]
4. Lopes, F.; Ruh, K.; Combs, D.K. Validation of an approach to predict total-tract fiber digestibility using a standardized in vitro technique for different diets fed to high-producing dairy cows. *J. Dairy Sci.* **2015**, *98*, 2596–2602. [CrossRef] [PubMed]

5. Mauricio, R.M.; Owen, E.; Mould, F.L.; Givens, I.; Theodorou, M.K.; France, J.; Davis, D.R.; Dhanoa, M.S. Comparison of bovine rumen liquor and bovine faeces as inoculum for an in vitro gas production technique for evaluating forages. *Anim. Feed Sci. Technol.* **2001**, *89*, 33–48. [CrossRef]
6. Hughes, M.; Mlambo, V.; Lallo, C.H.O.; Jennings, P.G.A. Potency of microbial inocula from bovine faeces and rumen fluid for in vitro digestion of different tropical forage substrates. *Grass Forage Sci.* **2012**, *67*, 263–273. [CrossRef]
7. Ramin, M.; Lerose, D.; Tagliapietra, F.; Huhtanen, P. Comparison of rumen fluid inoculum vs. faecal inoculum on predicted methane production using a fully automated in vitro gas production system. *Livest. Sci.* **2015**, *181*, 65–71. [CrossRef]
8. Akhter, S.; Owen, E.; Theodorou, M.K.; Butler, E.A.; Minson, D.J. Bovine faeces as a source of micro-organisms for the in vitro digestibility assay of forages. *Grass Forage Sci.* **1999**, *54*, 219–226. [CrossRef]
9. Kim, M.; Kim, J.; Kuehn, L.A.; Bono, J.L.; Berry, E.D.; Kalchayanand, N.; Freetly, H.C.; Benson, A.K.; Wells, J.E. Investigation of bacterial diversity in the feces of cattle fed different diets. *J. Anim. Sci.* **2014**, *92*, 683–694. [CrossRef]
10. Menke, K.H.; Steingass, H. Estimation of the energetic feed value obtained from chemical analysis and in vitro gas production using rumen fluid. *Anim. Res. Dev.* **1988**, *28*, 7–55.
11. Zicarelli, F.; Calabro, S.; Cutrignelli, M.I.; Infascelli, F.; Tudisco, R.; Bovera, F.; Piccolo, V. In vitro fermentation characteristics of diets with different forage/concentrate ratios: Comparison of rumen and faecal inocula. *J. Sci. Food Agric.* **2011**, *91*, 1213–1221. [CrossRef] [PubMed]
12. Spanghero, M.; Chiaravalli, M.; Colombini, S.; Fabro, C.; Froldi, F.; Mason, F.; Moschini, M.; Sarnataro, C.; Schiavon, S.; Tagliapietra, F. Rumen inoculum collected from cows at slaughter or from a continuous fermenter and preserved in warm, refrigerated, chilled or freeze-dried environments for in vitro tests. *Animals* **2019**, *9*, 815. [CrossRef] [PubMed]
13. Ankom Technology. In Vitro True Digestibility Using the DAISY Incubator. 2005. Available online: http://www.ankom.com/media/documents/IVDMD_0805_D200.pdf (accessed on 13 September 2010).
14. Pandian, C.S.; Reddy, T.J.; Sivaiah, K.; Blümmel, M.; Ramana Reddy, Y.R. Faecal matter as inoculum for in vitro gas production technique. *Anim. Nutr. Feed Technol.* **2016**, *16*, 271. [CrossRef]
15. Jiao, J.; Lu, Q.; Tan, Z.; Guan, L.; Zhou, C.; Tang, S.; Han, X. In vitro evaluation of effects of gut region and fiber structure on the intestinal dominant bacterial diversity and functional bacterial species. *Anaerobe* **2014**, *28*, 168–177. [CrossRef]
16. El-Meadaway, A.; Mir, Z.; Mir, P.S.; Zaman, M.S.; Yanke, L.J. Relative efficacy of inocula from rumen fluid or faecal solution for determining in vitro digestibility and gas production. *Can. J. Anim. Sci.* **1998**, *78*, 673–679. [CrossRef]
17. Guzmán, M.L.; Sager, R.L. Ruminant Fecal Inolucum for In Vitro Feed Digestibility Analysis. 2016. Available online: https://unsl.academia.edu/RicardoSager (accessed on 6 November 2019).
18. Van Vliet, P.C.J.; Reijs, J.W.; Bloem, J.; Dijkstra, J.; De Goede, R.G.M. Effects of cow diet on the microbial community and organic matter and nitrogen content of feces. *J. Dairy Sci.* **2007**, *90*, 5146–5158. [CrossRef]
19. Lettat, A.; Hassanat, F.; Benchaar, C. Corn silage in dairy cow diets to reduce ruminal methanogenesis: Effects on the rumen metabolically active microbial communities. *J. Dairy Sci.* **2013**, *96*, 5237–5248. [CrossRef]
20. Callaway, T.R.; Dowd, S.E.; Edrington, T.S.; Anderson, R.C.; Krueger, N.; Bauer, N.; Kononoff, P.J.; Nisbet, D.J. Evaluation of bacterial diversity in the rumen and feces of cattle fed different levels of dried distillers grains plus solubles using bacterial tag-encoded FLX amplicon pyrosequencing. *J. Anim. Sci.* **2010**, *88*, 3977–3983. [CrossRef]
21. Shanks, O.C.; Kelty, C.A.; Archibeque, S.; Jenkins, M.; Newton, R.J.; McLellan, S.L.; Huse, S.M.; Sogin, M.L. Community structures of fecal bacteria in cattle from different animal feeding operations. *Appl. Environ. Microbiol.* **2011**, *77*, 2992–3001. [CrossRef]
22. Lengowski, M.B.; Zuber, K.H.; Witzig, M.; Möhring, J.; Boguhn, J.; Rodehutscord, M. Changes in rumen microbial community composition during adaption to an in vitro system and the impact of different forages. *PLoS ONE* **2016**, *11*, e0150115. [CrossRef]
23. Kozakai, K.; Nakamura, T.; Kobayashi, Y.; Tanigawa, T.; Osaka, I.; Kawamoto, S.; Hara, S. Effect of mechanical processing of corn silage on in vitro ruminal fermentation, and in situ bacterial colonization and dry matter degradation. *Can. J. Anim. Sci.* **2007**, *87*, 259–267. [CrossRef]
24. Wilson, J.R. Cell wall characteristics in relation to forage digestion byruminants. *J. Agric. Sci.* **1994**, *122*, 173–182. [CrossRef]

25. Lengowski, M.B.; Witzig, M.; Möhring, J.; Seyfang, G.M.; Rodehutscord, M. Effects of corn silage and grass silage in ruminant rations on diurnal changes of microbial populations in the rumen of dairy cows. *Anaerobe* **2016**, *42*, 6–16. [CrossRef]
26. McDonald, J.E.; Rooks, D.J.; McCarthy, A.J. Methods for the isolation of cellulose-degrading microorganisms. *Methods Enzymol.* **2012**, *510*, 349–374. [PubMed]
27. Palmonari, A.; Gallo, A.; Fustini, M.; Canestrari, G.; Masoero, F.; Sniffen, C.J.; Formigoni, A. Estimation of the indigestible fiber in different forage types. *J. Anim. Sci.* **2016**, *94*, 248–254. [CrossRef] [PubMed]
28. Krizsan, S.J.; Huhtanen, P. Effect of diet composition and incubation time on feed indigestible neutral detergent fiber concentration in dairy cows. *J. Dairy Sci.* **2013**, *96*, 1715–1726. [CrossRef]
29. Raffrenato, E.; Ross, D.A.; Van Amburgh, M.E. Development of an in vitro method to determine rumen undigested aNDFom for use in feed evaluation. *J. Dairy Sci.* **2018**, *101*, 9888–9900. [CrossRef]
30. Aiple, K.P.; Steingass, H.; Menke, K.H. Suitability of a buffered faecal suspension as the inoculum in the Hohenheim gas test: 1. Modification of the method and its ability in the prediction of organic matter digestibility and metabolizable energy content of ruminant feeds compared with rumen fluid as inoculum. *J. Anim. Physiol. Anim. Nutr.* **1992**, *67*, 57–66.
31. Frey, J.C.; Pell, A.N.; Berthiaume, R.; Lapierre, H.; Lee, S.; Ha, J.K.; Angert, E.R. Comparative studies of microbial populations in the rumen, duodenum, ileum and faeces of lactating dairy cows. *J. Appl. Microbiol.* **2010**, *108*, 1982–1993. [CrossRef]
32. Raffrenato, E.; Erasmus, L.J. Variability of indigestible NDF in C3 and C4 forages and implications on the resulting feed energy values and potential microbial protein synthesis in dairy cattle. *S. Afr. J. Anim. Sci.* **2013**, *43*, 93–97.
33. Colombini, S.; Galassi, G.; Crovetto, G.M.; Rapetti, L. Milk production, nitrogen balance, and fiber digestibility prediction of corn, whole plant grain sorghum, and forage sorghum silages in the dairy cow. *J. Dairy Sci.* **2012**, *95*, 4457–4467. [CrossRef] [PubMed]
34. Mould, F.L.; Kliem, K.E.; Morgan, R.; Mauricio, R.M. In vitro microbial inoculum: A review of its function and properties. *Anim. Feed Sci. Technol.* **2005**, *123*, 31–50. [CrossRef]

7

Antimicrobial Effects of Black Soldier Fly and Yellow Mealworm Fats and their Impact on Gut Microbiota of Growing Rabbits

Sihem Dabbou [1,2], Ilario Ferrocino [3], Laura Gasco [3,*], Achille Schiavone [4], Angela Trocino [5], Gerolamo Xiccato [6], Ana C. Barroeta [7], Sandra Maione [4], Dominga Soglia [4], Ilaria Biasato [3], Luca Cocolin [3], Francesco Gai [8] and Daniele Michele Nucera [3]

1 Center Agriculture Food Environment (C3A), University of Trento, Via E. Mach 1, 38010 San Michele all'Adige, Italy; sihem.dabbou@unitn.it

2 Research and Innovation Centre, Fondazione Edmund Mach, 38010 San Michele all'Adige, Italy

3 Department of Agricultural, Forest and Food Sciences, University of Torino, Largo Paolo Braccini 2, 10095 Grugliasco, Italy; ilario.ferrocino@unito.it (I.F.); ilaria.biasato@unito.it (I.B.); lucasimone.cocolin@unito.it (L.C.); daniele.nucera@unito.it (D.M.N.)

4 Department of Veterinary Science, University of Turin, Largo P. Braccini 2, 10095 Grugliasco, Italy; achille.schiavone@unito.it (A.S.); sandra.maione@unito.it (S.M.); dominga.soglia@unito.it (D.S.)

5 Department of Comparative Biomedicine and Food Science, University of Padova, Viale dell'Università 16, Legnaro, 35020 Padova, Italy; angela.trocino@unipd.it

6 Department of Agronomy, Food, Natural Resources, Animal, and Environment, University of Padova, Viale dell'Università 16, 35020 Legnaro, Padova, Italy; gerolamo.xiccato@unipd.it

7 Nutrition and Animal Welfare Service, Department of Animal and Food Science, Faculty of Veterinary Medicine, Autonomous University of Barcelona, Bellaterra, 08193 Barcelona, Spain; Ana.Barroeta@uab.cat

8 Institute of Sciences of Food Production, National Research Council, Largo Paolo Braccini 2, 10095 Grugliasco, Italy; francesco.gai@ispa.cnr.it

* Correspondence: laura.gasco@unito.it

Simple Summary: The use of insect lipids as an alternative ingredient is an emergent topic in animal nutrition due to their antimicrobial activities. The present study evaluated the in vitro antimicrobial activities of two insect fats (black soldier fly (*Hermetia illucens* (HI) fat and yellow mealworm (*Tenebrio molitor* (TM) fat) and their effect as a total substitute for dietary soybean oil in cecal fermentation and gut microbiota of growing rabbits. The obtained results showed the potential of HI and TM fats as an antibacterial feed ingredient with a positive influence on the rabbit cecal microbiota. HI and TM fats therefore may be a sustainable lipid alternative to soybean oil in rabbit nutrition with possible interesting applications in the feed industry.

Abstract: This study aimed to evaluate the in vitro antimicrobial activities of two types of insect fats extracted from black soldier fly larvae (HI, *Hermetia illucens* L.) and yellow mealworm larvae (TM, *Tenebrio molitor* L.) and their effects as dietary replacement of soybean oil (S) on cecal fermentation pattern, and fecal and cecal microbiota in rabbits. A total of 120 weaned rabbits were randomly allotted to three dietary treatments (40 rabbits/group) —a control diet (C diet) containing 1.5% of S and two experimental diets (HI diet (HID) and TM diet (TMD)), where S was totally substituted by HI or TM fats during the whole trial that lasted 41 days. Regarding the in vitro antimicrobial activities, HI and TM fats did not show any effects on *Salmonella* growth. *Yersinia enterocolitica* showed significantly lower growth when challenged with HI fats than the controls. The insect fat supplementation in rabbit diets increased the contents of the cecal volatile fatty acids when compared to the control group. A metataxonomic approach was adopted to investigate the shift in the microbial composition as a function of the dietary insect fat supplementation. The microbiota did not show a clear separation as a function of the inclusion, even if a specific microbial signature was observed. Indeed, HI and TM fat supplementation enriched the presence of *Akkermansia* that was found to be correlated with

NH3-N concentration. An increase in *Ruminococcus*, which can improve the immune response of the host, was also observed. This study confirms the potential of HI and TM fats as antibacterial feed ingredients with a positive influence on the rabbit cecal microbiota, thus supporting the possibility of including HI and TM fats in rabbit diets.

Keywords: insect fat; *Hermetia illucens*; *Tenebrio molitor*; gut microbiota; antimicrobial effect; rabbit feeding

---

1. Introduction

In rabbit production, a high mortality, which can reach 80%, due to gastrointestinal disorders and epizootic enteropathy is the major health concern [1]. The rabbit's digestive health and physiology, as well as the immune system, are based on its abundant cecal microbiota [2,3]. The intestinal microbial community is abundant, since it contains about 100 to 1000 billion microorganisms per gram of digesta. Its diversity and complexity is very high, with about a thousand different species. Bacteria predominate with a 10^{11} to 10^{12} bacteria/g cecal content [4]. Cecal microbiota of rabbit species is dominated by phylum Firmicutes (90%), while the other phyla, conventionally found in the digestive ecosystems of mammals, are in the minority (10%). Lachnospiraceae and Ruminococcaceae are the dominant families of the cecal ecosystem (40% and 30%) followed by Bacteroidaceae and Rikenellaceae (less than 3%) [4]. The gut microbiota, as well as the factors affecting its composition, are considered important aspects to maintaining digestive health and therefore to enhancing rabbit production [2,5]. Diet is one of the main factors influencing the microbiota and the co-occurrence patterns of the cecal bacterial community [2,5]. Specifically, dietary fat intake can modulate gut microbiota [6,7]. It has been reported that some fatty acids (FA) and in particular medium chain fatty acids (MCFA) act as bacteriostatic (growth inhibiting) or bactericidal (killing) molecules [8,9]. Then, in addition to modulating the bacterial community, the amount and type of dietary fat can have an effect on the immune function both at systemic and intestinal levels [5]. Data regarding the effect of fat type and level in rabbit diet on gut health are still very limited and lead to contradictory results [5,10].

The well-known antimicrobial resistance caused by the misuse of antibiotics drugs in animal production, and the EU ban on their in-feed use (Regulation EC/1831/2003) has led to an increase in the incidence of livestock disease and economic damage. For several years, research has dedicated great efforts to investigating alternatives able to sustain production without causing an increase of antimicrobial resistance and several natural products have been investigated [11,12].

Recently, insects have been receiving considerable attention as novel alternative feed ingredients because of their excellent nutritive properties [13] and their potential effect on animal health [14,15]. It has been shown that insects can modulate the intestinal microbiota with positive effects on poultry growth, health, and resistance against pathogens [15–19]. Lipids are a main component of insects (30%–40% of the dry matter, DM) [20] and, once extracted during the insect larvae processing, they can be used as an interesting feed ingredient in animal farming [14,15,21–26]

Black soldier fly (*Hermetia illucens*, HI) and yellow mealworm (*Tenebrio molitor*, TM) fats are characterized by a different fatty acid (FA) profile. The HI fat is rich in saturated FAs (SFAs) and in MCFAs, mainly lauric acid (C12:0; 48% of the total FA profile) [14], which has an antimicrobial effect on a wide range of microbes [27]. On the other hand, the TM fat is considered as a source of n-6 poly-unsaturated FAs (PUFAs), with high linoleic acid content (9% of the total FA profile) [14]. MCFAs are effectively absorbed and metabolized on the gastro-intestinal tract and known for their physiological and antibacterial effects on Gram positive bacteria [28–30]. Furthermore, MCFAs can exert beneficial effects on intestinal health and microbial growth inhibition [31], as well as a favorable impact on the digestive health of the growing rabbit [32]. Both the short chain FAs (SCFAs; propionic acid and butyric acid) and the MCFAs (caproic acid and caprylic acid) have a direct in vitro antimicrobial activity against *Salmonella typhimurium* [33]. Matsue et al. [27] demonstrated that lauric acid has a high antimicrobial activity against pathogenic *Bacteroides* and *Clostridium* species, and consequently can modulate intestinal health. Significant in vitro gut antimicrobial effects against D-*Streptococci* and *Lactobacilli* have also been attributed to HI prepupae fat [30]. In a recent study, Sypniewski et al. [15] showed that the addition of HI fat to replace soybean oil (S) in turkey diets significantly reduced the presence of potentially-pathogenic microbes and decreased gastrointestinal-tract (GIT) inflammation by modulating the level of IL-6 and TNF-α.

In the light of what was reported above, the aim of the present study was to evaluate the in vitro antimicrobial activities of HI and TM fats and their effects on cecal traits and gut microbiota of growing rabbits.

2. Materials and Methods

2.1. Animal Ethics Statement

The study was performed in accordance with the guidelines of the European and the Italian laws (European Directive 86 609/EEC-Italian law D.L. 116/92), and was approved by the Bioethical Committee of the University of Torino (Italy) (Ref. 386638, 4/12/2017).

2.2. In Vitro Analyses for Antimicrobial Activity of Insect Fats

Bacterial strains were selected considering their impact on rabbits and on public health [34–37]. *Salmonella tiphymurium* (CIP 60.62T), *Salmonella enteritidis* (CRBIP 19.329), *Yersinia enterocolitica* (CIP 101.776), *Pasteurella multocida* (CIP 100.610) and *Listeria monocytogenes* (CIP 82.110T) were all purchased from Institute Pasteur (Paris, France). After overnight incubation following producers' instructions, strains were mixed with TM and HI fats [14] in order to reach a final concentration of 3 Log CFU/mL, verified by immediate plate streaks. Briefly, the broths for growth were prepared as follows: 200 μL of brain heart infusion broth (BHI, Oxoid, Fisher Scientific, Rodano, Milano, Italy) were added to 250 μL of insect fat and then to 50 μL of fresh bacterial overnight culture previously quantified by measuring the optical density at 600 nm (Prixma UV/VIS 5200, Fulltech Instruments, Rome, Italy) and diluted to 4 Log CFU/ml. Only for *Pasteurella multocida*, instead of BHI, tryptic soy broth (TSB, Oxoid, Fisher Scientific, Rodano, Milano, Italy) was used. These mixes were incubated 24 hours at 37 °C with horizontal shaking (RPR horizontal rotator, LE8S, Fisher Scientific, Rodano, Milano, Italy). At regular timings (every 2 hours), three aliquots for each mix were streaked on tryptic soy agar (TSA, Oxoid, Fisher Scientific, Rodano, Milano, Italy) (only *Pasteurella multocida*) or on brain heart infusion agar (BHA, Oxoid, Fisher Scientific, Rodano, Milano, Italy). Moreover, control tests were prepared—450 μL of BHI were added with 50 μL of quantified overnight cultures to reach a final concentration of 3 Log CFU/mL of broth (A) and a mix of soybean oil (S) and bacterial broth, prepared as described above for insect fats (B). Each control and strain/fat combination were analyzed in triplicates for a total of 84 samples per bacterial species—12 samples per 6-time intervals.

2.3. Inclusion of Hermetia illucens and Tenebrio molitor Fats in the Diet of Growing Rabbits

2.3.1. Experimental Design

For this experiment, three dietary treatments were tested in 120 growing rabbits (40 rabbits/diet) from 36 to 78 days of age—a control diet (C) containing 1.5% soybean oil (S), and two diets (HID and TMD) with total replacement of S with HI and TM larvae fat, respectively. TM and HI fats were obtained from commercial companies (Ynsect, Evry, France and Hermetia Deutschland GmbH & Co. KG, Baruth/Mark, Germany) by a mechanical process using high pressure and without solvents. The three diets contained an average of 16.6% DM of crude protein (CP) and 18.6 MJ/kg DM of gross energy. Detailed information about the rabbit farming conditions and live performance are provided by Gasco et al. [14]. Ingredients and chemical composition of the diets are shown in Table 1. Briefly, no overall significant differences among experimental groups were observed for growth performance during the trial.

Table 1. Ingredients (% as fed) and chemical composition (% DM) of experimental diets. (modified from Gasco et al. [14]).

	Experimental Diets		
Ingredients	**C**	**HID**	**TMD**
Dehydrated alfalfa meal (17 g CP/100 g)	32	32	32
Alfalfa hay	7.5	7.5	7.5
Wheat bran	23.5	23.5	23.5
Barley meal	10	10	10
Dried sugar beet pulp	16	16	16
Soybean meal (44 g CP/100 g)	7	7	7
Soybean oil	1.5	-	-
Hermetia illucens fat	-	1.5	-
Tenebrio molitor fat	-	-	1.5
Cane molasses	1.2	1.2	1.2
Dicalcium phosphate	0.3	0.3	0.3
Salt	0.4	0.4	0.4
L–methionine (98 g methionine/100 g)	0.1	0.1	0.1
Vitamin–mineral premix[a]	0.5	0.5	0.5
Chemical Composition			
Dry matter, %	89.4	89.2	89.6
Ash, % DM	8.58	7.77	7.75
Crude protein, % DM	17.0	16.8	16.3
Ether extract, % DM	4.22	3.92	3.87
Neutral detergent fiber (aNDF), % DM	40.2	41.7	40.5
Acid detergent fiber (ADF), % DM	21.7	23.0	22.8
Acid detergent lignin (ADL), % DM	4.81	5.09	5.02
Gross energy, MJ/kg DM	18.50	18.50	18.62

C: control diet with soybean oil; HID: diet with *Hermetia illucens* fat; TMD: diet with *Tenebrio molitor* fat; CP: crude protein; DM: dry matter. [a]Supplementation per kilogram of feed: vitamin A, 16,000 IU; vitamin D3, 1600 IU; vitamin E acetate, 30 mg; vitamin B1, 0.8 mg; vitamin B6, 1.65 mg; niacin, 40 mg; folic acid, 1 mg; Mn, 30 mg; Fe, 116 mg; Cu, 12.5 mg; Zn, 60 mg; Co, 0.45 mg; Ca, 1.3 mg; Se, 0.3 mg.

2.3.2. Fatty Acid Profiles of Insect Lipids and Experimental Diets

The FA profiles of insect lipids and feeds were determined as reported by Gasco et al. [14]. The fatty acid methyl esters (FAME) were separated, identified, and quantified. The results were expressed as g/kg of FAME (Table 2).

Table 2. Fatty acid profiles of the dietary fats and experimental diets (g kg-1 of total FAME) (modified from Gasco et al. [14]).

Fatty acids	Dietary Fats			Experimental Diets		
	S	**HI**	**TM**	**C**	**HID**	**TMD**
C12:0	0.2	480	2.3	0.5	203	3.0
C14:0	0.5	103	22.2	0.9	44.7	13.3
C16:0	104	127	176	121	161	184
C18:0	44.3	19.0	23.1	28.4	20.8	22.2
BCFAs	0.1	2.9	0.8	2.4	4.1	3.4
C16:1 n-7	0.9	32.0	16.6	0.12	19.9	10.4
C18:1 n-9	230	91.1	378	201	127	273

Table 2. *Cont.*

Fatty acids	Dietary Fats			Experimental Diets		
	S	HI	TM	C	HID	TMD
C18:2 n-6	515	90.0	332	521	310	389
C18:3 n-3	70.3	10.1	18.0	74.3	62.8	55.1
SFA [1]	158	748	231	165	454	240
UFA [1]	842	252	769	835	546	760
MUFA [1]	254	141	411	236	169	309
PUFA [1]	588	111	358	599	377	451
$\sum$n3	70.5	11.7	18.3	74.7	62.8	55.1
$\sum$n6	516	91.1	333	523	311	391

S: soybean oil; HI: *Hermetia illucens* fat; TM: *Tenebrio molitor* fat; C: control diet with soybean oil; HID: diet with *Hermetia illucens* fat; TMD: diet with *Tenebrio molitor* fat; FAME: fatty acid methyl ester; BCFAs: branched-chain fatty acids; SFA: saturated fatty acid; UFA: unsaturated fatty acid; MUFA: monounsaturated fatty acid; PUFA: polyunsaturated fatty acid. [1] included minor FAs.

2.3.3. Cecal Fermentation Traits

At slaughtering (78 days of age), a total of 30 rabbits (10 animals per treatment) were randomly selected and eviscerated. The pH, cecal ammonia, and volatile FA (VFA) profiles of the cecal contents were determined as reported by Tazzoli et al. [38].

2.3.4. DNA Extraction and 16S rRNA High-Throughput Amplicon Target Sequencing

In order to observe the development and dynamics of bacterial communities, hard feces were collected from 10 rabbits per group at 36, 51, and 77 days of age, whereas the cecal contents (n = 10 per group) were taken during the slaughtering at 78 days of age. Samples from the same group and the same collection site and day were pooled together in sterilized polyethylene bags, and immediately stored at −80 °C until examination. DNA from feces and cecal samples were extracted by using a commercial kit (QIAamp Fast DNA stool Mini Kit, QIAGEN®, Hilden, Germany) following the instructions reported by the manufacturer with slightly modification. The DNA were quantified by NanoDropTM 2000 Spectrophotometer (Fisher Scientific, Rodano, Milano, Italy) and standardized at 5 ng/uL and used a template in the PCR amplifying the V3–V4 region of the 16S rRNA gene using the primers and protocols described by Klindworth et al. [39]. The PCR amplicons were cleaned according to Illumina (San Diego, CA, USA) guidelines. The sequencing was performed with a MiSeq Illumina instrument with V3 chemistry and generated 250 bp paired-end reads according to the manufacturer's instructions.

2.3.5. Bioinformatics and Statistical Analysis

After paired-end sequencing, raw reads were analyzed as previously reported by Biasato et al. [17]. Sequences were first joined using FLASH software [40] with default parameters and filtered for low quality bases (at Phred < Q20) by QIIME 1.9.0 software [41]. Reads shorter than 300 bp were discarded by using Prinseq software [42]. The USEARCH software version 8.1 was used for chimera filtering and operational taxonomic units (OTUs) were clustered at 97% of similarity threshold by UCLUST algorithms [43]. Representative sequences of each cluster were mapped against the Greengenes 16S rRNA gene database version 2013 for taxonomic assignment. In order to avoid bias due to the different sequencing depth, OTU tables were rarefied at 3996 sequences/sample. tables display the higher taxonomy resolution that was reached. When the taxonomy assignment was not able to reach the genus level, the family or phyla were displayed. Alpha diversity indices were calculated using the diversity function of the vegan package [44]. Weighted and unweight UniFrac distance matrix and OTU tables were used to find differences between samples by anosim and adonis statistical tests through the function vegan in R environment. The pairwise Wilcoxon test was used as appropriate to determine significant differences in alpha diversity or OTU abundance. A generalized linear model was used in order to test the importance of continuous or discrete variables available (sampling time and insect inclusion) on the relative abundance of bacterial genera or family. The interactions between the levels of the fixed factors were evaluated by pairwise comparisons. Not-normally distributed variables were presented as median (range interquartile) and box plots represented the interquartile range between the first and the third quartiles, with the error bars showing the lowest and the highest value. Pairwise Spearman's non-parametric correlations were used to study the relationships between the relative abundance of microbial taxa in cecal samples and VFA variables. The correlation plots were visualized in R using the coplot package of R.

The statistical analysis for data related to in vitro antimicrobial activity and cecal traits was performed using IBM SPSS Statistics V25.0.0 software (IBM, Armonk, NY, USA) by means of ANOVA, followed, if significant, by Duncan test post-hoc. Regarding the antibacterial activities of insect fats, bacterial counts was evaluated at each time point and the average results compared across the different conditions—pure bacterial broth, bacterial broth mixed with soybean oil, bacterial broth mixed with TM fat, and bacterial broth mixed with HI fat. Significance was declared at $p \leq 0.05$. A statistical trend was considered for $p \leq 0.10$. Sequences data were deposited on NCBI database under the bioproject number PRJNA645756.

3. Results

3.1. *In Vitro Antimicrobial Activities of Insect Fats*

Overall, the bacterial counts, when quantified broths were challenged with insect fats, were lower than the controls for three out of five pathogenic species tested, whereas only the two strains of *Salmonella* did not show any significant difference in counts when compared to control tubes counts (therefore these data are not reported in Table 3). Considering the other species, for all of them results highlighted that HI fat caused a delay in bacterial growth, implying a bacteriostatic effect (Table 3). After 24 hours of incubation, counts of *Pasteurella moultocida* broths challenged with HI fat showed a mean log difference of −4.48 and −4.76 when compared to control broths with soybean oil (control B) and no oil addition (control A), respectively. Similar results were observed for *Yersinia enterocolitica* showing a mean log difference values of −5.9 and −5.97, with control B and A, respectively. Finally, *Listeria monocytogenes* counts also led to a similar pattern of results, showing mean log difference values of −5.11 and −5.15 when comparing counts of broths challenged with HI fat and the controls B and A. All these differences were statistically significant (Table 3). On the other hand, results related to TM fat showed that only *Pasteurella multocida* was effectively inhibited in growth—the mean log difference between controls and TM-challenged broths showed values of −2.64 and −2.92, respectively, for control B and control A. These values were lower than what reported above for broth challenged with HI fat (Table 3). Finally, differences in bacterial counts between HI fat challenged broths and controls were statistically meaningful from the 4th hour of incubation only in the case of *Yersinia enterocolitica*, whereas for the other species the bacteriostatic effects were achieved only from the 8th hour of incubation. TM-fat-challenged broths showed significant effect only in *Pasteurella mutocida* after 8 hours of incubation and this was maintained until the end of the experiment (see details in Table 3).

Table 3. Distribution of average cell counts (Log (CFU)/per mL) over time for the three bacterial species for which inhibition was recorded (means ± SEM; n = 3).

Growth Conditions	T4	T6	T8	T10	T12	T24
			Pasteurella multocida			
Control (A)	4.09 (0.19)	4.14 (0.06)	4.76 (0.32)—A	4.80 (0.34)—A	5.69 (0.49)—A	6.82 (0.44)—A
Control (B)	4.11 (0.17)	3.96 (0.12)	4.17 (0.26)—A	4.60 (0.40)—A	5.44 (0.57)—A	6.54 (0.39)—A
TSB + TM fat	3.95 (0.44)	3.75 (0.39)	3.08 (0.35)—B	3.21 (0.30)—B	2.91 (0.86)—B	3.90 (1.21)—B
TSB + HI fat	3.92 (0.20)	3.48 (0.17)	3.61 (0.19)—B	2.80 (0.59)—B	2.67 (0.49)—B	2.06 (0.36)—C
ANOVA	N.S	N.S	F = 4.99; p = 0.01	F = 5.07; p = 0.01	F = 6.36; p < 0.01	F = 13.75; p < 0.01
			Yersinia enterocolitica			
Control (A)	4.36 (0.16)—A	5.19 (0.28)—A	6. 76 (0.61)—A	7.32 (0.37)—A	8.18 (0.37)—A	9.95 (0.30)—A
Control (B)	4.10 (0.21)—A	5.05 (0.37)—A	5.91 (0.41)—A	6.83 (0.48)—A	7.35 (0.68)—A	9.88 (0.33)—A
BHI + TM fat	4.05 (0.31)—A	4.39 (0.52)—AB	4.98 (0.74)—AB	5.04 (0.94)—AB	5.93 (1.41)—AB	8.09 (2.13)—A
BHI + HI fat	3.34 (0.10)—B	2.69 (0.64)—B	3.01 (1.20)—B	3.02 (1.32)—B	2.94 (1.24)—B	3.98 (2.18)—B
ANOVA	F = 3.28; p = 0.05*	F = 6.38; p < 0.01	F = 4.95; p = 0.01	F = 7.71; p < 0.01	F = 6.46; p < 0.01	F = 5.42; p < 0.01
			Listeria monocytogenes			
Control (A)	4.58 (0.35)	5.58 (0.39)	6.47 (0.41)—A	7.41 (0.41)—A	8.28 (0.46)—A	10.00 (0.26)—A
Control (B)	4.52 (0.32)	5.45 (0.42)	6.26 (0.54)—A	6.94 (0.61)—A	7.95 (0.59)—A	9.96 (0.28)—A
BHI + TM fat	4.20 (0.25)	4.59 (0.79)	5.98 (0.46)—A	6.76 (0.70)—A	7.63 (0.61)—A	9.48 (0.03)—A
BHI + HI fat	4.00 (0.92)	3.66 (0.84)	4.1 (0.63)—B	4.48 (0.61)—B	4.51 (0.62)—B	4.85 (0.97)—B
ANOVA	N.S	N.S	F = 3.30; p = 0.05**	F = 3.98; p = 0.03	F = 6.85; p < 0.01	F = 29.92; p < 0.01

HI: *Hermetia illucens*; TM: *Tenebrio molitor*; T: incubation time in hours; TSB: tryptic soy broth; BHI: brain heart infusion; N.S.: non-significant; SEM: standard error of the means. *Significant results (without approximation $p = 0.046$); **results in trend of significance ($p = 0.051$). In the same columns, different letters identify significantly-different results at post-hoc test (Duncan). Control (A) indicates the broth culture without oil/fat in it and control (B) indicates a broth culture grown in BHI/tryptic soy agar and soybean oil.

3.2. Cecal Fermentation Traits

The cecal fermentation traits are reported in Table 4. The supplementation of insect fat increased the total VFA contents when compared to the control group (85.3 vs. 83.9 vs. 72.4 mmol/L; $p < 0.05$), whereas the pH, the cecal ammonia-N content, and the molar proportion of the different VFAs and their ratio were not influenced by the supplementation of insect fats ($p > 0.05$).

Table 4. Effect of lipid sources on cecal traits and fermentation parameters (n = 10 rabbits/group).

	C	HID	TMD	SEM	*p*-Value
pH	6.1	5.9	5.9	0.01	0.15
N–NH3 (mmol/L)	2.2	3.0	3.1	0.23	0.25
Total VFA (mmol/L)	72.4b	85.3a	83.9a	2.31	0.03
Acetic acid (C2; mmol/100 mmol VFA)	77.8	78.1	76.6	0.43	0.30
Propionic acid (C3; mmol/100 mmol VFA)	5.3	5.0	5.4	0.19	0.71
Butyric acid (C4; mmol/100 mmol VFA)	16.2	16.1	17.2	0.42	0.53
Valeric acid (C5; mmol/100 mmol VFA)	0.5	0.4	0.5	0.03	0.59
Caproic acid (C6; mmol/100 mmol VFA)	0.3	0.3	0.3	0.03	0.45
C2/C3 ratio	15.2	16.0	14.6	0.54	0.58
C2/C4 ratio	4.9	5.0	4.6	0.16	0.63

C: control diet with soybean oil; HID: diet with *Hermetia illucens* fat; TMD: diet with *Tenebrio molitor* fat; SEM: standard error of the means; N–NH3: ammonia-N; VFA: volatile fatty acid.

3.3. Cecal and Fecal Microbiota Characterization

The total number of high-quality paired-end sequences obtained from 16S rRNA sequencing reached 13,448,661 raw reads. After the filtering, 7,801,336 reads passed the filters applied through QIIME, with a median value of 59,114 ± 52,946 reads/sample, and a mean sequence length of 441 bp. The rarefaction analysis and Good's coverage, expressed as a median percentage (93%), also indicated satisfactory coverage for all the samples. We applied a generalized linear model (GLM) for the alpha-diversity in the fecal samples in order to test the effect of the treatment across time. The Chao1 index increased ($p < 0.01$) in the TM and HI groups when compared to the C group ($p < 0.05$), while the Shannon index was affected by the sampling time only. However, the number of observed species significantly increased in the HI group in comparison with the other diets ($p < 0.05$) (Figure 1).

No significant differences were observed when comparing the alpha diversity index as a function of the different diets in the cecal samples. Going more deeply in the microbiota comparison, adonis and analysis of similarity (ANOSIM) statistical tests, based on the weighted and unweighted UniFrac distance matrix showed significant differences among samples as a function of the sampling time ($p < 0.05$). However, the dietary inclusion of insect fat did not show any significant effect in the microbiota composition of the cecal samples.

Figure 2 summarizes the distribution of the microbiota across time and samples and displays a simple microbiota composition dominated by the presence of *Bacteroides*, Clostridiales, Lachnospiraceae, Ruminococcaceae and *Ruminococcus*. Comparing the relative abundance of the main OTUs across the fecal samples grouped as a function of the dietary supplementation by the GLM function, it was possible to observe that HI inclusion level increased the relative abundance of *Akkermansia* ($p < 0.05$) and *Ruminococcus* ($p < 0.01$) and reduced the presence of *Citrobacter* ($p < 0.05$).

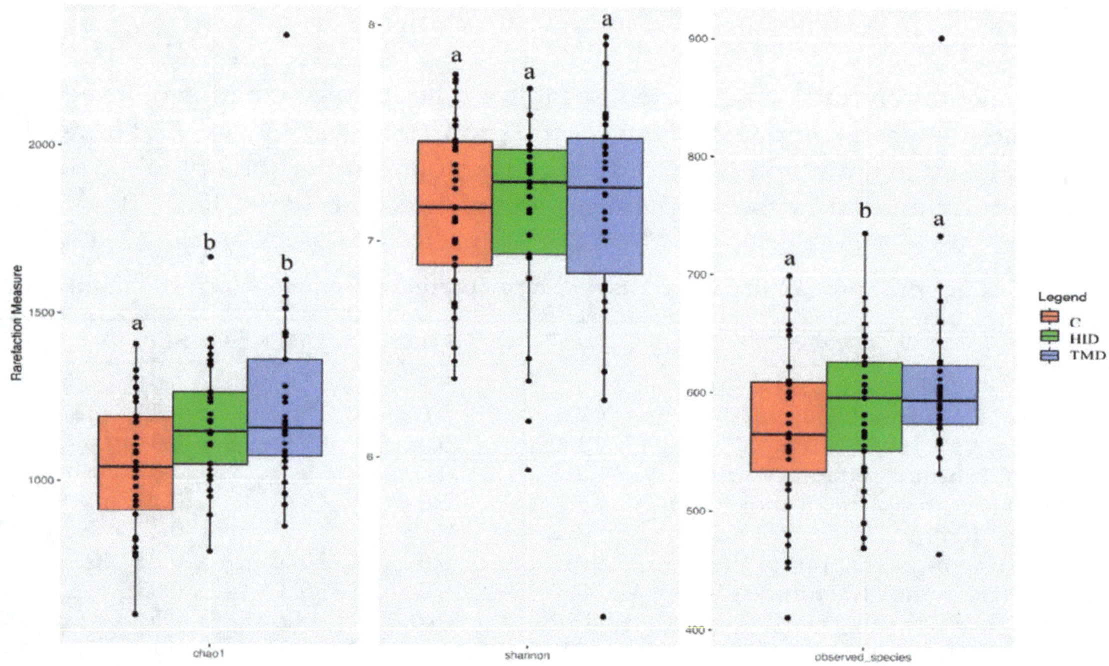

Figure 1. Boxplots showing the alpha diversity rarefaction index across fecal samples of rabbits fed with TMD (blue box), HID (green box), or C (red box) diets; C: control diet with soybean oil; HID: diet with *Hermetia illucens* fat; TMD: diet with *Tenebrio molitor* fat; Boxes represent the interquartile range (IQR) between the first and third quartile and the line inside represents the median (2nd quartile). Whiskers denote the lowest and the highest values within 1.56 IQR from the first and third quartile, respectively. Different letters in each box indicate significant difference.

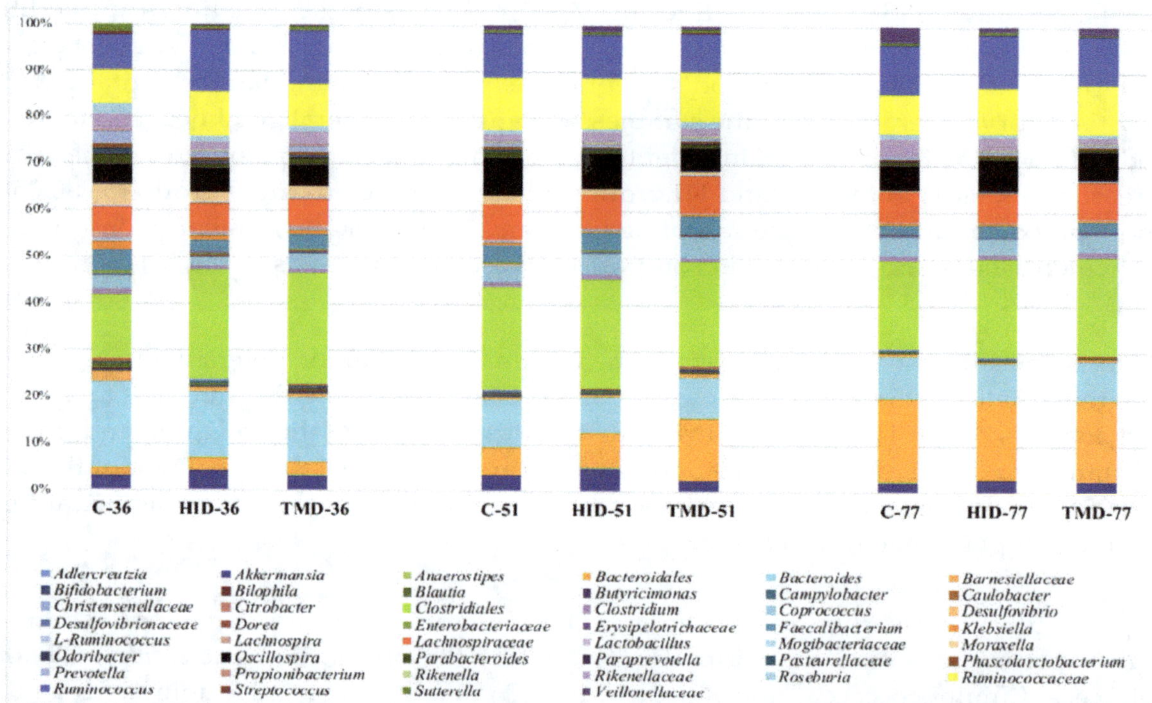

Figure 2. Taxonomic groups detected in fecal samples by means of 16S amplicon target sequencing. Only operational taxonomic units (OTUs) with an incidence above 0.2% in at least two samples are shown. The samples are labelled according to the type of fat supplementation (C: control diet with soybean oil; HID: diet with *Hermetia illucens* fat; TMD: diet with *Tenebrio molitor* fat) and sampling time (36, 51, or 77 days).

Both HID and TMD increased the relative abundance of Clostridiales ($p < 0.01$) and Desulfovibrionaceae ($p < 0.01$), while reducing the relative abundance of *Lachnospira* ($p < 0.05$) and *Phascolarctobacterium* ($p < 0.01$) when compared to the C diet (Figure 3).

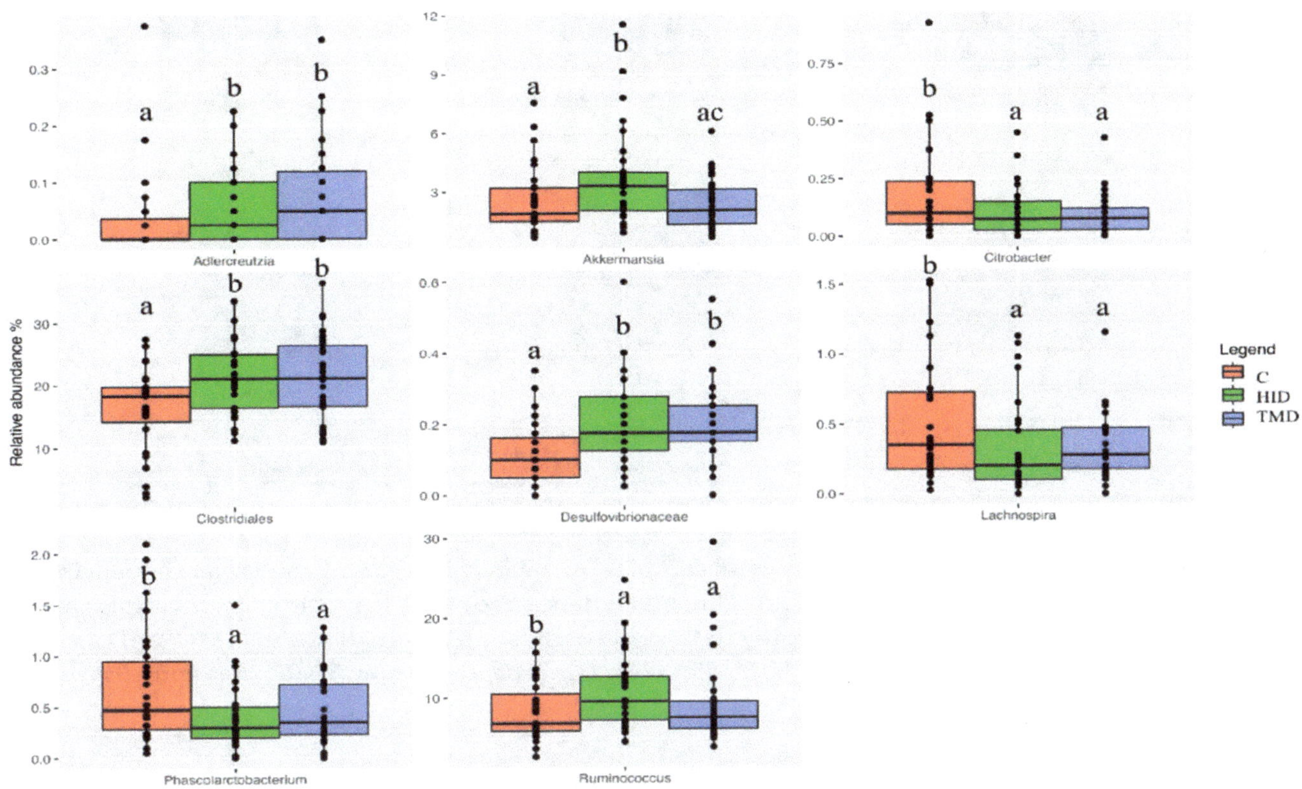

Figure 3. Boxplots showing the relative abundance at genus or family level of operational taxonomic units (OTUs) differential abundant based on the generalized linear model (GLM) test in fecal samples of rabbits fed with TMD (blue box), HID (green box), or C diets (red box). Different letters in each box indicate significant difference; C: control diet with soybean oil; HID: diet with *Hermetia illucens* fat; TMD: diet with *Tenebrio molitor* fat.

Taking into account the cecal samples (Figure 4), no differences were observed regarding the microbial composition and distribution. However, the pairwise comparisons using the Wilcoxon rank sum test showed that TM inclusion reduced the relative abundance of *Klebsiella, Lachnospira, Parabacteroides,* and *Odoribacter* compared to the C and HID (Figure 5, $p < 0.05$).

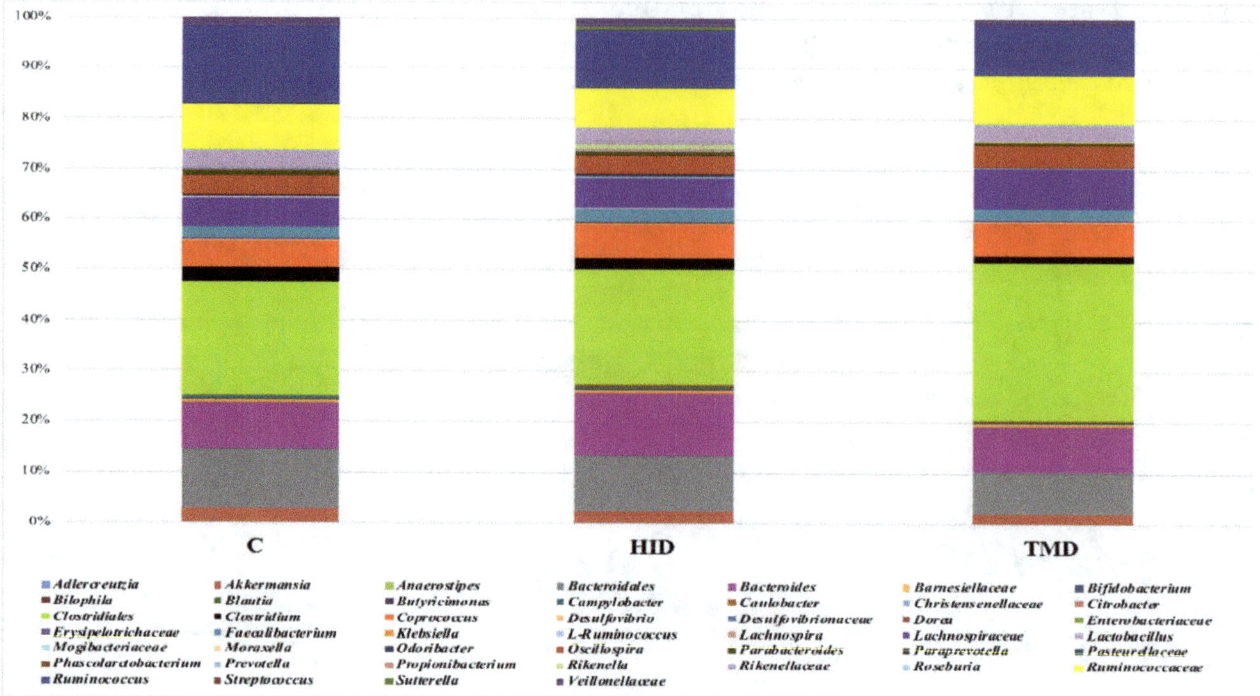

Figure 4. Taxonomic groups detected in cecal samples by means of 16S amplicon target sequencing. Only operational taxonomic units (OTUs) with an incidence above 0.2% in at least two samples are shown. The samples are labelled according to the type of fat supplementation, i.e., TMD, HID, or C diets; C: control diet with soybean oil; HID: diet with *Hermetia illucens* fat; TMD: diet with *Tenebrio molitor* fat.

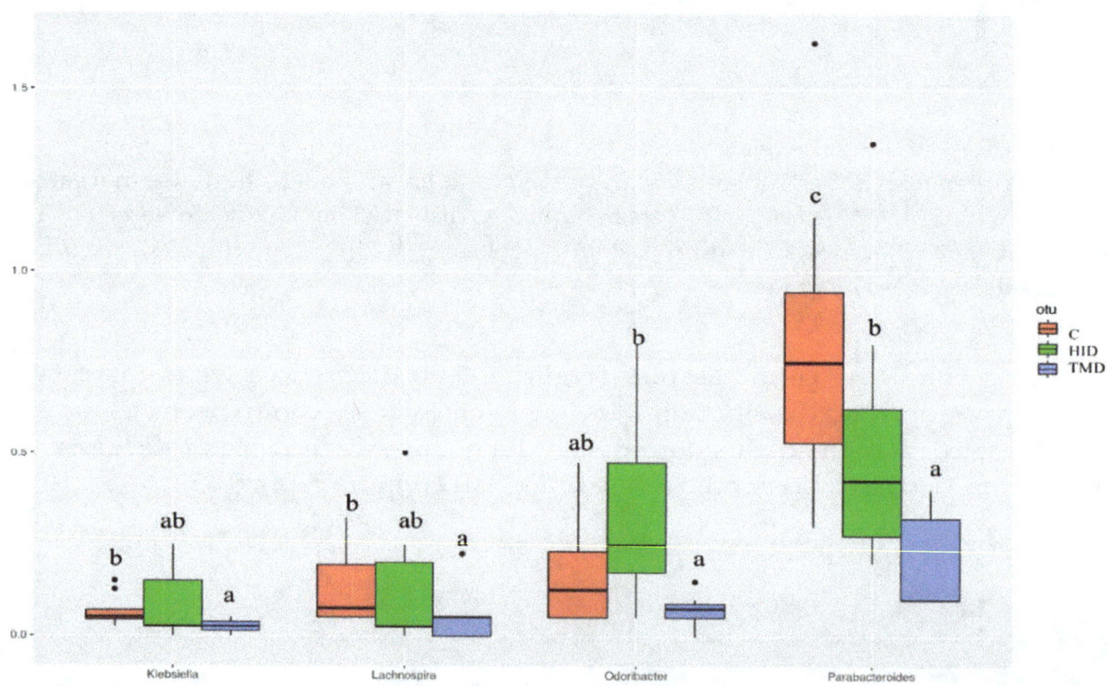

Figure 5. Boxplots showing the relative abundance at genus or family level of the operational taxonomic units (OTUs) differential abundance based on Wilcoxon rank sum test in cecal samples of rabbits fed with TM (blue box), HI (green box), or C (red box) diets. Different letters in each box indicate significant difference; C: control diet with soybean oil; HID: diet with *Hermetia illucens* fat; TMD: diet with *Tenebrio molitor* fat.

The correlations between VFA contents and microbiota are summarized in Figure 6. In detail, strong positive correlations between *Lachnospira* and propanoic acid, *Akkermansia*, Clostridiales, and NH3-N, and *Phascolarctobacterium* and acetic, propanoic, and caproic acids were detected ($p < 0.05$).

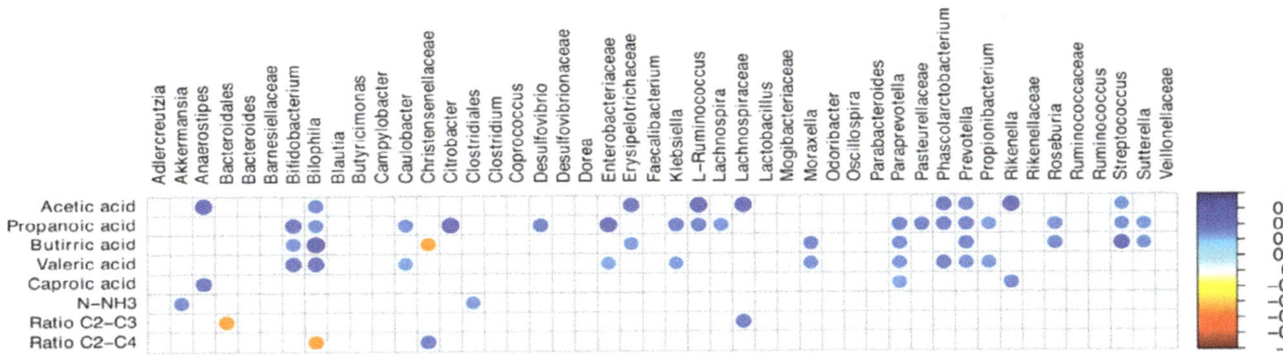

Figure 6. Spearman's rank correlation matrix of cecal operational taxonomic unit (OTU) abundance and the volatile fatty acid profile. The colors of the scale bar denote the nature of the correlation, with 1 indicating a positive correlation (blue) and -1 indicating a negative correlation (dark red), between the two datasets; N–NH3: ammonia-N; C2: Acetic acid; C3: Propanoic acid; C4: Butyric acid.

4. Discussion

The current study was conducted to evaluate the in vitro antimicrobial properties of HI and TM fats and their effect as an alternative lipid sources in rabbit diets. The possible utilization of insect fats in animal diets has been investigated recently, and so far, a few papers focusing on antimicrobial activities in vitro or in vivo are available, but no data are present in the literature on pathogen growth rate in presence of insect fats. Mustafa et al. [45] showed that the oil extracted from the melon bug (*Aspongopus vidulatus*) was able to inhibit the growth of bacterial species by using the agar well diffusion method, with only Gram positive bacteria (*Staphylococcus, Bacillus,* and *Enterococcus*) being susceptible to crude oil extracts. *Salmonella paratyphi* was also tested and no inhibition was recorded, similarly to what was observed in the present research with other serovars. A recent study of Spranghers et al. [30] on the in vitro effects of fats extracted from HI (from prepupae fed to weaned piglets) pointed out that D-*Streptococci* and *Lactobacilli* were the only bacterial populations that reduced their load after being challenged with insect fats, whereas no effects were recorded on Gram negative bacteria. The data presented in our paper suggest that fats extracted from TM and HI are able to delay bacterial growth of both Gram positive and Gram negative pathogenic bacteria, even if the susceptibility changes with the considered species. Interestingly, no bactericidal effect was observed, thus indicating the possibility of bacterial cells repairing damage that may be induced from FAs or monoglycerides [8]. The antimicrobial effect of TM fat was lower than that of HI fat. These results may be related to the different concentrations of SFA. Indeed, the HI fat used in this study was composed by 79% of SFA (25% in TM fat), and the major component was lauric acid, already reported as very effective against many bacterial species [8,30]. In a systematic study of the in vitro antimicrobial activities of FAs, monoglycerides, and diglycerides, Kabara et al. [46] showed that lauric acid was the most active FA against gram-positive bacteria. On the other hand, the antibacterial effects of unsaturated fatty acids (UFA) reported in literature [8] were not detected in this study, probably in relation to the prolonged incubation time and the temperature of 37 °C, which may have been responsible for the reduction/impairment of the activity of these molecules that characterize TM fat (75% of the FA vs. 21% of HI fat). This hypothesis may also explain why soybean oil did not show any activity, considering its higher level in monounsaturated fatty acids (MUFAs) and PUFAs among all fats used in this work.

The present observations highlighted the possibility of using insect fats in feed formulation, considering that they may be important for controlling growth of important microbial pathogens such as *Listeria monocytogenes, Yersinia enterocolitica,* and *Pasteurella multocida* that may be important pathogens for rabbits [34,37] or part of the gut flora potentially contaminating rabbit meat during slaughtering [35,36]. Reducing microbial loads in the rabbit gut, apart from animal welfare and safety implications, may also be important for food safety management during slaughtering operations. Studies performed on piglets fed with diets rich in FAs showed a decreased load of microbes in the stomach and gut, mostly if undesired species were considered (i.e., *streptococci*), thus promoting gut heath and growth performances [30,47]. However, as already emphasized by Spranghers et al. [30], more studies need to be performed in vivo, in order to assess the activity of fat in the gut, considering that the bacterial activities and the digestive enzymatic systems of the rabbit (i.e., lipases) may neutralize the FAs, therefore limiting their activity. Furthermore, the presence of FAs as antimicrobial molecules in feed may prevent the use of classical antibiotics in meat production, limiting the diffusion of microbes harboring antimicrobial resistance genes. In fact, antimicrobial resistance mechanisms were rarely observed in microbial species challenged with FAs [47], and to date no genetic modifications induced by exposing bacterial cells to antimicrobial FAs have been shown [8,48].

As far as the in vivo trial is concerned, there was a lack of differences among groups in cecal fermentation traits in our study. Cecal pH and VFA content are the main variables characterizing the extent and the pattern of cecal fermentation, thus constituting an indirect estimate and an important tool to qualitatively evaluate the cecal microbial activity. The dietary HI or TM fat inclusion led to a greater total VFA content in the cecum than that of control diet, thus potentially enhancing the gut with a modification of the fermentation patterns and the composition of the cecal microflora. Peeters et al. [49] previously observed that a high concentration of total cecal VFA in rabbits had a protective effect against enteropathogenic *Escherichia coli* infection. However, the molar proportion of different VFA was not affected by the total replacement of soybean oil with HI and TM larvae fats.

The present study is the first to investigate gut microbiota of growing rabbits fed diets supplemented with insect fats. As reported previously, HI fat was rich in MCFAs which have a higher antimicrobial activity and thus can stimulate gastrointestinal tract health through inhibition of potentially-pathogenic bacteria [31,33]. MCFAs, and in particular lauric acid, are presumably absorbed in the upper digestive tract, and might be useful in protecting against microbial infection, modulating inflammation, healing wounds, and controlling the balance and distribution of bacteria in gut microbiota [50]. The present study is the first to investigate gut microbiota of growing rabbits fed diets supplemented with insect fats. The results of this study revealed an enrichment of different taxa according to the dietary treatment and a similar microbial diversity and richness between feces and cecum samples. Cecal microbiota is a primary determinant for rabbit health, whereas the fecal microbiota provides an accurate method for studying the evolution of rabbit gut microbiota from weaning to slaughtering [51,52]. Investigating the differences between rabbits fed the C and the HI and TM fat diets, in the current study, no differences were found with regard to alpha-diversity measures in cecal and fecal samples. However, we observed a higher microbial diversity in the HI and TM groups. High levels of diversity generally help intestinal microbiota to determine effective colonization resistance against potential invading pathogens and to modulate animal reaction after a stress-environment [53]. Based on such considerations, the above-mentioned findings are indicative of a positive HI and TM fat-related effect on the gut microbiota of rabbits. *Firmicutes, Bacteroidetes,* and *Verrucomicrobia* represented the dominant bacterial phyla in the control and insect-fat-fed rabbits of the present study. These findings overall agree with previous research that identified *Firmicutes* and *Bacteroides* as the main bacterial phyla in the gut microbiota of rabbits [3,52,54,55]. In relation to the genera composition, the *Bacteroides*, Bacteroidales, Clostridiales, Lachnospiraceae, Ruminococcaceae, and *Ruminococcus*, mainly colonized the cecal and fecal microbiota of the rabbits fed soybean oil or insect fats in the current study. These findings are also in agreement with previous studies, which observed *Bacteroides* [52,55], Clostridiales [54], and *Ruminococcus* [53,56] as being the main bacterial genera in cecal and fecal microbiota of rabbits and being capable of degrading the polysaccharide and amino acid fermentation to produce VFAs. These findings could be the reason for its overrepresentation in cecum where it is supposed to play an active role [52,57]. A previous study reported that the presence of this taxa was positively correlated with the feed conversion rate [58].

Regarding the microbial composition, we did not observe any strong effects as a consequence of the dietary inclusion of HI and TM fats. However, a signature in the microbial population was observed. The fat of TM reduced some taxa such as *Klebsiella, Lachnospira, Parabacteroides,* and *Odoribacter*. On the other hand, the dietary supplementation of HI and TM fats enriched the presence of Clostridiales, Desulfovibrionaceae, *Ruminococcus,* and *Akkermansia*, which are the main taxa in the gut microbiota of rabbits [51]. It is well reported that *Akkermansia* can be considered a new-generation probiotic, able to degrade the mucin in the gut with the production of beneficial molecules like SCFAs thus exerting a significant improvement in the gut barrier and in the maintenance of intestinal health [59–61]. In addition, it was suggested that *Akkermansia* have an important role in the hydrolysis of various dietary polysaccharides, contributing to increase cellulose digestibility as well as methane metabolism [62–64]. *Akkermansia* would be also involved in carbohydrate digestion and in immune protection against inflammation [54]. The increase of this taxon related with the insect fat inclusion suggests an optimal gut environment in our rabbits, even if this observation needs further investigation to be confirmed. In addition, a strong positive correlation between *Akkermansia* and NH3-N was observed. This result confirmed that gut microbiota is responsible for a variety of metabolic activities including production of biologically-active substances. The Ruminococcaceae family is considered as an important producer of short-chain FAs (mainly butyrate, acetic, and succinic acids) through glucose metabolism and cellulose digestion [65,66]. It was reported that members of Ruminococcaceae are important components of the beneficial microbiota of several herbivores [67,68]. Their presence is related to an improvement of the immune system of the host via intestinal mucus degradation and a prevention of acidosis via lactate degradation [69]. Ruminococcaceae has been associated with antibiotic biosynthesis in a metatranscriptomic study on the human gut microbiota, suggesting a role in gut health [70]. The supplementation of HI fat also increased the presence of *Ruminococcus* (belonging to Lachnospiraceae family), that are butyrate-producing bacteria. This ability was also confirmed by the strong positive correlation between L-*Ruminococcus*, acetic, and propionic acid. The identification of short-chain FA-producing bacteria allows us to hypothesize a similar way of action for HI fat in the rabbit's gut. Therefore, the increase in the above-mentioned bacterial taxa by dietary HI fat may have helped the rabbits to maintain a healthy gut. The microbial signature driven by the inclusion of insect fat increases the presence of *Bacteroides, Clostridium, Akkermansia,* and *Ruminococcus* and suggests that dietary HI and TM fats may exert a positive influence on the cecal microbiota of rabbits. It should be pointed out that rabbits fed HI and TM fats showed no significant alterations at histopathological level and no differences in growth performance [14]. However, since a clear cause–effect relationship between diversity and composition of cecal microbiota and rabbit performance has not yet been confirmed, the gut microbiota findings need to be contextualized to underline the effect of insect fats in rabbit diets.

5. Conclusions

In conclusion, the results of the present study provide new information about the in vitro antibacterial properties and the in vivo effects at the gut level of the replacement of soybean oil with HI and TM larvae fat in rabbit diets. The in vitro activities of HI or TM fats against *Pasteurella, Yersinia,* and known pathogens of the rabbit gut, indicate a potential for impairing their growth in vivo in rabbits. Furthermore, the dietary inclusion of HI and TM fats stimulate VFA production at the cecum and may positively modulate the cecal and fecal microbiota of growing rabbits. However, further research is needed to confirm the antimicrobial potentiality of insect fats in rabbit feeding.

Author Contributions: Conceptualization, S.D., L.G., A.S., A.T., G.X., and F.G.; data curation, I.F., D.M.N., A.T., G.X., S.D., S.M., D.S., and A.C.B.; funding acquisition L.G.; methodology, S.D., D.M.N., I.F., F.G., I.B., L.C., and L.G.; writing—original draft preparation, S.D., I.F., and D.M.N.; writing—review and editing, all the authors. All authors have read and agreed to the published version of the manuscript.

Acknowledgments: For Schiavone A. this work is the result of the visit (21142/IV/19) funded by the Fundación Séneca-Agencia de Ciencia y Tecnología de la Región de Murcia (Spain) in connection with the "Jiménez De La Espada" Regional Programme for Mobility, Collaboration and Knowledge Exchange (Spain).

References

1. Licois, D.; Wyers, M.; Coudert, P. Epizootic Rabbit Enteropathy: Experimental transmission and clinical characterization. *Vet. Res.* **2005**, *36*, 601–613. [CrossRef] [PubMed]
2. Combes, S.; Fortun-Lamothe, L.; Cauquil, L.; Gidenne, T. Engineering the rabbit digestive ecosystem to improve digestive health and efficacy. *Animal* **2013**, *7*, 1429–1439. [CrossRef] [PubMed]
3. Vántus, V.B.; Kovács, M.; Zsolnai, A. The rabbit cecal microbiota: Development, composition and its role in the prevention of digestive diseases—A review on recent literature in the light of molecular genetic methods. *Acta Agrar. Kaposváriensis* **2014**, *18*, 55–65.
4. Fortun-Lamothe, L.; Theau-Clément, M.; Combes, S.; Allain, D.; Lebas, F.; Le Normand, B.; Gidenne, T. Physiologie. In *Le Lapin de la Biologie à L'élevage*, 1st ed.; Gidenne, T., Ed.; Quae: Versailles, France, 2015; pp. 48–49. ISBN 9782759224166.
5. Fortun-Lamothe, L.; Boullier, S. A review on the interactions between gut microflora and digestive mucosal immunity. Possible ways to improve the health of rabbits. *Livest. Sci.* **2007**, *107*, 1–18. [CrossRef]
6. Kim, S.J.; Kim, S.E.; Kim, A.R.; Kang, S.; Park, M.Y.; Sung, M.K. Dietary fat intake and age modulate the composition of the gut microbiota and colonic inflammation in C57BL/6J mice. *BMC Microbiol.* **2019**, *19*, 193. [CrossRef]
7. Yang, Q.; Liang, Q.; Balakrishnan, B.; Belobrajdic, D.P.; Feng, Q.J.; Zhang, W. Role of dietary nutrients in the modulation of gut microbiota: A narrative review. *Nutrients* **2020**, *12*, 381. [CrossRef]
8. Yoon, B.K.; Jackman, J.A.; Valle-González, E.R.; Cho, N.J. Antibacterial free fatty acids and monoglycerides: Biological activities, experimental testing, and therapeutic applications. *Int. J. Mol. Sci.* **2018**, *19*, 1114. [CrossRef]
9. Xiccato, G. Fat Digestion. In *Nutrition of the Rabbit*, 3rd ed.; De Blas, C., Wiseman, J., Eds.; CAB International: Wallingford, UK, 2020; pp. 58–68. ISBN 9781789241273. [CrossRef]
10. Gidenne, T.; Lebas, F.; Licois, D.; García, J. Nutrition and Feeding Strategy: Impact on Health Status. In *Nutrition of the Rabbit*, 3rd ed.; De Blas, C., Wiseman, J., Eds.; CAB International: Wallingford, UK, 2020; pp. 193–221. ISBN 9781789241273. [CrossRef]
11. Christaki, E.; Marcou, M.; Tofarides, A. Antimicrobial resistance in Bacteria: Mechanisms, evolution, and persistence: Review. *J. Mol. Evol.* **2020**, *88*, 26–40. [CrossRef]
12. Girard, M.; Bee, G. Invited review: Tannins as a potential alternative to antibiotics to prevent coliform diarrhea in weaned pigs. *Animal* **2020**, *14*, 95–107. [CrossRef]
13. Gasco, L.; Biasato, I.; Dabbou, S.; Schiavone, A.; Gai, F. Animals fed insect-based diets: State-of-the-art on digestibility, performance and product quality. *Animals* **2019**, *9*, 170. [CrossRef]
14. Gasco, L.; Dabbou, S.; Trocino, A.; Xiccato, G.; Capucchio, M.T.; Biasato, I.; Dezzutto, D.; Birolo, M.; Meneguz, M.; Schiavone, A.; et al. Effect of dietary supplementation with insect fats on growth performance, digestive efficiency and health of rabbits. *J. Anim. Sci. Biotechnol.* **2019**, *10*, 4. [CrossRef] [PubMed]
15. Sypniewski, J.; Kierończyk, B.; Benzertiha, A.; Mikołajczak, Z.; Pruszyńska Oszmałek, E.; Kołodziejski, P.; Sassek, M.; Rawski, M.; Czekała, W.; Józefiak, D. Replacement of soybean oil by *Hermetia illucens* fat in turkey nutrition: Effect on performance, digestibility, microbial community, immune and physiological status and final product quality. *Br. Poult. Sci.* **2020**, *61*, 294–302. [CrossRef]
16. Borrelli, L.; Coretti, L.; Dipineto, L.; Bovera, F.; Menna, F.; Chiariotti, L.; Nizza, A.; Lembo, F.; Fioretti, A. Insect based diet, a promising nutritional source, modulates gut microbiota composition and SCFAs production in laying hens. *Sci. Rep.* **2017**, *7*, 16269. [CrossRef]

17. Biasato, I.; Ferrocino, I.; Biasibetti, E.; Grego, E.; Dabbou, S.; Sereno, A.; Gai, F.; Gasco, L.; Schiavone, A.; Cocolin, L.; et al. Modulation of intestinal microbiota, morphology and mucin composition by dietary insect meal inclusion in free-range chickens. *BMC Vet. Res.* **2018**, *14*, 383. [CrossRef]
18. Biasato, I.; Ferrocino, I.; Grego, E.; Dabbou, S.; Gai, F.; Gasco, L.; Cocolin, L.; Capucchio, M.T.; Schiavone, A. Gut microbiota and mucin composition in female broiler chickens fed diets including yellow mealworm (*Tenebrio molitor*, L.). *Animals*, 2019; 9, 213. [CrossRef]
19. Biasato, I.; Ferrocino, I.; Dabbou, S.; Evangelista, R.; Gai, F.; Gasco, L. Black soldier fly and gut health in broiler chickens: Insights into the relationship between caecal microbiota and intestinal mucin composition. *J. Anim. Sci. Biotechnol.* **2020**, *11*, 11. [CrossRef] [PubMed]
20. Berezina, N. Insects: Novel source of lipids for a fan of applications. *Oil Seeds Fats Crop. Lipids.* **2017**, *24*, D402. [CrossRef]
21. Schiavone, A.; Cullere, M.; De Marco, M.; Meneguz, M.; Biasato, I.; Bergagna, S.; Dezzutto, D.; Gai, F.; Dabbou, S.; Gasco, L.; et al. Partial or total replacement of soybean oil by black soldier larvae (*Hermetia illucens* L.) fat in broiler diets: Effect on growth performances, feed-choice, blood traits, carcass characteristics and meat quality. *Ital. J. Anim. Sci.* **2017**, *16*, 93–100. [CrossRef]
22. Schiavone, A.; Dabbou, S.; De Marco, M.; Cullere, M.; Biasato, I.; Biasibetti, E.; Capucchio, M.T.; Bergagna, S.; Dezzutto, D.; Meneguz, M.; et al. Black soldier fly (*Hermetia illucens* L.) larva fat inclusion in finisher broiler chicken diet as an alternative fat source. *Animal* **2018**, *12*, 2032–2039. [CrossRef]
23. Dalle Zotte, A.; Cullere, M.; Martins, C.; Alves, S.P.; Freire, J.P.B.; Falcão-e-Cunha, L.; Bessa, R.J.B. Incorporation of Black Soldier Fly (*Hermetia illucens* L.) larvae fat or extruded linseed in diets of growing rabbits and their effects on meat quality traits including detailed fatty acid composition. *Meat Sci.* **2018**, *146*, 50–56. [CrossRef]
24. Kierónczyk, B.; Rawski, M.; Józefiak, A.; Mazurkiewicz, J.; Świątkiewicz, S.; Siwek, M.; Bednarczyk, M.; Szumacher-Strabel, M.; Cieślak, A.; Benzertiha, A.; et al. Effects of replacing soybean oil with selected insect fats on broilers. *Anim. Feed Sci. Technol.* **2018**, *240*, 170–183. [CrossRef]
25. Benzertiha, A.; Kierónczyk, B.; Rawski, M.; Kołodziejski, P.; Bryszak, M.; Józefiak, D. Insect oil as an alternative to palm oil and poultry fat in broiler chicken nutrition. *Animals* **2019**, *9*, 116. [CrossRef]
26. Gasco, L.; Dabbou, S.; Gai, F.; Brugiapaglia, A.; Schiavone, A.; Birolo, M.; Xiccato, G.; Trocino, A. Quality and consumer acceptance of meat from rabbits fed diets in which soybean oil is replaced with black soldier fly and yellow mealworm fats. *Animals* **2019**, *9*, 629. [CrossRef]
27. Matsue, M.; Mori, Y.; Nagase, S.; Sugiyama, Y.; Hirano, R.; Ogai, K.; Ogura, K.; Kurihara, S.; Okamoto, S. Measuring the antimicrobial activity of lauric acid against various bacteria in human gut microbiota using a new method. *Cell Transplant.* **2019**, *28*, 1528–1541. [CrossRef]
28. Skřivanová, E.; Marounek, M.; Benda, V.; Brezina, P. Susceptibility of *Escherichia coli, Salmonella* s.p. and *Clostridium perfringens* to organic acids and monolaurin. *Vet. Med.* **2006**, *51*, 81–88. [CrossRef]
29. Zeitz, J.O.; Fennhoff, J.; Kluge, H.; Stangl, G.I.; Eder, K. Effects of dietary fats rich in lauric and myristic acid on performance, intestinal morphology, gut microbes, and meat quality in broilers. *Poult. Sci.* **2015**, *94*, 2404–2413. [CrossRef] [PubMed]
30. Spranghers, T.; Michiels, J.; Vrancx, J.; Ovyn, A.; Eeckhout, M.; De Clercq, P.; De Smet, S. Gut antimicrobial effects and nutritional value of black soldier fly (*Hermetia illucens* L.) prepupae for weaned piglets. *Anim. Feed Sci. Tech.* **2018**, *235*, 33–42. [CrossRef]
31. Baltić, B.; Starčević, M.; Đorđević, J.; Mrdović, B.; Marković, R. Importance of medium chain fatty acids in animal nutrition. *IOP Conf. Ser. Earth Environ. Sci.* **2017**, *85*, 012048. [CrossRef]
32. Skřivanová, V.; Marounek, M. A note on the effect of triacylglycerols of caprylic and capric fatty acid on performance, mortality, and digestibility of nutrients in young rabbits. *Anim. Feed Sci. Tech.* **2006**, *127*, 161–168. [CrossRef]
33. Boyen, F.; Haesebrouck, F.; Vanparys, A.; Volf, J.; Mahu, M.; Van Immerseel, F.; Rychlik, I.; Dewulf, J.; Ducatelle, R.; Pasmans, F. Coated fatty acids alter virulence properties of *Salmonella Typhimurium* and decrease intestinal colonization of pigs. *Vet. Microbiol.* **2008**, *132*, 319–327. [CrossRef]
34. Rodriguez-Calleja, J.M.; Garcia-Lopez, I.; Garcia Lopez, M.L.; Santos, J.A.; Otero, A. Rabbit meats as a source of bacterial foodborne pathogens. *J. Food Prot.* **2006**, *69*, 1106–1112. [CrossRef]
35. De Cesare, A.; Parisi, A.; Mioni, R.; Comin, D.; Lucchi, A.; Manfreda, G. *Listeria monocytogenes* circulating in rabbit meat products and slaughterhouses in Italy: Prevalence data and comparison among typing results. *Foodborne Pathog. Dis.* **2017**, *14*, 167–176. [CrossRef] [PubMed]

36. Kylie, K.; McEwen, S.A.; Boerlin, P.; Reid-Smith, M.J.; Weese, J.S.; Turner, P.V. Prevalence of antimicrobial resistance in fecal *Escherichia coli* and *Salmonella enterica* in Canadian commercial meat, companion, laboratory, and shelter rabbits (*Oryctolagus cuniculus*) and its association with routine antimicrobial use in commercial meat rabbits. *Prev. Vet. Med.* **2017**, *147*, 53–57. [PubMed]
37. Massacci, F.R.; Magistrali, C.F.; Cucco, L.; Curcio, L.; Bano, L.; Mangili, P.M.; Scoccia, E.; Bisgaard, M.; Aalbæk, B.; Christensen, H. Characterization of *Pasteurella multocida* involved in rabbit infections. *Vet. Microbiol.* **2018**, *213*, 66–72. [CrossRef] [PubMed]
38. Tazzoli, M.; Trocino, A.; Birolo, M.; Radaelli, G.; Xiccato, G. Optimizing feed efficiency and nitrogen excretion in growing rabbits by increasing dietary energy with high-starch, high-soluble fibre, low-insoluble fibre supply at low protein levels. *Livest. Sci.* **2015**, *172*, 59–68. [CrossRef]
39. Klindworth, A.; Pruesse, E.; Schweer, T.; Peplies, J.; Quast, C.; Horn, M.; Glöckner, F.O. Evaluation of general 16S ribosomal RNA gene PCR primers for classical and next-generation sequencing-based diversity studies. *Nucleic Acids Res.* **2013**, *41*, e1. [CrossRef]
40. Magoč, T.; Salzberg, S.L. FLASH: Fast length adjustment of short reads to improve genome assemblies. *Bioinformatics* **2011**, *27*, 2957–2963. [CrossRef]
41. Caporaso, J.G.; Kuczynski, J.; Stombaugh, J.; Bittinger, K.; Bushman, F.D.; Costello, E.K.; Fierer, N.; Peña, A.G.; Goodrich, J.K.; Gordon, J.I.; et al. QIIME allows analysis of high-throughput community sequencing data. *Nat. Methods* **2010**, *7*, 335–336. [CrossRef]
42. Schmieder, R.; Edwards, R. Quality control and preprocessing of metagenomic datasets. *Bioinformatics* **2011**, *27*, 863–864. [CrossRef]
43. Edgar, R.C. Search and clustering orders of magnitude faster than BLAST. *Bioinformatics* **2010**, *26*, 2460–2461. [CrossRef]
44. Dixon, A.L.; Liang, L.; Moffatt, M.F.; Chen, W.; Heath, S.; Wong, K.C.C.; Taylor, J.; Burnett, E.; Gut, I.; Farrall, M.; et al. A genome-wide association study of global gene expression. *Nat. Genet.* **2007**, *39*, 1202–1207. [CrossRef]
45. Mustafa, N.E.M.; Mariod, A.A.; Matthäus, B. Antibacterial activity of *Aspongopus Viduatus* (melon bug) oil. *J. Food Saf.* **2008**, *28*, 577–586. [CrossRef]
46. Kabara, J.J.; Swieczkowski, D.M.; Conley, D.J.; Truant, J.P. Fatty Acids and Derivatives as Antimicrobial Agents. *Antimicrob. Agents Chemother.* **1972**, *2*, 23–28. [CrossRef] [PubMed]
47. Desbois, A.P. Potential applications of antimicrobial fatty acids in medicine, agriculture and other industries. *Recent Pat. Antiinfect. Drug Discov.* **2012**, *7*, 111–122. [CrossRef] [PubMed]
48. Davies, J.; Davies, D. Origins and evolution of antibiotic resistance. *Microbiol. Mol. Biol. Rev.* **2010**, *74*, 417–433. [CrossRef] [PubMed]
49. Peeters, J.E.; Maertens, L.; Orsenigo, R.; Colin, M. Influence of dietary beet pulp on caecal VFA, experimental colibacillosis and iota-enterotoxaemia in rabbits. *Anim. Feed Sci. Tech.* **1995**, *51*, 123–139. [CrossRef]
50. Dayrit, F.M. The properties of lauric acid and their significance in coconut oil. *J. Am. Oil Chem. Soc.* **2015**, *92*, 1–15. [CrossRef]
51. Kylie, K.; Weese, J.S.; Turner, P.V. Comparison of the fecal microbiota of domestic commercial meat, laboratory, companion, and shelter rabbits (*Oryctolagus cuniculi*). *BMC Vet. Res.* **2018**, *14*, 143. [CrossRef]
52. Fang, S.; Chen, X.; Zhou, L.; Wang, C.; Chen, Q.; Lin, R.; Xiao, T.; Gan, Q. Faecal microbiota and functional capacity associated with weaning weight in meat rabbits. *Microb. Biotechnol.* **2019**, *12*, 1441–1452. [CrossRef]
53. Sudo, N. Role of gut microbiota in brain function and stress-related pathology. *Biosci. Microbiota. Food. Health* **2019**, *38*, 75–80. [CrossRef]
54. Velasco-Galilea, M.; Piles, M.; Vinas, M.; Rafel, O.; Gonzalez-Rodriguez, O.; Guivernau, M.; Sánchez, J.P. Rabbit microbiota changes throughout the intestinal tract. *Front. Microbiol.* **2018**, *9*, 2144. [CrossRef]
55. Chen, S.Y.; Deng, F.; Jia, X.; Liu, H.; Zhang, G.W.; Lai, S.J. Gut microbiota profiling with differential tolerance against the reduced dietary fibre level in rabbit. *Sci. Rep.* **2019**, *9*, 1–9. [CrossRef] [PubMed]
56. Dabbou, S.; Ferrocino, I.; Kovitvadhi, A.; Dabbou, S.; Bergagna, S.; Dezzuto, D.; Schiavone, A.; Cocolin, L.; Gai, F.; Santoro, V.; et al. Bilberry pomace in rabbit nutrition: Effects on growth performance, apparent digestibility, caecal traits, bacterial community and antioxidant status. *Animal* **2019**, *13*, 53–63. [CrossRef] [PubMed]
57. Dai, Z.L.; Wu, G.; Zhu, W.Y. Amino acid metabolism in intestinal bacteria: Links between gut ecology and host health. *Front. Biosci.* **2011**, *16*, 1768–1786. [CrossRef] [PubMed]

58. North, M.K.; Dalle Zotte, A.; Hoffman, L. Composition of rabbit caecal microbiota and the effects of dietary quercetin supplementation and sex thereupon. *World Rabbit Sci.* **2019**, *27*, 185–198. [CrossRef]
59. Dao, M.C.; Everard, A.; Aron-Wisnewsky, J.; Sokolovska, N.; Prifti, E.; Verger, E.O.; Kayser, B.D.; Levenez, F.; Chilloux, J.; Hoyles, L.; et al. *Akkermansia muciniphila* and improved metabolic health during a dietary intervention in obesity: Relationship with gut microbiome richness and ecology. *Gut* **2016**, *65*, 426–436. [CrossRef]
60. Gómez-Gallego, C.; Pohl, S.; Salminen, S.; De Vos, W.M.; Kneifel, W. *Akkermansia muciniphila*: A novel functional microbe with probiotic properties. *Benef. Microbes* **2016**, *7*, 571–584. [CrossRef]
61. Zhai, Q.; Feng, S.; Arjan, N.; Chen, W. A next generation probiotic, *Akkermansia muciniphila*. *Crit. Rev. Food Sci. Nutr.* **2019**, *59*, 3227–3236. [CrossRef]
62. Dunfield, P.F.; Yuryev, A.; Senin, P.; Smirnova, A.V.; Stott, M.B.; Hou, S.; Ly, B.; Saw, J.H.; Zhou, Z.; Ren, Y.; et al. Methane oxidation by an extremely acidophilic bacterium of the phylum Verrucomicrobia. *Nature* **2007**, *450*, 879–882. [CrossRef]
63. Martinez-Garcia, M.; Brazel, D.M.; Swan, B.K.; Arnosti, C.; Chain, P.S.G.; Reitenga, K.G.; Xie, G.; Poulton, N.J.; Gomez, M.L.; Masland, D.E.D.; et al. Capturing single cell genomes of active polysaccharide degraders: An unexpected contribution of verrucomicrobia. *PLoS ONE* **2012**, *7*, e35314. [CrossRef]
64. Wertz, J.T.; Kim, E.; Breznak, J.A.; Schmidt, T.M.; Rodrigues, J.L. Genomic and physiological characterization of the *Verrucomicrobia* isolate *Diplosphaera colitermitum* gen. Nov., sp. nov., reveals microaerophily and nitrogen fixation genes. *Appl. Environ. Microbiol.* **2012**, *78*, 1544–1555. [CrossRef]
65. Liu, C.; Finegold, S.M.; Song, Y.; Lawson, P.A. Reclassification of *Clostridium coccoides, Ruminococcus hansenii, Ruminococcus hydrogenotrophicus, Ruminococcus luti, Ruminococcus productus* and Ruminococcus *schinkii* as *Blautia coccoides* gen. nov., comb. nov., *Blautia hansenii* comb. nov., *Blautia hydrogenotrophica* comb. nov., *Blautia luti* comb. nov., *Blautia producta* comb. nov., *Blautia schinkii* comb. nov. and description of *Blautia wexlerae* sp. nov., isolated from human faeces. *Int. J. Sys. Evol. Microbiol.* **2008**, *58*, 1896–1902.
66. Chen, J.; Zhang, H.; Wu, X.; Shang, S.; Yan, J.; Chen, Y.; Zhang, H.; Tang, X. Characterization of the gut microbiota in the golden takin (*Budorcas taxicolor bedfordi*). *AMB Express* **2017**, *7*, 1–10. [CrossRef] [PubMed]
67. Gulino, L.M.; Ouwerkerk, D.; Kang, A.Y.; Maguire, A.J.; Kienzle, M.; Klieve, A.V. Shedding light on the microbial community of the macropod foregut using 454-amplicon pyrosequencing. *PLoS ONE* **2013**, *8*, e61463. [CrossRef]
68. Tims, S.; Derom, C.; Jonkers, D.M.; Vlietinck, R.; Saris, W.H.; Kleerebezem, M.; De Vos, W.M.; Zoetendal, E.G. Microbiota conservation and BMI signatures in adult monozygotic twins. *ISME J.* **2013**, *7*, 707–717. [CrossRef] [PubMed]
69. Clark, A.; Mach, N. Exercise-induced stress behavior, gut-microbiota-brain axis and diet: A systematic review for athletes. *J. Int. Soc. Sports Nutr.* **2016**, *13*, 43. [CrossRef] [PubMed]
70. Gosalbes, M.J.; Durbán, A.; Pignatelli, M.; Abellan, J.J.; Jiménez-Hernández, N.; Pérez-Cobas, A.E.; Latorre, A.; Moya, A. Metatranscriptomic approach to analyze the functional human gut microbiota. *PLoS ONE* **2011**, *6*, e17447. [CrossRef]

8

Comparative Grain Yield, Straw Yield, Chemical Composition, Carbohydrate and Protein Fractions, In Vitro Digestibility and Rumen Degradability of Four Common Vetch Varieties Grown on the Qinghai-Tibetan Plateau

Yafeng Huang, Fangfang Zhou and Zhibiao Nan *

State Key Laboratory of Grassland Agro-ecosystems, Key Laboratory of Grassland Livestock Industry Innovation of Ministry of Agriculture and Rural Affairs, College of Pastoral Agriculture Science and Technology, Lanzhou University, Lanzhou 730020, China

* Correspondence: zhibiao@lzu.edu.cn;

Simple Summary: The common vetch (*Vicia sativa* L.) is an important legume crop of mixed crop-livestock systems that provides high-quality grains used as food/feed and straw used as ruminant feed. The objective of this study was to determine the variability in grain yield, straw yield, straw chemical composition, carbohydrate and protein fractions, in vitro gas production, and in situ ruminal degradability of four different varieties of common vetch grown on the Qinghai-Tibetan Plateau. The results showed that grain yield, straw yield, and straw nutrient value varied significantly among the four varieties. Overall, the findings indicated that in terms of straw yield and nutritive quality, variety Lanjian No. 1 has the greatest potential as a crop for supplementing ruminant diets in the smallholder mixed crop–livestock systems on the Qinghai-Tibetan Plateau.

Abstract: Four varieties of common vetch, including three improved varieties (Lanjian No. 1, Lanjian No. 2, and Lanjian No. 3) and one local variety (333A), were evaluated for varietal variations in grain yield, straw yield and straw quality attributes on the Qinghai-Tibetan Plateau. Crops were harvested at pod maturity to determine grain yield, straw yield, harvest index, and potential utility index (PUI). Straw quality was determined by measuring chemical composition, carbohydrate and protein fractions, in vitro gas production and in situ ruminal degradability. Results showed a significant effect ($p < 0.01$) of variety on the grain yield [875.2–1255 kg dry matter (DM)/ha], straw yield (3154–5556 kg DM/ha), harvest index (15.6–28.7%) and PUI (53.3–63.2%). Variety also had a significant effect on chemical composition, carbohydrate and protein fractions ($p < 0.05$) except non-structural carbohydrates and rapidly degradable sugars. Significant differences ($p < 0.05$) were observed among the varieties in potential gas production [188–234 mL/g DM], in vitro organic matter (OM) digestibility (43.7–54.2% of OM), and metabolizable energy (6.40–7.92 MJ/kg DM) of straw. Significant differences ($p < 0.001$) were also observed among the varieties in rapidly degradable DM fraction and effective DM degradability of straw; however, no difference was observed in other DM degradation parameters and neutral detergent fiber degradation parameters. In conclusion, based on straw yield and quality, Lanjian No. 1 has the greatest potential among the tested varieties as a crop for supplementing ruminant diets for smallholder farmers on the Qinghai-Tibetan plateau.

Keywords: common vetch; straw; nutritive value; varietal effect; ruminants

1. Introduction

Ruminants provides the majority of food worldwide, contributing approximately 25.9% of global meat production and nearly 100% of global milk production [1,2]. The demand for cattle and sheep meat has been growing at a global growth rate of 2.0% per year from 2016 and is predicted to continue till 2026 [3]. The Qinghai-Tibetan plateau supports approximately 41 million sheep and over 14.3 million cattle, one of the largest livestock systems in Asia [4,5]. The plateau is also the headwater region of most of Asia's five major rivers, including Yangtze, Yellow, Mekong, Ganges, and Indus rivers [4]. Thus, sustainable management of the Plateau is important for the livelihood of over 9.8 million nomadic populations and for protecting these crucial river systems.

Unfortunately, harsh climatic conditions (e.g., low temperatures) and short growing season on the Qinghai-Tibetan plateau limit crop production [4–6]. Combined with grassland degradation, the annual fodder gap in this region is expected to reach 2 million tonnes, which implies that each year 2.7 million sheep units with insufficient feedstuffs will result in 20–30% liveweight loss during winter and early spring [5,7]. An effective remedy to stabilize feed supply is to incorporate nitrogen-fixing annual cool-season feed legumes such as common vetch (*Vicia sativa* L.) into fallow lands that will improve soil properties and lead to increased grain and straw production [8].

The use of common vetch in the mixed crop-livestock systems has expanded greatly in the last decade [4,9–11]. Several studies showed high generation of crop residues after harvesting common vetch that has substantial potential to be used for feeding ruminant livestock [9,12]. Previous studies also reported that the nutritive quality of common vetch straw is relatively high, containing an average of 9.41% crude protein (CP), 55.2% in vitro organic matter digestibility (IVOMD) and 7.3 MJ/kg metabolizable energy (ME) [9,13]. This makes common vetch straw a good feed supplement for ruminants, which are offered low-quality cereal straw-based diets in many small-holder crop-livestock systems [9,14]. In addition, the fatty acid composition of grains has increased the use of common vetch as a feed for ruminants [15]. Although common vetch is important as a feed for ruminants, limited information is available on the varietal variations in straw composition (carbohydrate and protein fractions), in vitro gas production, and in situ ruminal degradability, because most of the earlier studies focused on grain yield, straw yield, and straw quality including only chemical composition, IVOMD and ME [9–13]. Russel et al. [16] suggested that carbohydrate and protein fractions can be used as a reliable indicator to accurately predict biological value and performance of feed in ruminants. Marcos et al. [17], and Blümmel and Ørskov [18] demonstrated that the chemical composition in combination with in vitro digestibility and in situ ruminal degradability can be used as crucial parameters to evaluate the nutritive value of feed.

Accordingly, the objective of this study was to evaluate grain and straw yields, as well as chemical composition, carbohydrate and protein fractions, in vitro gas production and in situ ruminal degradability of straw of four different common vetch varieties grown on the Qinghai-Tibetan Plateau.

2. Materials and Methods

2.1. Location, Experimental Design, and Sampling

A detailed description of the location, experimental design and sampling is in the companion paper on the nutritive value of common vetch grain [4] and are summarized here.

This experiment was conducted during the cropping season of 2015 at the Xiahe Experimental Station of Lanzhou University, Gansu, China (35°45′ N, 102°34′ E; altitude 2880 m). The region under this study is situated at the eastern margin of the Qinghai-Tibetan Plateau. The location received an annual average precipitation of 452 mm (80% in May and September) and recorded a mean annual air temperature of 3.5 °C (1984–2014, 31 years). The previous crop was rape (*Brassica campestris* L.).

Three improved varieties (Lanjian No. 1, Lanjian No. 2, Lanjian No. 3) and one local variety (333A) were utilized in the field study. These improved varieties, developed by the Common Vetch Breeding Program of Lanzhou University, are well-adapted and extensively grown by smallholder farmers in the region surrounding the research site. Agronomic characteristics of these tested varieties are shown in Table 1. All four varieties were planted under the same agronomic conditions following a completely randomized design with four replicates. The individual plot area was 40 m^2 (8 m × 5 m) with a row spacing of 20 cm, and grains were sown by hand at a depth of 3–5 cm at a density of 150 viable grains m-2. Grains were sown (6 May 2015) before inoculation with rhizobium (CCBAU01069, China Agricultural University, Beijing, China), which was recommended based on the symbiont performance of these varieties [19]. Irrigation and fertilizers were not applied after sowing, and weeds in each plot were adequately controlled manually.

Table 1. Agronomic characteristics of the common vetch varieties utilized in this study.

Agronomic Characteristic	333A	Lanjian No.1	Lanjian No.2	Lanjian No.3
Days to mature (day)	134	145	132	124
1000 grains weight (g)	54	79	71	76
Plant height (cm)	92	106	80	69
Altitude (m.a.s.l)	–	<3000	<3500	<4000
Year of release	1987	2014	2015	2011

Source: Ministry of Agriculture, Beijing, China.

Harvesting of the common vetch plants was done at the pod maturity (6, 15, 15, and 26 September for Lanjian No. 3, 333A, Lanjian No. 2 and Lanjian No. 1, respectively). The crops were manually harvested from two representative subplots (1 m × 1 m) of each plot for each variety and threshed to obtain grain and straw samples. Harvest index was calculated as follows: Harvest index (%) = Grain yield × 100 / (Grain yield + Straw yield). All samples were oven-dried at 65 °C for 48 h and ground to pass through a 2-mm sieve for analysis of in situ ruminal incubation and to pass through a 1-mm sieve for chemical analysis and in vitro gas production measurement. All procedures involved animals were approved by the Animal Ethics Committee of Lanzhou University (protocol AEC-LZU-2016-01).

2.2. *In Vitro Gas Production*

Four adult Dorper rams (approximately 33-month-old; 58.4 ± 1.24 kg body weight) fitted with flexible rumen cannulas were used as donors of ruminal fluid. The rams were kept in individual stalls and had free access to fresh water and mineral/vitamin licks. The rams were daily fed 1.2 kg of 550 g/kg DM sheepgrass [*Leymus chinensis*, (Trin.) Tzvel], 140 g/kg DM soybean meal, 294 g/kg DM maize (*Zea mays* L.) seed, 8.6 g/kg DM calcium hydrophosphate, 5.0 g/kg DM salt and 2.4 g/kg DM mineral-vitamin mix at maintenance energy level in equal portions at 08:00 and 16:30 hours. In vitro gas production (GP) was measured as described by Blümmel and Ørskov [18]. Rumen fluids were obtained before morning feeding and strained through four layers of cheesecloth into a preheated, insulated bottle. Briefly, approximately 200 mg DM of each sample (in duplicate) was weighed into calibrated glass syringes (100 mL). Each syringe was preheated at 39 °C before injecting 30 mL rumen fluid/buffer mixture [20]. Then, the syringes were placed vertically in a water bath at 39 °C with three syringes without a sample used as blank. The volume of GP was manually recorded after 0, 4, 6, 8, 12, 24, 36, 48 and 72 h of incubation and was blank corrected. The incubation run was repeated three times. All operations involving rumen fluid were conducted under a continuous flush of CO_2 to ensure anaerobic conditions.

The GP data were fitted with time using the following equation [17], as follows: $y = b \times [1 - e^{-c\,(t\,-\,lag)}]$, where y = the volume of GP at time t; b = the potential GP (mL/g DM); c (h^{-1}) = the fractional rate of GP (h^{-1}); lag = the initial delay in the onset of gas production (h); t = incubation time (h). Parameters b, c and lag were determined by an iterative least square method using the NLIN procedure of SAS 9.2 (SAS Inst. Inc., Cary, NC). The IVOMD [% organic matter (OM)] and ME (MJ/kg DM) were estimated using the equations of Menke and Steingass [21] as: IVOMD = 14.88 + 0.889 GP + 0.45 CP + 0.0651 ASH; ME = 2.20 + 0.136 GP + 0.057 CP + 0.0029 EE^2, where GP = the net gas production after

24 h of incubation (mL/200mg DM), ASH = the ash content (% DM), CP = the crude protein content (% DM), and EE = the ether extract content (% DM).

2.3. *In Situ Ruminal Incubation*

In situ ruminal degradability of DM and neutral detergent fiber (NDF) of straw samples was determined using the nylon bag technique described by Nandra et al. [22]. Briefly, approximately 5.0 g of each sample (in duplicate) was weighted into nylon bags (9 × 5 cm; 50-μm pore size) and incubated in the ventral sacs of the rumen of the same four rams used for the production of ruminal fluid in Section 2.2. The bags were inserted into the rumen for 0 (control), 4, 8, 12, 24, 48 and 72 h. Following removal, the bags were briefly washed under cold running water and frozen (–20°C) until further analysis. All bags were defrosted, manually washed in cold tap water until the water was clear, oven-dried at 60°C for 48 h, and weighed. The dried undigested residues of replicates per same time within sheep were pooled to measure DM and NDF. Straw DM and NDF disappearance rates were estimated from the difference in straw weight before and after incubation. The kinetic parameters of DM and NDF degradation were determined using the exponential equation described by Ørskov and McDonald [23]. The effective degradability of DM (EDDM) and NDF (EDNDF) were calculated as $ED = A + [(B \times C)/(C + k)]$, where A = the soluble fraction, B = the potentially degradable fraction, C = the rate of degradation of fraction B, k = the rumen outflow rate ($0.031\ h^{-1}$) [4].

2.4. *Laboratory Analysis*

Determination of DM (ID 930.15), nitrogen (N; ID 988.05), ether extract (EE; ID 920.85), ash (ID 938.08), acid detergent fiber (ADF; ID 973.18) and acid detergent lignin (ADL; ID 973.18) were analyzed following the methods of the Association of Official Agricultural Chemists [24]. The CP content was calculated by multiplying the nitrogen value by 6.25. The NDF content was determined following the method by Van Soest et al. [25] using heat-stable α-amylase and sodium sulfite. Contents of NDF and ADF were expressed inclusive of residual ash. Acid detergent insoluble protein (ADIP), neutral detergent insoluble protein (NDIP) were measured by Kjeldahl analysis of the ADF and NDF bag residues, respectively, using the procedure described by Licitra et al. [26].

Carbohydrate fractions of straw samples from the four common vetch varieties were determined as proposed by the Cornell Net Carbohydrate and Protein system (CNCPS) [27]. The system divides carbohydrate into four fractions in terms of their degradation rate as follow: C_A, rapidly degradable sugars; CB_1, intermediately degradable pectin and starch; CB_2, slowly degradable cell wall; and C_C, undegradable/lignin-bound cell wall. Total carbohydrates (TCHO) content was calculated as $TCHO = 100 - (CP + EE + Ash)$ [28]. Non-structural carbohydrates (NSC) and structural carbohydrates (SC) were estimated using the equations given by Caballero et al. [29] as: $NSC = TCHO - SC$ and $SC = NDF - NDIP$. Starch content was determined by enzymatic hydrolysis of α-linked glucose polymers [30].

The CP of straw samples was fractionated into five different fractions according to CNCPS as described by Licitra et al. [26] and Sniffen et al. [27]. These fractions include: fraction P_A, non-protein nitrogen (NPN), calculated as the difference between total N and true protein N, analyzed using sulphuric acid (0.5 M) and sodium tungstate (0.30 M); fraction P_{B1}, buffer-soluble protein, estimated by subtracting buffer-insoluble protein precipitated with freshly prepared (1 g/10 mL) sodium azide and borate-phosphate buffer (pH 6.7–6.8) solution from true protein; fraction P_{B2}, neutral detergent- soluble protein, calculated by subtracting NDIP from buffer-insoluble protein; fraction P_{B3}, acid detergent-soluble protein, calculated by subtracting ADIP from NDIP; and fraction P_C, ADIP, is indigestible protein. All measurements were performed in duplicate and appropriate chemical standards were included in each analytical run.

2.5. Calculations and Statistical Analyses

Potential utility index (PUI) was estimated from the amount of utilizable portion of the total biomass yield for grain and straw regardless of the economic value as described by Alkhtib et al. [31]:

$$\text{PUI (\%)} = 100 \times [\text{grain yield} + 0.01 \times \text{IVOMD (\%)} \times \text{straw yield}] / \text{total biomass yield}$$

Data collected were subjected to one-way ANOVA using SPSS software (Version 21.0. IBM Corporation, Armonk, NY, USA). The fermentation parameters were subjected to separate analysis of variance with varieties as fixed effect and the incubation run as random effect. Differences between means were compared using the Duncan significant difference test at $p < 0.05$.

3. Results

3.1. Grain Yield, Straw Yield, and PUI

Table 2 shows a significant effect of variety on grain yield and straw yield ($p < 0.01$). The grain yield ranged from 875.2 to 1255 kg DM/ha with an average value of 1088 kg DM/ha. The grain yield of 333A and Lanjian No. 1 was significantly less than that of other improved varieties. The straw yield varied from 3154 to 5556 kg DM/ha. The least straw yield was observed for Lanjian No. 3 and the greatest for Lanjian No. 1.

Variety had a significant ($p < 0.001$) effect on harvest index and PUI. Harvest index varied from 15.6 to 28.7%. Harvest index of Lanjian No. 2 was less than that of Lanjian No. 3, but greater than that of the other varieties ($p < 0.01$). The PUI varied from 53.3 to 63.2%. The PUI of the local variety was less than the improved varieties, which had similar PUI (average 62.1%).

Table 2. Influence of variety on grain yield, straw yield, harvest index, and potential utility index (PUI) of four common vetch varieties.

Dependent Variable	333A	Lanjian No. 1	Lanjian No. 2	Lanjian No. 3	SEM [1]	*p*-Value
Grain yield (kg DM/ha)	875.2 ^{b}	1024 ^{b}	1196 ^{a}	1255 ^{a}	45.1	0.001
Straw yield (kg DM/ha)	4319 ^{b}	5556 ^{a}	4158 ^{bc}	3154 ^{c}	267.6	0.003
Harvest index (%)	17.0 ^{c}	15.6 ^{c}	22.6 ^{b}	28.7 ^{a}	1.39	<0.001
PUI (%)	53.3 ^{b}	61.4 ^{a}	62.8 ^{a}	63.2 ^{a}	0.660	<0.001

^{a,b} Within a raw, different letters represent the significant differences at p-value < 0.05. [1] DM, dry matter; SEM, standard error of the mean.

3.2. Chemical Composition

Table 3 shows a significant effect of variety on DM, ash, CP and EE contents of the straw samples ($p < 0.01$). The DM content among varieties varied from 90.1 to 90.5% and was the greatest in Lanjian No. 2. The ash content varied from 10.2 to 13.5% DM. The least ash content was for 333A and the greatest for Lanjian No. 1. The CP content varied from 9.76 to 13.8% DM. The local variety had significantly less CP content compared to the improved varieties. There were significant differences in CP content among the improved varieties, which was greater for Lanjian No. 1. The EE content varied from 0.459 to 1.11% DM. The EE content of 333A was similar to that of Lanjian No. 3 but was significantly greater than that of Lanjian No. 1 and Lanjian No. 2.

Significant differences ($p < 0.05$) were observed in the cell wall contents of different varieties. The NDF, ADF, and ADL contents varied from 45.0 to 54.1% DM, 27.4 to 33.2% DM, and 6.08 to 9.56% DM, respectively. The NDF, ADF and ADL contents were the greatest in 333A and the least in Lanjian No. 1. Hemicellulose and cellulose contents varied from 17.6 to 21.4% DM and 21.4 to 23.7% DM, respectively. The hemicellulose and cellulose contents of variety 333A were higher than those of Lanjian No. 1, but similar to those of other varieties. Phosphorus content varied from 0.185 to 0.296% DM. The phosphorus content was considerably greater ($p < 0.05$) in Lanjian No. 3 than in Lanjian No. 2

and 333A, with Lanjian No. 1 being intermediate. Calcium content was also significantly influenced by variety ($p < 0.001$). It varied from 1.00 to 1.54% DM and was the greatest in Lanjian No. 1.

Table 3. Influence of variety on chemical composition (% dry matter unless stated otherwise) of straw in four common vetch varieties.

Dependent Variable	333A	Lanjian No. 1	Lanjian No. 2	Lanjian No. 3	SEM [1]	*p*-Value
Dry matter %	90.2 [bc]	90.4 [ab]	90.5 [a]	90.1 [c]	0.055	0.009
Ash	10.2 [c]	13.5 [a]	11.9 [b]	11.0 [bc]	0.393	0.003
Crude protein	9.76 [c]	13.8 [a]	12.4 [b]	11.5 [b]	0.424	<0.001
Ether extract	1.06 [a]	0.459 [b]	0.515 [b]	1.11 [a]	0.088	<0.001
Neutral detergent fiber	54.1 [a]	45.0 [c]	49.4 [b]	51.0 [b]	0.915	<0.001
Acid detergent fiber	33.2 [a]	27.4 [c]	30.3 [b]	29.6 [b]	0.603	<0.001
Acid detergent lignin	9.56 [a]	6.08 [c]	6.53 [bc]	7.05 [b]	0.369	<0.001
Hemicellulose	20.9 [a]	17.6 [b]	19.2 [ab]	21.4 [a]	0.537	0.025
Cellulose	23.6 [a]	21.4 [b]	23.7 [a]	22.5 [ab]	0.353	0.039
Phosphorus	0.185 [b]	0.249 [ab]	0.206 [b]	0.296 [a]	0.0149	0.019
Calcium	1.13 [bc]	1.54 [a]	1.00 [c]	1.32 [ab]	0.061	<0.001

[a,b] Within a raw, different letters represent the significant differences at p-value < 0.05. [1] SEM, standard error of the mean.

3.3. Carbohydrate and Protein Fractions

As shown in Table 4, the TCHO content was significantly different ($p < 0.001$) among varieties and varied from 72.2 to 79.0% DM. The greatest TCHO content was observed in 333A and the least in Lanjian No. 1. The NSC content varied from 29.3 to 31.2% DM with no difference ($p > 0.05$) among varieties. The SC content varied from 41.1 to 49.7% DM. The variation in SC content was the greatest in 333A and the least in Lanjian No. 1. Significant differences ($p < 0.05$) were observed in CHO fractions except C_A fraction (Table 4). The C_{B1} fraction was greatest ($p < 0.001$) in Lanjian No. 1 (26.0% CHO), intermediate in Lanjian No. 2 (22.9% CHO), and least in Lanjian No. 3 (20.1% CHO) and 333A (17.7% CHO). The C_{B2} fraction varied from 33.9 to 39.5% CHO. Variety 333A had significantly less ($p < 0.05$) C_{B2} fraction than Lanjian No. 2 and Lanjian No. 3 but similar to Lanjian No. 1. The Cc fraction varied from 20.2 to 29.0% CHO. The C_C fraction of variety 333A was significantly greater ($p < 0.001$) compared to the improved varieties, which had similar C_C fraction.

Table 4. Influence of variety on carbohydrate and protein fractions of straw in four common vetch varieties.

Dependent Variable	333A	Lanjian No. 1	Lanjian No. 2	Lanjian No. 3	SEM [1]	*p*-Value
Carbohydrates (% dry matter)						
TCHO	79.0 [a]	72.2 [c]	75.2 [b]	76.4 [b]	0.665	<0.001
NSC	29.3	31.2	29.8	29.6	0.474	0.575
SC	49.7 [a]	41.1 [c]	45.4 [b]	46.7 [b]	0.886	<0.001
Carbohydrate fractions (% CHO)						
C_A	19.4	17.1	18.1	18.7	0.645	0.435
C_{B1}	17.7 [c]	26.0 [a]	22.9 [b]	20.1 [c]	0.885	<0.001
C_{B2}	33.9 [b]	36.7 [ab]	39.5 [a]	39.0 [a]	0.835	0.042
C_C	29.0 [a]	20.2 [b]	20.8 [b]	22.1 [b]	0.998	<0.001
Protein fractions (% CP)						
P_A	7.16 [c]	11.5 [a]	9.43 [b]	8.55 [bc]	0.499	0.004
P_{B1}	37.3 [a]	25.8 [c]	30.1 [bc]	31.8 [b]	1.31	0.004
P_{B2}	10.8 [c]	34.0 [a]	27.9 [b]	22.3 [b]	2.36	<0.001
P_{B3}	23.6 [a]	13.3 [c]	14.8 [c]	18.4 [b]	1.10	<0.001
P_C	21.1 [a]	15.4 [c]	17.8 [b]	18.8 [b]	0.628	0.001

[a,b] Within a raw, different letters represent the significant differences at p-value < 0.05. [1] C_A, rapidly degradable sugars; C_{B1}, intermediately degradable pectin and starch; C_{B2}, slowly degradable cell wall; C_C, unavailable/lignin bound cell wall; CP, crude protein; TCHO, total carbohydrates; NSC, non-structural carbohydrates; P_A, non-protein nitrogen; P_{B1}, buffer soluble protein; P_{B2}, neutral detergent soluble protein; P_{B3}, acid detergent soluble protein; P_C, indigestible protein; SC, structural carbohydrates; SEM, standard error of the mean.

Variety had a significant influence on the protein fractions of straw ($p < 0.01$; Table 4). The P_A fraction varied from 7.16 to 11.5% CP, and the least value was recorded in 333A and the greatest in Lanjian No. 1. The P_{B1} fraction varied from 25.8 to 37.3% CP; the greatest value was recorded in 333A. The P_{B2} fraction varied from 10.8 to 34.0% CP. The variation in P_{B2} fraction was the least in 333A and the greatest in Lanjian No. 1. The P_{B3} fraction varied from 13.3 to 23.6% CP. The P_{B3} fraction was significantly greater for 333A than for Lanjian No. 3, and significantly greater for Lanjian No. 3 than the other varieties. The P_C fraction varied from 15.4 to 21.1% CP. The P_c fraction was greatest in 333A and least in Lanjian No. 1.

3.4. In Vitro Gas Production

Table 5 shows the gas production parameters, IVOMD and ME of straw in four common vetch varieties. The potential gas production differed considerably ($p < 0.001$) between the varieties and varied from 188 to 234 mL/g DM. The potential gas production of Lanjian No. 3 was greater than that of 333A; however, it was less than that of the other varieties. The fractional rate of GP was not affected by variety ($p > 0.05$), averaging 0.0631 h^{-1}. The *lag* value was similar ($p > 0.05$) for the four varieties (average 0.633 h).

Variety had a significant influence on the IVOMD content of straw ($p < 0.001$; Table 5). Straw IVOMD varied from 43.7 to 54.2% OM, and the varieties were ranked in order Lanjian No. 1 > Lanjian No. 2 > Lanjian No. 3 > 333A. Significant difference in the ME content was also observed among the different varieties ($p < 0.001$), and it varied from 6.40 to 7.92 MJ/kg DM. The least value was recorded for 333A and the greatest for Lanjian No. 1 and Lanjian No. 2.

Table 5. Influence of variety on gas production parameters, in vitro organic matter digestibility (IVOMD) and metabolizable energy (ME) of straw in four common vetch varieties.

Dependent Variable	333A	Lanjian No. 1	Lanjian No. 2	Lanjian No. 3	SEM [1]	*p*-Value
b (mL/g DM)	188 [c]	234 [a]	228 [a]	208 [b]	3.39	<0.001
c (h^{-1})	0.0608	0.0650	0.0649	0.0617	0.000740	0.334
lag (h)	0.608	0.656	0.656	0.612	0.00981	0.131
IVOMD (% OM)	43.7 [d]	54.2 [a]	52.0 [b]	48.4 [c]	0.676	<0.001
ME (MJ/kg DM)	6.40 [c]	7.92 [a]	7.61 [a]	7.08 [b]	0.0995	<0.001

[a,b] Within a raw, different letters represent the significant differences at *p*-value < 0.05. [1] *b*, potential gas production; *c*, fractional rate of gas production; DM, dry matter; *lag*, initial delay in the onset of gas production; OM, organic matter; SEM, standard error of the mean.

3.5. In Situ Ruminal Degradability

The soluble DM fraction was significantly different ($p < 0.001$) among varieties and varied from 22.8 to 28.4% DM (Table 6). The soluble DM fraction of 333A was similar to that of Lanjian No. 3, but less than that of Lanjian No. 1 and Lanjian No. 2. The potentially degradable DM fraction and rate of DM degradation were similar ($p > 0.05$) for the four varieties and recorded an average of 43.2% DM and 0.0442 h^{-1}, respectively. The EDDM value was significantly different ($p < 0.001$) among the four varieties. Average EDDM value was 51.0% DM and ranged from 46.7 to 55.2% DM. Variety 333A had significantly less EDDM value compared to Lanjian No. 1 and Lanjian No. 2, but it was similar to Lanjian No. 3.

The NDF degradation profiles in situ are given in Table 5. Soluble NDF fraction, potentially degradable NDF fraction, and rate of NDF degradation were not influenced ($p > 0.05$) by variety and recorded an average of 11.5% NDF, 46.7% NDF and 0.0374 h^{-1}, respectively. No difference was observed in EDNDF value among varieties, but there was a trend toward greater EDNDF for Lanjian No. 1 ($p = 0.078$).

Table 6. Influence of variety on in situ ruminal degradation kinetics of dry matter (DM) and neutral detergent fiber (NDF) of straw in four common vetch varieties.

Dependent Variable	333A	Lanjian No. 1	Lanjian No. 2	Lanjian No. 3	SEM [1]	*p*-Value
DM						
A (% DM)	22.8 [b]	28.4 [a]	26.8 [a]	24.4 [b]	0.630	<0.001
B (% DM)	41.1	45.1	43.9	42.6	0.581	0.066
C (h^{-1})	0.0430	0.0453	0.0444	0.0442	0.0005	0.377
EDDM (% DM)	46.7 [c]	55.2 [a]	52.7 [ab]	49.5 [bc]	0.953	<0.001
NDF						
A (% NDF)	10.6	12.2	12.0	11.2	0.535	0.735
B (% NDF)	45.3	48.8	46.7	46.0	0.536	0.101
C (h^{-1})	0.0366	0.0388	0.0371	0.0373	0.0004	0.142
EDNDF (% NDF)	35.1	39.3	37.4	36.2	0.618	0.078

[a,b] Within a raw, different letters represent the significant differences at *p*-value < 0.05. [1] *A*, soluble fraction; *B*, potentially degradable fraction; *C*, rate of degradation of fraction B; EDDM, effective dry matter degradability; EDNDF, effective neutral detergent fiber degradability; SEM, standard error of the mean.

4. Discussion

4.1. Grain Yield, Straw Yield, and PUI

In smallholder crop-livestock systems, improvement in crop straw yield implies increase in milk and meat production from ruminants [6,32]. Substantial variability differences among varieties were found for grain and straw yields, partly due to differences in days to pod maturity and harvest index as suggested by Abd El-Moneim [8], Larbi et al. [9], and Kafilzadeh and Maleki [32]. Our findings are consistent with the earlier reports on common vetch [9] as well as faba bean (*Vicia faba* L.) [31], lentil (*Lens culinaris*) [32] and chickpea (*Cicer arietinum*) [33]. The grain yield recorded in this study (875.2–1255 kg DM/ha) is within the reported range (287–1783 kg DM/ha) [9], but less than ranges reported value of 1340–2240 kg DM/ha [12]. Meanwhile, the straw yield (3154–5556 kg DM/ha) is greater than that reported by Larbi et al. [9] (629–2226 kg DM/ha), but slightly less than that reported by Albayrak et al. [12] (4620–7320 kg DM/ha). The yields vary between studies [9,12] as consequence of the differences in the varieties, agronomic practices and growing conditions (e.g., soil type and climate) [8,12,34]. The grain yield of the improved varieties was significantly greater than that of the local variety in this study. Similar results were reported for other leguminous crops such as faba bean [31] and lentil [35]. The local variety demonstrated inferior PUI compared to the improved varieties. This is consistent with the findings in faba bean [31]; however, contrary to the findings of Tolera et al. [34] who observed less PUI for the improved varieties compared with the local varieties.

4.2. Chemical Composition

Higher CP and lower cell wall contents (NDF, ADF, cellulose and ADL) can be used as indicators of good feed quality [28]. Large varietal differences in straw chemical composition observed in our study is in agreement with the earlier findings in common vetch [9,11,13], as well as faba bean [31], lentil [32] and chickpea [33]. The range of CP content observed in this study (9.76–13.8% DM) was more than the threshold content (8.0% DM CP) required for optimum activity of rumen microorganisms in ruminants [36]. The NDF content among the varieties varied from 45.0 to 54.1% DM with a mean value of 49.9% DM. Van Soest [37] indicated that NDF content over 65% leads to effects on voluntary intake and production by ruminants. This makes common vetch straw a good source of CP supplements for ruminants in smallholder crop-livestock/agro-pastoral systems. Common vetch straw NDF and ADF contents of 52.2 and 36.1% DM reported by Makkar et al. [13] are consistent with our results; however, the CP contents (6.2% DM) is less than our results. Phosphorus and calcium contents observed in this study are similar to those reported by Abreu and Bruno-Soares [38]. The differences in chemical composition of straw between studies may be due to varietal variability, differences in growing condition (e.g., soil type and climate), or differences in harvesting and postharvest handling practices.

4.3. Carbohydrate and Protein Fractions

The significant varietal differences in carbohydrate fractions of common vetch straw are in agreement with reported for sorghum (*Sorghum bicolor* [L.] Moench) [28], timothy (*Phleum pratense* L.) [39] and wheat (*Triticum aestivum* L.) [40]. The values for TCHO, NSC and SC obtained in this study are comparable with the previous studies on other legume crops such as *Trifolium alaxendrinum* [41]. Among the CHO fractions, C_{B2} fraction was the highest in the straw of common vetch varieties analyzed. Others have reported similar results in sorghum, berseem (*Trifolium alexandrium*) and cowpea (*Vigna sinensis*) [28,41]. However, there is limited data available on the CHO fractions of common vetch straw. The pattern of CHO fractions observed in our study for common vetch straw is comparable with the earlier reports on other forage crops [41–43].

Wide range of protein fractions observed in the varieties is consistent with the findings in sorghum [28] and wheat [40]. Swarna et al. [44] indicated that P_{B2} and P_{B3} fractions represent a bypass protein of forage, while P_c fraction represents the non-degraded fraction. Compared to other varieties, Lanjian No. 1 had lower P_c fraction and higher P_{B2} + P_{B3} fractions. These observations on protein fractions suggest that straw of variety Lanjian No. 1 could be used as the better nitrogen source for ruminants. There is limited data available on the contents of protein fractions of common vetch. The pattern of protein fractions revealed here is similar to the reports on lucerne (*Medicago sativa* L.) [43] and black gram [*Vigna mungo* (L.) Hepper] [44].

4.4. In Vitro Gas Production

The gas produced by in vitro fermentation reflects the degree of feed fermentation and digestibility [34]. In this study, we observed significant differences among varieties in potential gas production, but not in gas production rate and *lag* time. Our results are partially in line with previous studies on other crops such as spineless cacti (*Opuntia* spp) [20] and chickpea [33]. Studies on gas production from common vetch straws are scarce, with the exception of the Spain studies [45]. López et al. [45] reported greater potential gas production, and less gas production rate and *lag* time in comparison with those observed in the current study. Cone and van Gelder [46] indicated that despite high degradability, feed with high CP typically produce less gas during fermentation as protein fermentation produces ammonia, which affects the carbonate buffer balance by neutralizing H^+ ions from volatile fatty acids without releasing carbon dioxide. The differences in gas production parameters between studies may be attributed to the differences in straw CP content, varieties, season, and location.

The IVOMD and ME contents of straw varied among the varieties, which is partly due to varietal variations in straw ADF and NDF and may be due to the proportions of straw morphological fractions, which were not measured in this study. In this study, varieties Lanjian No. 1 and Lanjian No. 2 recorded straw CP and IVOMD as high as 8.0% DM and 50.0% OM, respectively, which suggests that the straw of these varieties may be effectively used as a CP supplement to ruminants fed low-quality cereal straw-based diets [9]. The straw IVOMD range in this study is comparable to the range reported earlier in other vetch varieties [9], while the ME range is slightly less than those reported for vetches and other legume straws [13]. The straw quality varied between studies possibly due to differences in varieties, cell wall lignification, leaf to stem ratio, and the stage at which the straw was harvested. Larbi et al. [9] and Makkar et al. [13] earlier reported that straw IVOMD and ME contents of common vetch are influenced by varieties, growing season, and stage of straw harvest.

4.5. In Situ Ruminal Degradability

Higher ruminal degradability of high-fiber forages is satisfying because it implies improved the nutrient availability to rumen microbes [47]. The observed differences in DM degradation profiles of the straw varieties may be related to their varietal traits reflected as substantial differences in morphological and chemical composition. Our results are consistent with those reported for other crops such as maize [47]. For DM degradation parameters, the value of *A* fraction observed in this study for common vetch straw is greater and the values of *B* and *C* fractions are similar to those reported by Bruno-Soares et al. [48]. In this study, the value of *A* fraction of Lanjian No. 1 was significantly greater than other varieties, while the values for *B* and *C* fractions were similar for the four varieties analyzed. The differences in EDDM between the varieties can be mainly attributed to the differences in *A* fraction and not to *B* and *C* fractions, which is consistent with the literature [47]. The EDDM reported for common vetch straw in our study are similar to that reported for chickpea straw (51.8% DM) [49], but greater than that reported for fenugreek (*Trigonella foenum-graecum*) straw (32.2% DM) [50]. The degradation profiles of NDF were not influenced by variety in our study, which is in agreement with the earlier reports on maize and chickpea [47,49]. Bruno-Soares et al. [48] reported less rapidly degradable NDF fraction and rate of NDF degradation, and similar potentially degradable NDF fraction in comparison with those observed in this study. The EDNDF of straw in this study was less than reported by Abbeddou et al. [51] for lentil straw (45.9% NDF), but greater than that reported by Mustafa et al. [50] for fenugreek straw (20.3% NDF). Different studies recorded different ruminal degradation kinetics due to differences in varieties/species, straw composition, and animal species [4,49–51].

5. Conclusions

The results of this study showed varietal differences in grain and straw yields, and straw nutrient value in common vetch. Evaluation of common vetch varieties showed that Lanjian No. 1 had straw yield, straw CP, non-protein nitrogen, neutral detergent soluble protein, IVOMD, ME, and EDDM greater than other varieties, despite its less grain yield. Variety Lanjian No. 3 demonstrated early maturity and greater grain yield; however, straw yield and quality were less than Lanjian No. 1. Variety Lanjian No. 2 had greater grain yield and moderate straw CP; however, straw IVOMD, ME, EDDM and EDNDF contents were comparable with Lanjian No. 1. Based on these results, variety Lanjian No. 1 is the best option among varieties examined for smallholder farmers on the Qinghai-Tibetan plateau.

Author Contributions: Y.H. and Z.N. conceived and designed the experiment, Y.H. and F.Z. performed the experiments, Y.H. analyzed the data, and Y.H. and Z.N. wrote the manuscript.

References

1. Food and Agriculture Organization of the United Nations Home Page. Available online: http://www.fao.org/docrep/015/am081m/am081m00.html (accessed on 1 November 2018).
2. Wathes, C.M.; Buller, H.; Maggs, H.; Campbell, M.L. Livestock production in the UK in the 21st century: A perfect storm averted? *Animals* **2013**, *3*, 574–583. [CrossRef] [PubMed]
3. Mottet, A.; Teillard, F.; Boettcher, P.; Dé Besi, G.; Besbes, B. Review: Domestic herbivores and food security: Current contribution, trends and challenges for a sustainable development. *Animal* **2018**, *12*, s188–s198. [CrossRef]
4. Huang, Y.F.; Li, R.; Coulter, J.A.; Zhang, Z.X.; Nan, Z.B. Comparative grain chemical composition, ruminal degradation in vivo, and intestinal digestibility in vitro of *Vicia sativa* L. varieties grown on the Tibetan Plateau. *Animals* **2019**, *9*, 212. [CrossRef] [PubMed]
5. Shang, Z.H.; Gibb, M.J.; Leiber, F.; Ismail, M.; Ding, L.M.; Guo, X.S.; Long, R.J. The sustainable development of grassland-livestock systems on the Tibetan plateau: Problems, strategies and prospects. *Rangel. J.* **2014**, *36*, 267–296. [CrossRef]
6. Nan, Z.B.; Abd El-Moneim, A.M.; Larbi, A.; Nie, B. Productivity of vetches (*Vicia* spp.) under alpine grassland conditions in China. *Trop. Grassl.* **2006**, *40*, 177–182.
7. Zhang, Q.P.; Bell, L.W.; Shen, Y.Y.; Whish, J.P.M. Indices of forage nutritional yield and water use efficiency amongst spring-sown annual forage crops in north-west China. *Eur. J. Agron.* **2018**, *93*, 1–10. [CrossRef]

8. Abd El Moneim, A.M. Agronomic potential of three vetches (*Vicia* spp.) under rainfed conditions. *J. Agron. Crop Sci.* **1993**, *170*, 113–120. [CrossRef]
9. Larbi, A.; Hassan, S.; Kattash, G.; El-Moneim, A.M.; Jammal, B.; Nabil, H.; Nakkul, H. Annual feed legume yield and quality in dryland environments in north-west Syria: 2. grain and straw yield and straw quality. *Anim. Feed Sci. Technol.* **2010**, *160*, 90–97. [CrossRef]
10. Huang, Y.F.; Gao, X.L.; Nan, Z.B.; Zhang, Z.X. Potential value of the common vetch (*Vicia sativa* L.) as an animal feedstuff: A review. *J. Anim. Physiol. Anim. Nutr.* **2017**, *101*, 807–823. [CrossRef] [PubMed]
11. Fortina, R.; Gasmi-Boubaker, A.; Lussiana, C.; Malfatto, V.; Tassone, S.; Renna, M. Nutritive value and energy content of the straw of selected *Vicia* L. taxa from Tunisia. *Ital. J. Anim. Sci.* **2015**, *14*, 280–284. [CrossRef]
12. Albayrak, S.; Sevimay, C.S.; Töngel, Ö. Effects of inoculation with rhizobium on seed yield and yield components of common vetch (*Vicia sativa* L.). *Turk. J. Agric. For.* **2006**, *30*, 31–37.
13. Makkar, H.P.S.; Goodchild, A.V.; Abd El-Moneim, A.M. Cell-constituents, tannin levels, by chemical and biological assays and nutritive value of some legume foliage and straw. *J. Food Agric.* **1996**, *71*, 129–136. [CrossRef]
14. Haddad, S.G.; Husein, M.Q. Nutritive value of lentil and vetch straws as compared with alfalfa hay and wheat straw for replacement ewe lambs. *Small Rumin. Res.* **2001**, *40*, 255–260. [CrossRef]
15. Renna, M.; Gasmi-Boubaker, A.; Lussiana, C.; Battaglini, L.M.; Belfayez, K.; Fortina, R. Fatty acid composition of the seed oil of selected *Vicia* L. taxa from Tunisia. *Ital. J. Anim. Sci.* **2014**, *13*, 308–316. [CrossRef]
16. Russel, J.B.; O'Connor, J.D.; Fox, D.G.; Van Soest, P.J.; Sniffen, C.J. A net carbohydrate and protein system for evaluating cattle diets. I. Ruminal fermentation. *J. Anim. Sci.* **1992**, *70*, 3551–3561. [CrossRef] [PubMed]
17. Marcos, C.N.; García-Rebollar, P.; de Blas, C.; Carro, M.D. Variability in the chemical composition and in vitro ruminal fermentation of olive cake by-products. *Animals* **2019**, *9*, 109. [CrossRef] [PubMed]
18. Blümmel, M.; Ørskov, E.R. Comparison of in vitro gas production and nylon bag degradability of roughage in predicting feed intake in cattle. *Anim. Feed Sci. Technol.* **1993**, *40*, 109–119. [CrossRef]
19. Fu, P. Effects of Inoculation with Rhizobium and Phosphate on Growth of Vetch. Master's Thesis, Lanzhou University, Lanzhou, China, 2015; pp. 1–52.
20. Batista, A.M.; Mustafa, A.F.; McAllister, T.; Wang, Y.X.; Soita, H.; McKinnon, J.J. Effects of variety on chemical composition, in situ nutrient disappearance and in vitro gas production of spineless cacti. *J. Sci. Food Agric.* **2003**, *83*, 440–445. [CrossRef]
21. Menke, K.H.; Steingass, H. Estimation of the energetic feed value obtained from chemical analysis and gas production using rumen fluid. *Anim. Res. Dev.* **1988**, *28*, 7–55.
22. Nandra, K.S.; Dobos, R.C.; Orchard, B.A.; Neutze, S.A.; Oddy, V.H.; Cullis, B.R.; Jones, A.W. The effect of animal species on in sacco degradation of dry matter and protein of feeds in the rumen. *Anim. Feed Sci. Technol.* **2000**, *83*, 273–285. [CrossRef]
23. Ørskov, E.R.; McDonald, I. The estimation of protein degradability in the rumen from incubation measurements weighted according to rate of passage. *J. Agric. Sci.* **1979**, *92*, 499–503. [CrossRef]
24. AOAC (Association of Official Analytical Chemists). *Official Methods of Analysis*, 15th ed.; AOAC International: Arlington, VA, USA, 1990; pp. 71–90.
25. Van Soest, P.J.; Robertson, J.B.; Lewis, B.A. Methods for dietary fibre, neutral detergent fibre, and non-starch polysaccharides in relation to animal nutrition. *J. Dairy Sci.* **1991**, *74*, 3583–3597. [CrossRef]
26. Licitra, G.; Hernandez, T.M.; Van Soest, P.J. Standardizations of procedures for nitrogen fractionation of ruminant feeds. *Anim. Feed Sci. Technol.* **1996**, *57*, 347–358. [CrossRef]
27. Sniffen, C.J.; O'Connor, J.D.; Van Soest, P.J.; Fox, D.G.; Russell, J.B. A net carbohydrate and protein system for evaluating cattle diets: II. Carbohydrate and protein availability. *J. Anim. Sci.* **1992**, *70*, 3562–3577. [CrossRef]
28. Singh, S.; Bhat, B.V.; Shukla, G.P.; Gaharana, D.; Anele, U.Y. Nutritional evaluation of different varieties of sorghum stovers in sheep. *Anim. Feed Sci. Technol.* **2017**, *227*, 42–51. [CrossRef]
29. Caballero, R.; Alzueta, C.; Ortiz, L.T.; Rodríguez, M.L.; Barro, C.; Rebolé, A. Carbohydrate and protein fractions of fresh and dried common vetch at three maturity stages. *Agron. J.* **2001**, *93*, 1006–1013. [CrossRef]
30. Rode, L.M.; Yang, W.Z.; Beauchemin, K.A. Fibrolytic enzyme supplements for dairy cows in early lactation. *J. Dairy Sci.* **1999**, *82*, 2121–2126. [CrossRef]
31. Alkhtib, A.S.; Wamatu, J.A.; Wegi, T.; Rischkowsky, B.A. Variation in the straw traits of morphological fractions of faba bean (*Vicia faba* L.) and implications for selecting for food-feed varieties. *Anim. Feed Sci. Technol.* **2016**, *222*, 122–131. [CrossRef]

32. Wamatu, J.; Mersha, A.; Tolera, A.; Beyan, M.; Alkhtib, A.; Eshete, M.; Ahmed, S.; Rischkowsky, B. Selecting for food-feed traits in early and late maturity lentil genotypes (*Lens culinaris*). *J. Exp. Biol. Agri. Sci.* **2017**, *5*, 697–705.
33. Kafilzadeh, F.; Maleki, E. Chemical composition, in vitro digestibility and gas production of straws from different varieties and accessions of chickpea. *J. Anim. Physiol. Anim. Nutr.* **2012**, *96*, 111–118. [CrossRef]
34. Tolera, A.; Tsegaye, B.; Berg, T. Effects of variety, cropping year, location and fertilizer application on nutritive value of durum wheat straw. *J. Anim. Physiol. Anim. Nutr.* **2008**, *92*, 121–130. [CrossRef]
35. Alkhtib, A.; Wamatu, J.; Ejeta, T.; Rischkowsky, B. Integrating the straw yield and quality into multi-dimensional improvement of lentil (*Lens culinaris*). *J. Sci. Food Agric.* **2017**, *97*, 4135–4141. [CrossRef]
36. Belachew, Z.; Yieshak, K.; Taye, T.; Janssens, G. Chemical composition and in sacco ruminal degradation of tropical trees rich in condensed tannins. *Czech J. Anim. Sci.* **2013**, *4*, 176–192. [CrossRef]
37. Van Soest, P.J. *Nutritional Ecology of the Ruminant*, 2nd ed.; Cornell University Press: Ithaca, NY, USA, 1994.
38. Abreu, J.M.F.; Bruno-Soares, A.M. Characterization and utilization of rice, legume and rape straws. *Toxicon* **1998**, *34*, 292–293.
39. Yu, P.; Christensen, D.A.; McKinnon, J.J.; Markert, J.D. Effect of variety and maturity stage on chemical composition carbohydrate and protein subfractions, in vitro rumen degradability and energy values of timothy and alfalfa. *Can. J. Anim. Sci.* **2003**, *83*, 279–290. [CrossRef]
40. Mustafa, A.F.; Christensen, D.A.; McKinnon, J.J.; Hucl, P.J. Chemical characterization of selected Western Canadian Spring and durum wheat cultivars using the Cornell net carbohydrate and protein analysis. *J. Sci. Food Agric.* **1999**, *79*, 1659–1665. [CrossRef]
41. Gupta, A.; Singh, S.; Kundu, S.S.; Jha, N. Evaluation of tropical feedstuffs for carbohydrate and protein fractions by CNCP system. *Indian J. Anim. Sci.* **2011**, *81*, 1154–1160.
42. Singh, S.; Kushwaha, B.P.; Nag, S.K.; Mishra, A.K.; Singh, A.; Anele, U.Y. In vitro ruminal fermentation, protein and carbohydrate fractionation, methane production and prediction of twelve commonly used Indian green forages. *Anim. Feed Sci. Technol.* **2012**, *178*, 2–11. [CrossRef]
43. Kamble, A.B.; Puniya, M.; Kundu, S.S.; Shelke, S.K.; Mohini, M. Evaluation of forages in terms of carbohydrate, nitrogen fractions and methane production. *Indian J. Anim. Nutr.* **2011**, *28*, 231–238.
44. Swarna, V.; Dhulipalla, S.K.; Elineni, R.R.; Dhulipalla, N.N. Evaluation of certain crop residues for carbohydrate and protein fractions by cornell net carbohydrate and protein system. *J. Adv. Vet. Anim. Res.* **2015**, *2*, 213–216. [CrossRef]
45. López, S.; Davies, D.R.; Giraldez, F.J.; Dhanoa, M.S.; Dijkstra, J.; France, J. Assessment of nutritive value of cereal and legume straws based on chemical composition and in vitro digestibility. *J. Sci. Food Agric.* **2005**, *85*, 1550–1557. [CrossRef]
46. Cone, J.W.; van Gelder, A.H. Influence of protein fermentation on gas production profiles. *Anim. Feed Sci. Technol.* **1999**, *76*, 251–264. [CrossRef]
47. Zeller, F.M.E.; Edmunds, B.L.; Schwarz, F.J. Effect of genotype on chemical composition, ruminal degradability and in vitro fermentation characteristics of maize residual plants. *J. Anim. Physiol. Anim. Nutr.* **2014**, *98*, 982–990. [CrossRef]
48. Bruno-Soares, A.M.; Abreu, J.M.F.; Guedes, A.; Dias-da-Silva, A. Chemical composition, DM and NDF degradation kinetics in rumen of seven legume straws. *Anim. Feed Sci. Technol.* **2000**, *83*, 75–80. [CrossRef]
49. Kılıçalp, N.; Hızlı, H.; Mart, D. Chemical composition and rumen degradation characteristics of different chickpea (*Cicer Arietinum* L.) lines straw. *Turk. J. Agric. Food Sci. Technol.* **2017**, *5*, 459–463. [CrossRef]
50. Mustafa, A.F.; Christensen, D.A.; McKinnon, J.J. In vitro and in situ evaluation of fenugreek (*Trigonel foenum-graecum*) hay and straw. *Can. J. Anim. Sci.* **1996**, *76*, 625–628. [CrossRef]
51. Abbeddou, S.; Rihawi, S.; Hess, H.D.; Iniguez, L.; Mayer, A.C.; Kreuzer, M. Nutritional composition of lentil straw, vetch hay, olive leaves, and saltbush leaves and their digestibility as measured in fat-tailed sheep. *Small Rumin. Res.* **2011**, *96*, 126–135. [CrossRef]

9

Effects of Gas Production Recording System and Pig Fecal Inoculum Volume on Kinetics and Variation of In Vitro Fermentation using Corn Distiller's Dried Grains with Solubles and Soybean Hulls

Jae-Cheol Jang [1], Zhikai Zeng [1], Gerald C. Shurson [1] and Pedro E. Urriola [1,2,*]

[1] Department of Animal Science, University of Minnesota, St. Paul, MN 55108, USA; jang0046@umn.edu (J.-C.J.); zzeng@umn.edu (Z.Z.); shurs001@umn.edu (G.C.S.)
[2] Department of Veterinary Population Medicine, University of Minnesota, St. Paul, MN 55108, USA
[*] Correspondence: urrio001@umn.edu

Simple Summary: Various in vitro methodologies have been developed and used to estimate the digestibility of feed ingredients, such as corn distillers dried grains with solubles (cDDGS) and soybean hulls (SBH) which contain high concentrations of dietary fiber. This study evaluated two in vitro gas production recording systems (manual vs. automated) and two initial fecal inoculum volumes (30 vs. 75 mL) on the parameters of in vitro fermentation of cDDGS and SBH. The results showed that the use of 75-mL inoculum volume with 0.5 g substrate tended to reduce the variation of measurements compared to the 30-mL inoculum volume with 0.2 g substrate regardless of the gas production recording system. These findings suggest that using larger inoculum volume with more substrate increases the precision of measurements. Furthermore, the automated system decreases labor for conducting the assay.

Abstract: An experiment was conducted to investigate the effect of inoculum volume (IV), substrate quantity, and the use of a manual or automated gas production (GP) recording system for in vitro determinations of fermentation of corn distillers dried grains with solubles (cDDGS) and soybean hulls (SBH). A 2 × 2 × 2 factorial arrangement of treatments was used and included the factors of (1) ingredients (cDDGS or SBH), (2) inoculum volume and substrate quantity (IV30 = 0.2 g substrate + 30 mL inoculum or IV75 = 0.5 g substrate + 75 mL inoculum), and (3) GP recording system (MRS = manual recording system or ARS = automated recording system). Feed ingredient samples were pre-treated with pepsin and pancreatin, and the hydrolyzed residues were subsequently incubated with fresh pig feces in a buffered mineral solution. The GP recording was monitored for 72 h, and the kinetics were estimated by fitting data using an exponential model. Compared with SBH, cDDGS yielded less ($p < 0.01$) maximal gas production (G_f), required more time ($p < 0.02$) to achieve half gas accumulation ($T/2$), and had less ($p < 0.01$) fractional rate of degradation (μ) and in vitro fermentability of dry matter (IVDMF). Using the ARS resulted in less IVDMF ($p < 0.01$) compared with MRS (79.0% vs. 81.2%, respectively). Interactions were observed between GP recording system and inoculum volume and substrate quantity for Gf ($p < 0.04$), μ ($p < 0.01$), and $T/2$ ($p < 0.04$) which implies that increasing inoculum volume and substrate quantity resulted in decreased Gf (332 mL/g from IV30 vs. 256 mL/g from IV75), μ (0.05 from IV30 vs. 0.04 from IV75), and $T/2$ (34 h for IV30 vs. 25 h for IV75) when recorded with ARS but not MRS. However, the recorded cumulative GP at 72 h was not influenced by the inoculum volume nor recording system. The precision of G_f (as measured by the coefficient of variation of G_f) tended to increase for IV30 compared with IV75 ($p < 0.10$), indicating that using larger inoculum volume and substrate quantity (IV75) reduced within batch variation in GP kinetics. Consequently, both systems showed comparable results in GP kinetics, but considering convenience and achievement of consistency, 75 mL of inoculum volume with 0.5 g substrate is recommended for ARS.

Keywords: corn distillers dried grains with solubles; gas collection technique; in vitro; pig fecal inoculum; soybean hulls

1. Introduction

The use of increasing amounts of dietary fiber (DF) in swine feeding programs contributes to various environmental [1], animal well-being [2], and sustainability [3] impacts. About 46.3 million metric tonnes of feed was fed to pigs in the United States in 2016, consisting of 16% corn distillers dried grains with solubles (cDDGS), and 15% total soybean products [4]. However, these ingredients contain higher amounts of DF and less starch compared with corn, resulting in a greater production of short-chain fatty acids (SCFA) by gut microbiota when pigs are fed diets containing cDDGS or soybean hulls (SBH) than corn. The SCFA affect the intestinal epithelial cells and affect the intestinal integrity by regulating ion absorption and gut motility [5].

Various in vitro methodologies have been developed and used to estimate the digestibility of various feed ingredients, including ingredients that contain high concentrations of DF. The most widely used procedure is a three-step in vitro assay that combines replicated enzymatic hydrolysis from the stomach through small intestine [6] with representative large intestine fermentation using swine feces as a living bacterial inoculum [7]. This procedure has been well accepted to estimate in vitro dry matter digestibility (IVDMD) in the large intestine and total gas production of various feed ingredients for swine [5,8–11]. An automated recording system (ARS) for gas production (GP) was introduced in the early 1990s to reduce the amount of labor, compared with the manual recording system (MRS) when evaluating diets and feed ingredients for ruminants [12]. The ARS technique measures the kinetics of microbial fermentation in an automated fashion by monitoring the gas pressure and ventilation process [12]. Several in vitro studies have investigated the advantages and disadvantages of using ARS to measure the gas production profile and fermentation kinetics in ruminant-based in vitro systems [13,14]. However, the type of feed ingredient, amount of fecal inoculum, quantity of substrate, and the type of recording system may affect the accuracy and precision of the parameters estimated.

Our current study was conducted to determine the effects of inoculum volume and recording system on in vitro gas production and the concentration of SCFA produced from the fermentation of cDDGS and SBH. This investigation was based on the hypothesis that the type of ingredient, volume of fecal inoculum, and amount of substrate in a bottle would affect the accuracy and precision of gas production parameter measurements, including the concentration of SCFA when using the ARS in a pig-based in vitro digestibility system.

2. Materials and Methods

2.1. Experimental Design, Feed Samples, and Enzymatic Hydrolysis

This experiment was conducted using a 2 × 2 × 2 factorial arrangement of treatments to examine the effects of feed ingredients (cDDGS or SBH), fecal inoculum volume (IV30 = 200 mg substrate + 30 mL inoculum or IV75 = 500 mg substrate + 75 mL inoculum), and GP recording system (MRS or ARS) on IVDMF and the production of SCFA. Hydrolyzed corn DDGS and SBH residues were obtained from the two-step procedure involving pepsin and pancreatin hydrolysis in our previous studies [8,15] that was developed by Boisen and Fernandez [6]. Briefly, 2 g of each cDDGS and SBH sample was weighed into a 500-mL Pyrex Erlenmeyer flask and incubated at 39 °C in a water bath. Then, 100 mL of phosphate buffer solution (0.1 M 7:1 KH_2PO_4:Na_2HPO_4, pH 6.0) and 40 mL 0.2 M HCl solution (pH 2.0) were added. The pH was adjusted to 2.0 by adding 1 M HCl or 1 M NaOH. The addition of 2 mL of 5 mg/mL chloramphenicol (C0378; Sigma-Aldrich Corp., St. Louis, MO, USA) solution (dissolved in ethanol) was added to prevent bacterial growth during hydrolysis. A volume of 4 mL of 100 mg/mL fresh porcine pepsin (P7000, 421 pepsin units / mg solids; Sigma-Aldrich Corp.)

solution (dissolved in 0.2 M HCl) was added to each bottle and incubated in a water bath at 39 °C for 2 h. All the flasks were shaken gently by hand for 5 s every 15 min. Subsequently, 40 mL of 0.2 M phosphate buffer (7:1 KH_2PO_4:Na_2HPO_4, pH 6.8) and 20 mL of 0.6 M NaOH were added to each flask. Finally, 4 mL of 100 mg/mL fresh porcine pancreatin (P1750, 4 times the specifications of the United States Pharmacopeia; Sigma-Aldrich Corp.) solution (dissolved in 0.2 M phosphate buffer) was added. The hydrolysis continued for 4 h under the same conditions as used for pepsin hydrolysis. Subsequent in vitro fermentation analysis was performed using these residues according to the procedure developed by Jha et al. [10,11].

2.2. Experimental Design and In Vitro Fermentation Procedures

Before fermentation, samples of cDDGS and SBH were hydrolyzed by enzymatic digestion with pepsin and pancreatin. The residues from enzymatic digestion were then subsequently pooled within each ingredient source for in vitro fermentation. Blank inocula without substrates were used as controls. The experimental scheme was as follows: 8 treatments × 3 replications + 4 blanks repeated over three batches. Briefly, either 0.2 g or 0.5 g of pooled hydrolyzed cDDGS and SBH samples (ground to 1 mm in particle size) was weighed and incubated in a buffer solution containing macro-and micro-minerals [16]. Feces were collected by rectal stimulation from one finishing pig per batch. Pigs were fed a conventional corn-soybean meal-based diet without antibiotics (Innovation Campus, Cargill Animal Nutrition, Elk River, MN, USA). Collected fecal samples were immediately placed in air-tight plastic syringes and kept in a water bath at 39 °C until incubation. The time from fecal collection until incubation was less than 1 h. In the laboratory, the inoculum was formulated by diluting blended feces in an inoculation solution composed of distilled water (474 mL/L), trace mineral solution (0.12 mL/L containing 132 g/L of $CaCl_2$, 100 g/L of $MnCl_3{\cdot}4H_2O$, 10 g/L of $CoCl_2{\cdot}6H_2O$, and 80 g/L of $FeCl_3{\cdot}6H_2O$), in vitro buffer solution (237 mL/L containing 4.0 g/L of NH_4HCO_3 and 35 g/L of $NaHCO_3$), macromineral solution (237 mL/L composed of 5.7 g/L of Na_2HPO_4, 6.2 g/L of KH_2PO_4, 0.583 g/L of $MgSO_4{\cdot}7H_2O$, and 2.22 g/L of NaCl), and resazurin (blue dye, 0.1% wt/vol solution; 1.22 mL/L) and filtered through four layers of cheesecloth. The final inoculum concentration was 0.05 g feces/mL of buffer. Either 30 mL or 75 mL of inoculum aliquots were respectively transferred into bottles containing 200 mg or 500 mg of the hydrolyzed sample substrates to provide an equal inoculum to substrate ratio (6.67 mL/mg) between the two systems. Carbon dioxide (CO_2) was provided to maintain an anaerobic environment during the entire inoculum preparation process.

The headspace gas pressure in the bottles was recorded using either MRS or ARS. The gas was measured manually at 11 time points post-inoculation using an inverted 25-mL burette with its stopcock end attached to the vacuum, and its open end submerged into a water bath (39 °C) in MRS. The ARS was designed to measure the kinetics of microbial fermentation by monitoring the gas pressure automatically every 5 min and recording remotely using a commercial apparatus (Ankom^{RF} Gas Production System, Ankom Technology, Macedon NY, USA) equipped with real-time sensors. The headspace volume was 57.5 mL in MRS and 257.5 mL in ARS. For the ARS system, accumulated gas in the headspace was automatically released when the pressure exceeded 35 psi. Recording of headspace pressure was terminated at 72 h post-incubation. At the end of the 72 h, the supernatant from each bottle was collected for SCFA analysis.

2.3. Chemical Analyses

Before liquid chromatography–mass spectrometry (LC-MS) analysis, samples of fermentation supernatants were derivatized with hydroquinone (HQ) for the determination of SCFA concentrations [17]. Briefly, two microliters of the extracted supernatant were mixed with 70 μL of acetonitrile (ACN) containing 7.5 μM acetic acid-d4, 10 μL dipridyl disulfide (DPDS), 10 μL triphenylphosphine (TPP), and 10 μL HQ. The mixture was incubated at 60 °C for 30 min, chilled on ice, and mixed with 100 μL H_2O. The vials were then centrifuged at 21,000 × *g* at 4 °C for 10 min. The processed HQ-reaction mixture from chemical derivatization of samples was injected into ultra-performance liquid chromatography (UPLC) system (Xevo-G2-S; Waters, Milford, MA, USA). The concentration of individual compounds was determined by calculating the ratio between the peak area of compounds and the peak area of internal standards. Acetic acid-d4 was used as an internal standard calibration curve for precise SCFA quantification. The acquired data were processed by software (QuanLynx, Waters, Milford, MA, USA).

2.4. Calculations

The in vitro fermentability of dry matter (IVDMF) during fecal inoculum fermentation was calculated as follows:

IVDMF, % = [(dry weight of the hydrolyzed residue − dry weight of the residue after fermentation)/dry weight of the hydrolyzed residue] × 100

After correction for the blank units, the recorded cumulative gas pressure (psi) was converted into mL of gas produced per g DM using Avogadro's law as follows:

Gas volume, mL = gas pressure × [V/RT] × 22.4 L/mol × 1000 mL/L,

where V denotes head space volume in the bottle (L), R was the gas constant 8.314472 L k Pa/K/mol, and T represents the temperature in Kelvin (273 °K + Celsius temperature in the bottle).

Gas accumulation curves (mL/g DM) recorded during the 72 h of fermentation were fitted by the following model developed by France et al. [18]:

$$G\ (\text{mL}/\text{g DM}) = 0,\ \text{if } 0 < t < L$$

$$G = G_f\ (1 - \exp\ (-[b\ (t - L) + c\ (\sqrt{t} - \sqrt{L})])),\ \text{if } t \geq L,$$

where G denotes the gas accumulation at a specific time (t), G_f (mL/g DM) was the maximum gas volume for $t = \infty$, and L (h) represents the lag time before the fermentation began and is determined by the initial delay until the onset of gas production occurs. In the present study, gas accumulation of the cDDGS treatment rapidly reached one-fourth of the maximum accumulation in 2 h, and the parameter L (h) was very close to 0, which resulted in the model failing to converge. Therefore, L (h) data were removed from the final model. The constants b (h^{-1}) and c ($h^{-1/2}$) determine the fractional rate of degradation of the substrate μ (h − 1), which is postulated to vary with time as follows:

$$\mu = b + c/(2\sqrt{t}),\ \text{if } t \geq L$$

Kinetics parameters of gas production (G_f, T/2, G72, and μ at T/2) were compared in the statistical analysis, with T/2 representing the time to half asymptote when $G = G_f/2$.

2.5. Statistical Analyses

The kinetics of gas production parameters were fitted based on the individual time series data and were analyzed using PROC NLIN of SAS version 9.4 (SAS Inst. Inc., Cary, NC, USA). The IVDMF, fitted gas production kinetic parameters, and the concentration of SCFA were analyzed using the GLIMMIX procedure of SAS version 9.4 (SAS Inst., Inc., Cary, NC, USA), with individual bottles considered as the experimental unit. The model included substrates (cDDGS and SBH), inoculum volume (30 mL and 75 mL), GP recording system (MRS and ARS), and their interactions (Substrate × Volume, Substrate × System, Volume × System, and Substrate × Volume × System) as the fixed factors and batches of samples as random factors. The average coefficient of variance (CV) was calculated based on the average values of kinetic parameters within each treatment using PROC GLM of SAS version 9.4 (SAS Inst., Inc., Cary, NC, USA). The least square means of individual treatments were separated by the Tukey method. Results were considered significant at $p \leq 0.05$ and trends at $0.05 < p \leq 0.10$.

3. Results and Discussion

3.1. Fermentation Kinetics and Metabolites

Soybean hulls yielded greater ($p < 0.01$) maximal gas production (G_f), required less time ($p < 0.02$) to achieve half gas accumulation ($T/2$), and had greater ($p < 0.01$) fractional rate of degradation (μ) and IVDMF compared to cDDGS (Table 1). Each of ingredients showed similar gas production curves regardless of gas recording system and inoculum volume (Figure 1). Results for IVDMF of cDDGS (69.2%) obtained in the current study were greater than that reported in previous studies (59.6% by Jha et al. [9]; 55.7% by Huang et al. [8]), but maximum gas volume (G_f) of cDDGS (200 mL/g DM) was comparable to those reported by Jha et al. [9] (200 mL/g DM) and Huang et al. [8] (208 mL/g DM). Different kinetics of GP between these two ingredients can be explained by their fiber composition. Soybean hulls contain about 5.5 times more soluble dietary fiber (SDF) than insoluble dietary fiber (IDF), whereas cDDGS contains 1.6 times more SDF than IDF [19]. It has been suggested that apparent ileal digestibility (AID) and apparent total tract digestibility (ATTD) of SDF are a result of greater fermentation compared with IDF in growing-finishing pigs [20]. Moreover, while SDF is mainly fermented in the proximal colon, IDF is fermented primarily in the distal colon [21], which is likely due to the hydrophobic and the crystalline characteristics of these types of DF [22]. Consequently, a greater SDF/IDF ratio in SBH may have resulted in a sharp increase in the fractional rate of degradation during earlier fermentation stage (<8 h) compared with cDDGS in the current study.

Gas production (GP) kinetics parameters were not different between the GP recording systems, whereas IVDMF was less ($p < 0.01$) when recorded in the ARS system compared with the MRS system (79.0% vs. 81.2%, respectively). Moreover, interactions were observed between GP recording system and inoculum volume and substrate quantity for G_f ($p < 0.04$) and μ ($p < 0.01$), and the time to half asymptote ($T/2$, $p < 0.04$). According to the meta-analysis on methodological factors influencing GP during in vitro rumen fermentation, the GP recording apparatus with venting system (i.e., ARS) resulted in greater gas production estimates compared to the MRS GP recording apparatus operating without venting system [23]. Furthermore, the absence of automatic ventilation system in MRS increased headspace pressure, so that it may have caused a partial dissolution of carbon dioxide (CO_2) in the inoculum, and subsequently resulted in the underestimation of GP as well as restricting microbial respiration. Results from the current experiment showed no difference in parameters of GP kinetics between the two systems. One possible explanation for the lack of differences may be due to the differences between the headspace volumes to fermentation inoculum ratio between the systems, which was 4.9 for ARS compared to 2.8 for MRS in the current experiment. This ARS ratio is greater than the ratio used in a previous in vitro study conducted using swine fecal inoculum with ARS (ratio: 3.2, Pastorelli et al. [24]). However, the optimal ratio between headspace and fermentation inoculum has not yet been established. The smaller ratio may result in greater underestimation of GP because of higher pressure [25], whereas the larger ratio may result in lower pressure and cause inhibition of microbial activity [13]. Therefore, based on the current results, it can be expected that relatively larger ratio between headspace volume to inoculum in ARS may interfere with the microbial fermentation in the bottle, resulting in decreased IVDMF, as well as increased within batch variation. However, further investigations are required to determine the optimal ratio between headspace and inoculum volume when using swine fecal inoculum in ARS.

Table 1. Parameters of the fitted kinetics and concentration of short-chain fatty acids (SCFA) after in vitro fermentation of corn distillers dried grains with solubles (cDDGS) and soybean hulls (SBH) using an automatic recording system (ARS) and manual recording system (MRS) using different fecal inoculum volumes (30 and 75 mL) obtained from pigs [1].

Item	ARS		MRS		SEM [2]	Substrates		SEM	p-Values [3]			
	30 mL	75 mL	30 mL	75 mL		cDDGS	SBH		Sub	Sys	Vol	Sys × Vol
Fermentation kinetics												
N [4]	8	8	8	8		20	20					
Gf [5]	332^{a}	256^{b}	281^{ab}	286^{ab}	37.4	200^{B}	362^{A}	34.3	0.011	0.971	0.334	0.047
μ [6]	0.04^{b}	0.05^{a}	0.04^{ab}	0.04^{ab}	0.015	0.02^{B}	0.05^{A}	0.014	0.012	0.446	0.544	0.012
T/2 [7]	34.34^{a}	25.07^{b}	27.16^{ab}	28.16^{ab}	7.832	32.02^{A}	26.10^{B}	7.794	0.028	0.984	0.977	0.041
G72 [8]	211	210	224	240	12.2	159^{B}	285^{A}	9.8	0.016	0.987	0.826	0.110
IVDMF [9]	79.2	79.0	80.8	81.0	1.10	69.2^{B}	90.9^{A}	0.93	0.012	0.012	0.731	0.463
Fermentation metabolites, mmol/g												
Acetic acid	4.78	4.19	4.32	4.53	0.504	3.71^{B}	5.20^{A}	0.504	0.009	0.906	0.346	0.429
Propionic acid	1.14	1.17	1.10	1.12	0.063	1.08^{B}	1.18^{A}	0.031	0.047	0.284	0.433	0.917
Butyric acid	0.58	0.71	0.54	0.58	0.055	0.64	0.57	0.054	0.175	0.130	0.271	0.414
Valeric acid	0.19	0.16	0.15	0.22	0.034	0.19	0.16	0.024	0.306	0.688	0.570	0.108
Total SCFA	6.69	6.23	6.10	6.46	0.584	5.64^{B}	7.11^{A}	0.413	0.001	0.757	0.428	0.485

^{A,B} Means within a row with different uppercase superscript letters differ significantly (main effect, $p < 0.05$). ^{a,b} Means within a row with different lowercase superscript letters differ significantly (Sys × Vol, $p < 0.05$). [1] The least squares mean value presented based on the three replications per treatment. [2] Standard error of the means. [3] Sub = feed ingredients as substrates; Sys = gas production recording system; Vol = inoculum volume. [4] Number of observations in fermentation. [5] Maximum gas volume (mL/g DM incubated). [6] Fractional rate of degradation (h – 1) at t = *T*/2. [7] *T*/2, half-time to asymptote (h). [8] Volume of gas production at 72 h. [9] In vitro dry matter fermentability.

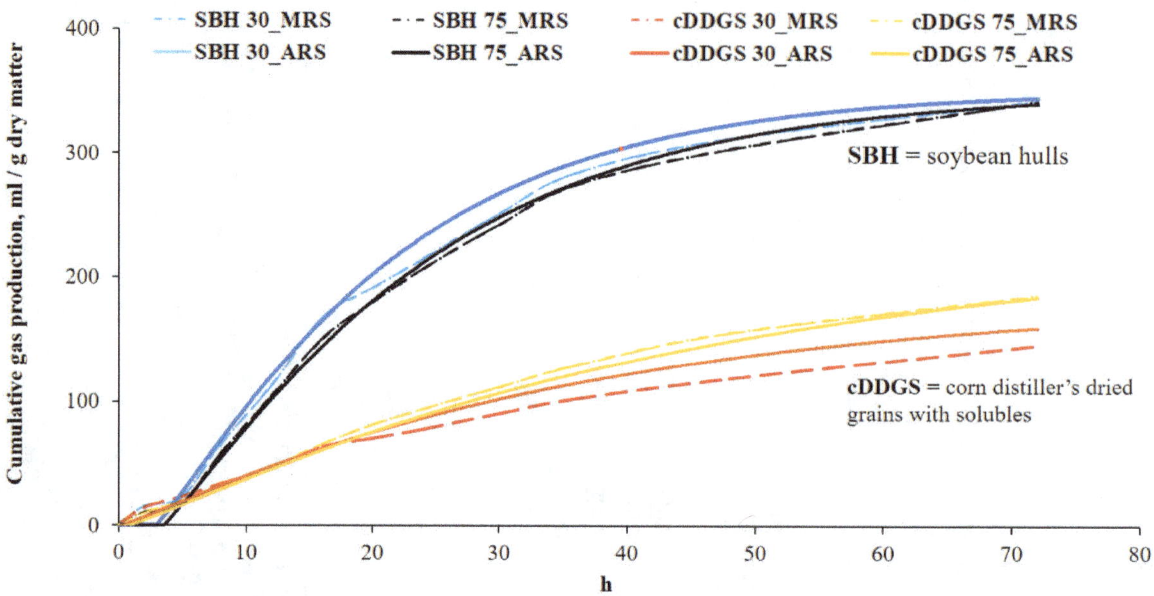

Figure 1. Gas accumulation curves of two ingredients (soybean hulls = SBH; and corn dried distiller's grains with solubles = DDGS) and inoculum volume (30 and 75 mL) incubated either in automatic gas production recording system (ARS) or manual gas production recording system (MRS) during 72 h.

Regardless of substrates, acetic acid was the most abundant SCFA produced during in vitro fermentation. The samples of SBH produced more acetic acid ($p < 0.01$), propionic acid ($p < 0.05$), and total SCFA ($p < 0.01$) compared with cDDGS (Table 1). These results are in agreement with those from a previous in vitro study by Jha and Leterme [26], indicating that both in vitro GP recording systems yielded accurate estimates of microbial fermentation. The greater SCFA production observed during fermentation of SBH compared with cDDGS can be attributed to the solubility of DF in the ingredient. Ferulic acid consists of cross-linked cell wall polysaccharides and other cell wall components such that it might be associated with fiber matrix rigidity [27]. Insoluble DF has been linked to decreased SCFA production resulting from slower fermentation rates compared to soluble DF, and insoluble DF contains 100 times greater ferulic acid content than soluble DF [28]. Thus, there was a greater amount of soluble dietary fiber available for microbiota fermentation in SBH, resulting in increased production of SCFA compared to cDDGS (Figure 1).

3.2. The Average Coefficient of Variance

The hypothesis of this study was that error frequency and severity would be relatively greater in MRS compared to ARS because of the intensive labor involved during the first phase of microbial fermentation. Although we observed no differences between the two GP recording systems for the coefficient of variation (CV) of GP kinetic parameters and IVDMF, the CV tended to be less ($p < 0.10$) when using the greater inoculum volume (IV75) compared to using the smaller inoculum volume (IV30, Table 2). Also, the CV tended to be less in cDDGS on time to half asymptote ($T/2$, $p < 0.07$) and IVDMF ($p < 0.09$) compared to SBH.

Comparison of results from our variability analysis to results from other studies is difficult because each study analyzed results using different mathematical methods. However, results from a previous study evaluating the repeatability and reproducibility of an ARS using rumen fluid from four laboratories indicated that fermentable organic matter had the greatest repeatability and reproducibility (0.2 to 1.9%, and 0.3 to 4.5%, respectively), followed by kinetic parameters (Gf = 1.1 to 2.5% in repeatability and 1.7 to 3.8% in reproducibility; $T/2$ = 4.3 to 13.2% in repeatability and 4.7 to 13.2% in reproducibility; μ= 8.2 to 12.8% in repeatability and 18.6 to 27.5% in reproducibility) [29]. This pattern was similar to the results obtained in the current study, indicating that the CV for IVDMF had the least variation, and kinetic parameters showed comparatively greater variation. Also, a similar CV pattern of kinetics was observed in a recent study using the ring test of in vitro GP recording systems conducted in four different laboratories in Europe (Denmark, United Kingdom, Spain, Italy), using the same wireless apparatus that we used in the current experiment [30]. These researchers also indicated that the least variation among parameters of GP kinetics observed were as follows: GP at 48 h (CV = 4.8%), *Gf* (CV = 6.4%), μ (CV = 11.4%), and *T*/2 (CV = 14.1%).

Table 2. Average coefficient of variation of the in vitro fermentation kinetic parameters for corn distillers dried grains with solubles (cDDGS) and soybean hulls (SBH) using an automatic recording system (ARS) and manual recording system (MRS) with two different pig fecal inoculum volumes (30 and 75 mL) [1].

Item	ARS		MRS		SEM [2]	Substrates		SEM	*p*-Values [3]			
	30 mL	75 mL	30 mL	75 mL		cDDGS	SBH		Sub	Sys	Vol	Sys × Vol
Gf [4]	29.2	23.7	31.5	13.9	12.53	22.7	26.4	8.16	0.757	0.758	0.100	0.762
μ [5]	43.7	45.7	50.6	38.5	14.49	31.2	58.0	6.06	0.068	0.985	0.688	0.941
T/2 [6]	32.2	48.9	55.6	27.0	16.06	29.1	52.8	9.20	0.122	0.958	0.716	0.590
G72 [7]	33.6	20.9	10.2	9.8	5.53	19.9	17.2	5.92	0.750	0.136	0.447	0.203
IVDMF [8]	4.35	3.15	3.85	4.85	0.03	5.02	3.08	0.81	0.121	0.588	0.929	0.764

[1] The least squares mean value presented based on the three replications per treatment. [2] Standard error of the means. [3] Sub = feed ingredients as substrates; Sys = gas production recording system; Vol = inoculum volume. [4] Maximum gas volume (mL/g DM incubated). [5] Fractional rate of degradation (h − 1) at t = *T*/2. [6] *T*/2, half-time to asymptote (h). [7] Termination gas volume (at 72 h). [8] In vitro dry matter fermentability.

Rymer et al. [31] indicated that the largest source of variation in the GP technique could be attributed to the source of inoculum and its microbial activity. In our study, we assumed that the fecal samples may vary among fecal donor age and may significantly increase the CV in the current experiment. Fecal sampling procedures were irregularly managed because of the bio-security of the company-owned research farm. There is a relatively large variation in age and body weight (60 to 100 kg) of fecal donors between batches. Kim et al. [32] indicated that pig microbial ecosystems in the GIT continue to change as pigs grow, and is influenced by various factors, including genetics, diet, and antibiotics. Therefore, our results reflect the fact that using fecal inocula from pigs of different ages leads to differences in microbial fermentability derived from different microbial communities within the batch of samples, resulting in increased variability of GP kinetic curves. The use of inoculum from one fecal donor per batch can be another factor. In our previous in vitro study, fecal samples were randomly collected from three out of five growing pigs for each batch of feed ingredient samples analyzed [8], resulting in CV's of kinetic parameters (*Gf*, and *T*/2) of 5.2 and 4.5% in SBH, respectively, and 9.8 and 18.5% in cDDGS, respectively, which were 5.18 and 2.81 times less than the CV's obtained from the current experiment. Rymer et al. [31] emphasized that fecal samples should be collected from several animals for in vitro fermentation analysis because each pig has different fecal microflora composition even though they are from the same genetic line and consume the same diet. Evidence from human studies has shown that using inoculum from at least three donors may enhance the predictive value of in vitro colonic fermentation [33]. Based on the results from the current experiment, we suggest collecting fecal samples from more than three pigs is necessary for improving the accuracy of pig in vitro fermentation assays of high fiber ingredients.

4. Conclusions

The results of this experiment demonstrate that both the GP recording systems (manual and automatic) were accurate at recording the gas production during in vitro fermentation similar to results reported in the literature for cDDGS and SBH. These results also suggest that there is an improvement in precision when larger volumes of fecal inocula are used if the ratio of substrate and headspace are kept in proportion between the GP recording systems.

Author Contributions: Z.Z. and P.E.U. designed the experiment. J.-C.J. and Z.Z. conducted the experimental work. All authors contributed to analyzing the data and writing of the manuscript.

Acknowledgments: The authors wish to thank Jinlong Zhu, Edward Zhaohui Yang, and Yuan-Tai Hung for laboratory work.

References

1. Aarnink, A.J.A.; Verstegen, M.W.A. Nutrition, key factor to reduce environmental load from pig production. *Livest. Sci.* **2007**, *109*, 194–203. [CrossRef]
2. De Leeuw, J.A.; Bolhuis, J.E.; Bosch, G.; Gerrits, W.J.J. Effects of dietary fibre on behaviour and satiety in pigs. *Proc. Nutr. Soc.* **2008**, *67*, 334–342. [CrossRef] [PubMed]
3. Basset-Mens, C.; van der Werf, H.M.G. Scenario-based environmental assessment of farming systems: The case of pig production in France. *Agric. Ecosyst. Environ.* **2005**, *105*, 127–144. [CrossRef]
4. United States Department of Agriculture (USDA). *Agricultural Statistics 2017*; Department of Agriculture, National Agricultural Statistics Service: Washington, DC, USA, 2017.
5. Kim, C.H.; Park, J.; Kim, M. Gut microbiota-derived short-chain fatty acids, T cells, and inflammation. *Immune Netw.* **2014**, *14*, 277–288. [CrossRef] [PubMed]
6. Boisen, S.; Fernández, J.A. Prediction of the total tract digestibility of energy in feedstuffs and pig diets by in vitro analyses. *Anim. Feed Sci. Technol.* **1997**, *68*, 277–286. [CrossRef]
7. Bindelle, J.; Buldgen, A.; Lambotte, D.; Wavreille, J.; Leterme, P. Effect of pig faecal donor and of pig diet composition on in vitro fermentation of sugar beet pulp. *Anim. Feed Sci. Technol.* **2007**, *132*, 212–226. [CrossRef]
8. Huang, Z.; Urriola, P.E.; Salfer, I.J.; Stern, M.D.; Shurson, G.C. Differences in in vitro hydrolysis and fermentation among and within high-fiber ingredients using a modified three-step procedure in growing pigs. *J. Anim. Sci.* **2017**, *95*, 5497–5506. [CrossRef]
9. Jha, R.; Woyengo, T.A.; Li, J.; Bedford, M.R.; Vasanthan, T.; Zijlstra, R.T. Enzymes enhance degradation of the fiber-starch-protein matrix of distillers dried grains with solubles as revealed by a porcine in vitro fermentation model and microscopy. *J. Anim. Sci.* **2015**, *93*, 1039–1051. [CrossRef]
10. Jha, R.; Bindelle, J.; Rossnagel, B.; Van Kessel, A.; Leterme, P. In vitro evaluation of the fermentation characteristics of the carbohydrate fractions of hulless barley and other cereals in the gastrointestinal tract of pigs. *Anim. Feed Sci. Technol.* **2011**, *163*, 185–193. [CrossRef]
11. Jha, R.; Bindelle, J.; Van Kessel, A.; Leterme, P. In vitro fibre fermentation of feed ingredients with varying fermentable carbohydrate and protein levels and protein synthesis by colonic bacteria isolated from pigs. *Anim. Feed Sci. Technol.* **2011**, *165*, 191–200. [CrossRef]
12. Davies, Z.S.; Mason, D.; Brooks, A.E.; Griffith, G.W.; Merry, R.J.; Theodorou, M.K. An automated system for measuring gas production from forages inoculated with rumen fluid and its use in determining the effect of enzymes on grass silage. *Anim. Feed Sci. Technol.* **2000**, *83*, 205–221. [CrossRef]
13. Tagliapietra, F.; Cattani, M.; Bailoni, L.; Schiavon, S. In vitro rumen fermentation: Effect of headspace pressure on the gas production kinetics of corn meal and meadow hay. *Anim. Feed Sci. Technol.* **2010**, *158*, 197–201. [CrossRef]
14. He, Z.X.; Zhao, Y.L.; McAllistera, T.A.; Yang, W.Z. Effect of in vitro techniques and exogenous feed enzymes on feed digestion. *Anim. Feed Sci. Technol.* **2016**, *213*, 148–152. [CrossRef]
15. Kerr, B.J.; Gabler, N.K.; Shurson, G.C. Formulating diets containing corn distillers dried grains with solubles on a net energy basis: Effects on pig performance and on energy and nutrient digestibility. *Prof. Anim. Sci.* **2015**, *31*, 497–503. [CrossRef]
16. Menke, K.H.; Steingass, H. Estimation of the energetic feed value obtained from chemical analysis and in vitro gas production using rumen fluid. *Anim. Res. Dev.* **1988**, *28*, 7–55.
17. Lu, Y.; Yao, D.; Chen, C. 2-hydrazinoquinoline as a derivatization agent for LC-MS-based metabolomic investigation of diabetic ketoacidosis. *Metabolites* **2013**, *3*, 993–1010. [CrossRef] [PubMed]
18. France, J.; Dhanoa, M.S.; Theodorou, M.K.; Lister, S.J.; Davies, D.R.; Isac, D. A model to interpret gas accumulation profiles associated with in vitro degradation of ruminant feeds. *J. Theor. Biol.* **1993**, *163*, 99–111. [CrossRef]

19. Jaworski, N.W.; Stein, H.H. Disappearance of nutrients and energy in the stomach and small intestine, cecum, and colon of pigs fed corn-soybean meal diets containing distillers dried grains with solubles, wheat middlings, or soybean hulls. *J. Anim. Sci.* **2017**, *95*, 727–739. [CrossRef]
20. Urriola, P.E.; Shurson, G.C.; Stein, H.H. Digestibility of dietary fiber in distillers coproducts fed to growing pigs. *J. Anim. Sci.* **2010**, *88*, 2373–2381. [CrossRef]
21. Jha, R.; Berrocoso, J.F.D. Dietary fiber and protein fermentation in the intestine of swine and their interactive effects on gut health and on the environment: A review. *Anim. Feed Sci. Technol.* **2016**, *212*, 18–26. [CrossRef]
22. Bach Knudsen, K.E.; Hansen, I. Gastrointestinal implications in pigs of wheat and oat fractions 1. Digestibility and bulking properties of polysaccharides and other major constituents. *Br. J. Nutr.* **1991**, *65*, 217–232. [CrossRef] [PubMed]
23. Maccarana, L.; Cattani, M.; Tagliapietra, F.; Schiavon, S.; Bailoni, L.; Mantovani, R. Methodological factors affecting gas and methane production during in vitro rumen fermentation evaluated by meta-analysis approach. *J. Anim. Sci. Biotechnol.* **2016**, *7*, 1–12. [CrossRef] [PubMed]
24. Pastorelli, G.; Faustini, M.; Attard, E. In vitro fermentation of feed ingredients by fresh or frozen pig fecal inocula. *Anim. Sci. J.* **2014**, *85*, 690–697. [CrossRef] [PubMed]
25. Cattani, M.; Tagliapietra, F.; Maccarana, L.; Hansen, H.H.; Bailoni, L.; Schiavon, S. Technical note: In vitro total gas and methane production measurements from closed or vented rumen batch culture systems. *J. Dairy Sci.* **2014**, *97*, 1736–1741. [CrossRef] [PubMed]
26. Jha, R.; Leterme, P. Feed ingredients differing in fermentable fibre and indigestible protein content affect fermentation metabolites and faecal nitrogen excretion in growing pigs. *Animal* **2012**, *6*, 603–611. [CrossRef] [PubMed]
27. Bunzel, M.; Ralph, J.; Marita, J.M.; Hatfield, R.D.; Steinhart, H. Diferulates as structural components in soluble and insoluble cereal dietary fibre. *J. Sci. Food Agric.* **2001**, *81*, 653–660. [CrossRef]
28. Pedersen, M.B.; Bunzel, M.; Schäfer, J.; Knudsen, K.E.B.; Sørensen, J.F.; Yu, S.; Lærke, H.N. Ferulic acid dehydrodimer and dehydrotrimer profiles of distiller's dried grains with solubles from different cereal species. *J. Agric. Food Chem.* **2015**, *63*, 2006–2012. [CrossRef]
29. Van Laar, H.; Van Straalen, W.M.; Van Gelder, A.H.; De Boever, J.L.; D'heer, B.; Vedder, H.; Kroes, R.; de Bot, P.; Van Hees, J.; Cone, J.W. Repeatability and reproducibility of an automated gas production technique. *Anim. Feed. Sci. Technol.* **2006**, *127*, 133–150. [CrossRef]
30. Cornou, C.; Storm, I.M.L.D.; Hindrichsen, I.K.; Worgan, H.; Bakewell, E.; Yáñez-Ruiz, D.R.; Abecia, L.; Tagliapietra, F.; Cattani, M.; Ritz, C.; et al. A ring test of a wireless in vitro gas production system. *Anim. Prod. Sci.* **2013**, *53*, 585–592. [CrossRef]
31. Rymer, C.; Huntington, J.A.; Williams, B.A.; Givens, D.I. In vitro cumulative gas production techniques: History, methodological considerations and challenges. *Anim. Feed. Sci. Technol.* **2005**, 9–30. [CrossRef]
32. Kim, H.B.; Borewicz, K.; White, B.A.; Singer, R.S.; Sreevatsan, S.; Tu, Z.J.; Isaacson, R.E. Longitudinal investigation of the age-related bacterial diversity in the feces of commercial pigs. *Vet. Microbiol.* **2011**, *153*, 124–133. [CrossRef] [PubMed]
33. McBurney, M.I.; Thompson, L.U. Effect of human faecal donor on in vitro fermentation variables. *Scand. J. Gastroenterol.* **1989**, *24*, 359–367. [CrossRef] [PubMed]

10

In Vitro Evaluation of Different Dietary Methane Mitigation Strategies

Juana C. Chagas, Mohammad Ramin and Sophie J. Krizsan *

Department of Agricultural Research for Northern Sweden, Swedish University of Agricultural Sciences (SLU), Skogsmarksgränd, 90183 Umeå, Sweden; juana.chagas@slu.se (J.C.C.); mohammad.ramin@slu.se (M.R.)

* Correspondence: sophie.krizsan@slu.se;

Simple Summary: Dietary methane mitigation strategies do not necessarily make food production from ruminants more energy-efficient, but reducing methane (CH_4) in the atmosphere immediately slows down global warming, helping to keep it within 2 °C above the pre-industrial baseline. There is no single most efficient strategy for mitigating enteric CH_4 production from domestic ruminants on forage-based diets. This study assessed a wide variety of dietary CH_4 mitigation strategies in the laboratory, to provide background for future studies with live animals on the efficiency and feasibility of dietary manipulation strategies to reduce CH_4 production. Among different chemical and plant-derived inhibitors and potential CH_4-reducing diets assessed, inclusion of the natural antimethanogenic macroalga *Asparagopsis taxiformis* showed the strongest, and dose-dependent, CH_4 mitigating effect, with the least impact on rumen fermentation parameters. Thus, applying *Asparagopsis taxiformis* at a low daily dose was the best potential dietary mitigation strategy tested, with promising long-term effects, and should be further studied in diets for lactating dairy cows.

Abstract: We assessed and ranked different dietary strategies for mitigating methane (CH_4) emissions and other fermentation parameters, using an automated gas system in two in vitro experiments. In experiment 1, a wide range of dietary CH_4 mitigation strategies was tested. In experiment 2, the two most promising CH_4 inhibitory compounds from experiment 1 were tested in a dose-response study. In experiment 1, the chemical compounds 2-nitroethanol, nitrate, propynoic acid, p-coumaric acid, bromoform, and *Asparagopsis taxiformis* (AT) decreased predicted in vivo CH_4 production (1.30, 21.3, 13.9, 24.2, 2.00, and 0.20 mL/g DM, respectively) compared with the control diet (38.7 mL/g DM). The 2-nitroethanol and AT treatments had lower molar proportions of acetate and higher molar proportions of propionate and butyrate compared with the control diet. In experiment 2, predicted in vivo CH_4 production decreased curvilinearly, molar proportions of acetate decreased, and propionate and butyrate proportions increased curvilinearly with increased levels of AT and 2-nitroethanol. Thus 2-nitroethanol and AT were the most efficient strategies to reduce CH_4 emissions in vitro, and AT inclusion additionally showed a strong dose-dependent CH_4 mitigating effect, with the least impact on rumen fermentation parameters.

Keywords: antimethanogenic; chemical inhibition; global warming; halogenated compound; macroalgae; methane production; methanogenic inhibitor; plant inhibitory compound

1. Introduction

The global population is growing and, although there is enough food in the world today, there are major differences in how people live. Meat and milk from ruminants are high-quality foods and a large proportion of their production is based on grass, but production is still resource-intensive. Future intensification of agriculture can reinforce negative effects such as greenhouse gas (GHG) emissions, the main contributor to climate change through global warming [1,2].

Methane (CH_4) is a powerful GHG that plays a key part in global climate change and concentrations have been rising rapidly in the atmosphere over the past decade. Recently published data based on radioactive carbon (C^{14}) content in CH_4 indicate that anthropogenic emissions of CH_4 in recent decades have been higher than previously estimated [3]. Satellite data [4] suggest that the increased global CH_4 emissions in the period 2005–2015 were mostly due to increased extraction of shale gas, and that the natural gas and oil industry contributes twice as much CH_4 emissions as animal agriculture.

Methanogenesis in the rumen is an essential metabolic process required to remove molecular hydrogen generated during fermentation. The production of CH_4 is influenced by animal species, age, management, and diet. The Rumen Census project sequenced a wide variety of rumen and camelid foregut microbial communities in many samples from a wide variety of animal species and countries, to identify factors such as diet, host species, or geography causing the greatest variation in CH_4 emissions [5]. The results showed that rumen archaeal diversity was similar irrespective of host or diet, and a core rumen bacterial population of 67% of the community occurred irrespective of host or diet. The main diversity changes in other bacteria present were caused by diet, and not host genetics [6]. Although dietary strategies for mitigating enteric CH_4 production in ruminants have been intensively studied, no single most efficient dietary strategy has been identified for dairy cows on forage-based diets.

Methane losses from typical dairy cow diets are 6–7% of gross energy intake, but losses are approximately 3% in feedlot situations, indicating that feeding high-concentrate diets can reduce CH_4 production [7]. However, recent data indicate that the effect on CH_4 production of including more grain in dairy cows diets is small [8] and use of this strategy can therefore be questioned. Use of antimethanogenics or plant inhibitory compounds in ruminant diets can also reduce GHG emissions, and has been suggested as an effective and feasible strategy in the livestock sector [9]. Dietary mitigation strategies do not necessarily make food production from ruminants more energy-efficient, but they reduce CH_4 emissions to the atmosphere and thus immediately slow down global warming [10], contributing to keep the planet within 2 °C of the pre-industrial baseline [11]. The use of CH_4 inhibitors might be the most immediate and efficient strategy to reduce CH_4 emissions from dairy cows. The most successful inhibitor suggested in vivo so far is 3-nitroxypropanol that has showed CH_4 reducing effects when provided to dairy cows in a low dose [12]. The tropical macroalgae *Asparagopsis Taxiformis* is a recent and natural supplement that has shown very promising CH_4 inhibitory effects in vitro [13].

In vitro gas production technique has been developed to evaluate factors influencing digestibility and fermentation kinetics from feeds. The technique has been used to estimate CH_4 emission with the advantage of screening large number of samples, providing large amount of data points, and allowing accurate predictions of in vivo CH_4 production [14].

This study assessed and ranked a wide variety of dietary CH_4 mitigation strategies using an automated gas in vitro system, in order to provide background for future in vivo evaluations of dietary manipulation strategies for efficiently reducing CH_4 production from domestic ruminants.

2. Materials and Methods

Two in vitro experiments were conducted to assess different dietary antimethanogenic compounds. In experiment 1, the dietary CH_4 mitigating strategies tested comprised six chemical inhibitory compounds at two levels, three plant-derived inhibitory treatments at two levels, five different potentially CH_4-reducing diets with the active ingredients in two levels except for one of the diets, and two typical grass silage fermentation acids at two levels to mimic different silage fermentation qualities. In experiment 2, the two most promising CH_4 inhibitory treatments from experiment 1 were tested in a dose-response experiment designed to represent a wide range of treatment levels.

2.1. Experimental Treatments

2.1.1. Experiment 1

All experimental diets were composed from a control diet that consisted of timothy grass (*Phleum pratense*), rolled barley (*Hordeum vulgare*), and rapeseed (*Brassica napus*) meal in a ratio of 545:363:92 g/kg diet dry matter (DM). The grass and rolled barley originated from Röbäcksdalens research farm in Umeå (63°45′ N, 20°17′ E), Sweden. The rapeseed meal was a commercial solvent-extracted and heat-moisture-treated protein supplement ExPro-00SF (Aarhus Karlshamn AB, Malmö, Sweden). All potential dietary CH_4 mitigating strategies tested in experiment 1 are listed in Table 1. The chemical compounds 2-nitroethanol (2-NE), propynoic acid, ferulic acid, p-coumaric acid, and bromoform (Sigma-Aldrich Sweden AB, Stockholm, Sweden) were added without replacing any DM of the control diet. Nitrate was added to the control diet to represent one level of 21 g NO_3/kg DM or 0.0890 g $Ca(NO_3)_2 \times 4H_2O$/g DM (Sigma-Aldrich Sweden AB, Stockholm, Sweden). The nitrate treatment was compared with a zero-nitrate treatment in which 0.0350 g urea/g DM and 0.051 g $CaCO_3$/g DM (J.T. Baker BV, Deventer, The Netherland) were added to the control diet to achieve an isonitrogenous and equivalent diet (159 g crude protein (CP)/kg DM). The plant-derived compounds rowan (*Sorbus aucuparia*) berries and the forb fireweed (*Chamerion angustifolium*) were added to replace grass and barley in the control diet, such that the ratio of forage:concentrate was kept constant relative to all other diets. These ingredients were collected in Umeå (63° N, 20° E), Sweden in October and July 2018, respectively. The red seaweed *Asparagopsis taxiformis* (AT) was added in such a small dose in both levels of the treatment that no replacement of control dietary ingredients was made. The seaweed was harvested in the Azores (38.6° N, 28° W), Portugal, in October 2018. Replacements in the potentially CH_4-reducing diets were also made so that the forage:concentrate ratio was kept constant relative to all other diets and to contain 160 g CP/kg diet DM. Rapeseed oil and oats (*Avena sativa*) were added to replace grass and barley on a DM basis. These ingredients were also collected in Umeå in July 2018. Dried distiller's grains (Agrodrank 90, Agroetanol, Östergötland, Sweden) replaced rapeseed meal in the control diet and was added to represent an increment of 20 g/kg DM in CP between the levels (CP concentration 160 g/kg DM and 180 g/kg DM, respectively). The CP concentration was made iso-nitrogenous to the control diet for the lowest level when dried distiller's grain replaced rapeseed meal.

Table 1. Experimental treatments evaluated in vitro in experiment 1 for methane (CH_4) mitigation potential.

Treatments	Levels	
Chemical compounds		
2-nitroethanol	5 m*M*	10 m*M*
Nitrate	None [1]	21 g/kg DM [2]
Propynoic acid	2 m*M*	4 m*M*
Ferulic acid	10 m*M*	20 m*M*
p-Coumaric acid	10 m*M*	20 m*M*
Bromoform	1.5 mg/g DM	3 mg/g DM
Plant-derived treatments		
Rowan berries	50 g/kg DM	100 g/kg DM
Fireweed	50 g/kg DM	100 g/kg DM
Asparagopsis taxiformis	10 g/kg OM	20 g/kg OM
Potentially CH_4-reducing treatments		
Rapeseed oil	40 g/kg DM	80 g/kg DM
Dried distiller's grain	90 g/kg DM	180 g/kg DM
Barley:oat	175:175 g/kg	0:350 g/kg
Maize silage:grass	275:275 g/kg [3]	545:0 g/kg [4]
Red clover:grass	275:275 g/kg	None
Lactic acid	60 g/kg DM	120 g/kg DM
Lactic acid + acetic acid	80 + 30 g DM	80 + 60 g DM

DM = dry matter; [1] 0.035 g of urea + 0.051 g of $CaCO_3$ on DM basis included in control diet in comparison with nitrate treatment; [2] 0.089% $Ca(NO_3)_2 \times 4H_2O$ on DM basis; [3] Urea was added to correct CP at 160 g/kg DM; [4] Urea was added to correct CP at 160 g/kg DM.

In the treatments where maize (*Zea mays*) silage replaced grass silage, urea was added to make diets isonitrogenous to the control diet. No correction of CP concentration was made in the diet when red clover (*Trifolium pratense*) replaced grass.

2.1.2. Experiment 2

In experiment 2, AT (0, 0.06, 0.13, 0.25, 0.5, and 1.0 g/kg of diet organic matter (OM)) and 2-NE (0, 0.3, 0.7, 1.3, 2.6, and 5.1 m*M*) were tested in a dose-response experiment comprising six different treatment levels. The same control diet as in experiment 1, of timothy, rolled barley, and rapeseed meal, was used in experiment 2.

2.2. *In Vitro Incubations*

The handling of animals in this experiment was approved by the Swedish Ethics Committee on Animal Research (Dnr A 32-16), represented by the Court of Appeal for Northern Norrland in Umeå, and the experiment was carried out in accordance with laws and regulations governing experiments performed with live animals in Sweden.

Two lactating Swedish Red cows, fed ad libitum on a diet of 600 g/kg grass silage and 400 g/kg concentrate on a DM basis (presenting chemical composition as 509 g/kg of DM, 425 g/kg NDF, and 171 g/kg CP), were used as donor animals of rumen inoculum for all incubations. The rumen fluid from each cow was filtered separately using a double layer of cheesecloth into Thermos flasks that were pre-warmed and flushed with carbon dioxide (CO_2) prior to collection. Rumen fluid was transported to the laboratory within 15 min. Equal amounts from each cow were immediately blended, strained through four layers of cheesecloth, and added to buffered mineral solution [15] including Peptone^{TM} (pancreatic digested casein; Merck, Darmstadt, Germany) at 39 °C under constant mixing and CO_2 flushing, to give a buffered rumen fluid solution with a rumen fluid:buffer ratio of 1:4 by volume.

Prior to each in vitro incubation, dietary ingredients were dried at 60 °C for 48 h and milled in a Retsch SM 2000 cutting mill (Retsch GmbH, Haan, Germany) to pass through a 1-mm screen. Then 1003 ± 38 mg of DM substrate were weighed into serum bottles flushed with CO_2, and 60 mL of the previously prepared buffered rumen fluid were added. All bottles were placed in a water bath and gently and continuously agitated at 39 °C during an incubation period of 48 h.

These procedures were repeated for six runs in total and all samples were incubated, with three replicates of each sample. All runs included triplicate bottles with blanks (i.e., bottles with 60 mL of buffered rumen fluid with no sample or treatment in), and samples were randomly allocated to the in vitro incubation bottles and never incubated in the same bottle in more than one run.

2.3. *In Vitro Gas Production Measurements and Sampling*

Gas production was measured using a fully automated system (Gas Production Recorder, GPR-2, Version 1.0 2015, Wageningen UR), with readings made every 12 min and corrected to the normal air pressure (101.3 kPa) [16].

Measurement of CH_4 in vitro was performed according to Ramin and Huhtanen [14] on gas samples withdrawn during the incubation period (0.2 mL) from each bottle at 2, 4, 8, 24, 32, and 48 h. Concentration of CH_4 was determined with a Varian Star 3400 CX gas chromatograph (Varian Analytical Instruments, Walnut Creek, CA, USA) equipped with a thermal conductivity detector.

Liquid samples of 0.6 mL were collected from the bottles at 8, 24, and 48 h of incubation and immediately stored at −20 °C until analysis of volatile fatty acids (VFA). Liquid samples for ammonia-nitrogen (NH_3-N) analysis were taken at 8 and 24 h of incubation, and also stored at −20 °C before further analysis. Liquid samples from the replicate treatments between runs were pooled before NH_3-N and VFA analysis.

After 48 h of incubation, all bottles were removed from the water bath and placed on ice to stop fermentation. The residue was used for in vitro determination of true organic matter digestibility (TOMD).

2.4. Chemical Analysis

The concentrations of DM and OM in the individual dietary ingredients were quantified by AOAC [17] method 930.15 and method 942.05, respectively. Concentrations of nitrogen were determined by Kjeldahl digestion of 1000 mg sample in 12 M sulfuric acid using Foss Tecator Kjeltabs Cu (Höganäs, Sweden) in a Block Digestion 28 system (SEAL Analytical Ltd., Mequon, WI, USA), followed by determination of total nitrogen by continuous flow analysis using an Auto Analyzer 3 (SEAL Analytical Ltd., Mequon, WI, USA). The samples were analyzed for neutral detergent fiber (NDF) using a heat-stable α-amylase [18] in an ANKOM200 Fiber Analyzer (Ankom Technology Corp., Macedon, NY, USA).

In vitro TOMD was determined for all samples in all runs by analyzing ash-free NDF concentrations in the residues using 07-11/5 Sefar Petex (Sefar AG, Heiden, Switzerland) in situ bags according to Krizsan et al. [19].

Individual VFA concentrations in rumen fluid samples were determined using a Waters Alliance 2795 UPLC system as described by Puhakka et al. [20], and NH_3-N concentration according to the method provided by SEAL Analytical (Method no. G-102-93 multitest MT7) using AutoAnalyzer 3.

Bromoform concentration in AT was analyzed according to Roque et al. [21] using an Agilent 7890B GC applied to Agilent 7000C triple quad Mass Spectrometer equipped with a ZB-5ms column (Agilent Technologies, Inc. Santa Clara, CA, USA).

2.5. Calculations

Mean blank gas production within run was subtracted from sample gas production. In vivo predicted CH_4 production was calculated as described by Ramin and Huhtanen [14] as:

$$CH_4 = 265 \times CH_4 \text{ concentration} + \text{total gas production} \times CH_4 \text{ concentration} \times 0.55$$

where total gas production is in mL/g sample, 265 is the total headspace volume (mL), and 0.55 is the ratio of CH_4 emissions in the outflow gas from the in vitro system. A mean retention time of 50 h (20 h in the first compartment and 30 h in the second compartment), corresponding to the maintenance level of feed intake, was used in model simulations.

Total VFA (TVFA) production was calculated as: the molar proportion of individual VFA were calculated related to TVFA.

$$\text{TVFA (mmol)} = (\sum \text{individual VFA concentration} - \text{mean of blank VFA}) \times 0.06 \text{ (amount of buffered rumen fluid)}$$

The molar proportion of individual VFA were calculated related to TVFA.

The in vitro TOMD was calculated as:

$$\text{TOMD (g/kg)} = \frac{\text{incubated OM (g)} - \text{NDF residue corrected for ash and blank (g)}}{1000 \times \text{incubated OM (g)}}$$

2.6. Statistical Analysis

Data on in vivo predicted CH_4 production and in vitro TOMD from Experiment 1 were analyzed using the MIXED procedure in SAS (SAS Institute Inc., Cary, NC, version 9.4), by a model correcting for random effect of bottle and fixed effect of run and treatment:

$$Y_{ijk} = \mu + T_i + R_j + B_k + e_{ijk}$$

where Y_{ijk} is dependent variable ijk, μ is overall mean, T_i is treatment i, R_j is run j, B_k is bottle k, and $e_{ijk} \sim N(0,\sigma_e^2)$ is the random residual error. Orthogonal contrasts were included for evaluation of control diet vs. treatment and of linear responses to level of treatment.

Data on measured VFA and NH_3-N concentrations from Experiment 1 were evaluated in a repeated measurements model using the Toeplitz function in the MIXED procedure in SAS (SAS Institute Inc., Cary, NC, USA, version 9.4) (level within treatment was used as subject). The model accounted for effects of treatment and time, and interactions between treatment and time:

$$y_{ij} = \mu + T_i + A_j + (TA)_{ij} + e_{ij}$$

where y_{ij} is the dependent variable ij, μ is overall mean, T_i is treatment i, A_j is time j, $(TA)_{ij}$ is interaction between treatment i and time j, and $e_{ij} \sim N(0,\sigma_e^2)$ is the random residual error.

Data on predicted in vivo CH_4 production, in vitro TOMD, total VFA (TVFA), and molar proportions of individual VFA and NH_3-N from experiment 2 were subjected to linear and quadratic regression analysis using the REG procedure in SAS (SAS Institute Inc., Cary, NC, USA, version 9.4). Best fit was judged from lowest root mean square error and highest adjusted R^2.

Effects were considered statistically significant at p-value ≤ 0.05.

3. Results

The chemical composition of the control diet and the potential CH_4 reducing diets is shown in Table 2. The AT bromoform concentration was 6.84 mg/g DM.

Table 2. Chemical composition (g/kg DM) of control and potential methane (CH_4) reducing diets evaluated in vitro in experiment 1.

Treatment	Level	Organic Matter	Crude Protein	Neutral Detergent Fiber
Control diet	—	944	160	387
Rapeseed oil	40 g/kg DM	906	154	372
Rapeseed oil	80 g/kg DM	869	149	356
Dried distiller's grain	90 g/kg DM	946	161	378
Dried distiller's grain	180 g/kg DM	946	181	366
Barley: oat	175:175 g/kg	944	165	385
Barley: oat	0:350 g/kg	944	170	383
Maize silage: grass	275:275 g/kg	954	160	355
Maize silage: grass	545:0 g/kg	963	160	323
Red clover: grass	275:275 g/kg	932	171	345
Lactic acid	60 g/kg DM	887	151	364
Lactic acid	120 g/kg DM	831	143	341
Lactic acid + acetic acid	80 + 30 g/kg DM	840	144	345
Lactic acid + acetic acid	80 + 60 g/kg DM	812	140	333

NDF = neutral detergent fibre.

3.1. Experiment 1

Predicted in vivo CH_4 production derived from analysis of 48 h gas in in vitro incubation of the control diet was 38.7 mL/g DM, in vitro TOMD was 867 g/kg, TVFA was 3.62 mmol, and molar proportion of acetate, butyrate, and propionate was 583, 125, and 237 mmol/mol, respectively. In comparison with the control diet the chemical compounds 2-NE, nitrate, propynoic acid, p-coumaric acid, bromoform, and the plant compound AT, decreased ($p \leq 0.01$) in vivo CH_4 predicted production (Table 3). Addition of 2-NE, bromoform, and AT gave the strongest inhibition ($p < 0.01$) of predicted in vivo CH_4 production among all experimental treatments (97%, 95%, and 99% reduction in the value for the control diet). The reduction in predicted in vivo CH_4 production achieved by the other compounds ranged between 38% and 64% of the value for the control diet. Surprisingly, none of the potential CH_4 reducing diets or lactic acid and acetic acid addition affected CH_4 production in this study ($p \geq 0.20$). In vitro TOMD was negatively affected by the chemical compounds p-coumaric acid and bromoform ($p < 0.01$), while rapeseed oil inclusion in the diet increased in vitro TOMD compared with the control diet ($p = 0.04$). Propynoic acid and bromoform decreased ($p \leq 0.01$) TVFA compared with the control diet. Several of the treatments altered the molar proportions of individual VFA. Acetate decreased ($p \leq 0.03$) on adding 2-NE, propynoic acid, p-coumaric acid, bromoform, AT, or lactic acid to the control diet. For all those treatments except p-coumaric acid and bromoform, there was a concomitant increase ($p \leq 0.05$) in molar proportions of propionic and butyric acid compared with the control diet. Results of nitrate vs. zero nitrate treatment were: TVFA 2.91 vs. 3.01 mol, acetate 597 vs. 604 mmol/mol propionate 250 vs. 227 mmol/mol and butyrate 87 vs. 123 mmol/mol.

The molar proportion of isobutyrate, isovalerate and valerate, and NH_3-N for the control diet and experimental treatments are given in, Table 4. The molar proportions of the branched-chain volatile fatty acids (BCVFA) were altered by many of the CH_4 mitigating strategies tested. Compared to the control diet, isobutyrate increased ($p \leq 0.01$) for p-coumaric acid treatment, while for bromoform treatment the molar proportion decreased ($p \leq 0.01$). The treatments, 2-NE, propynoic, p-coumaric, ferulic acid, AT, lactic acid, and lactic acid + acetic acid, increased ($p \leq 0.04$), and bromoform decreased ($p \leq 0.01$) the molar proportion of isovalerate compared to the control diet. Propynoic acid decreased ($p \leq 0.05$) while bromoform and AT increased ($p \leq 0.05$) molar proportion of valerate. Results of nitrate vs. zero nitrate treatment were: isobutyrate 6.63 vs. 9.96 mmol/mol, isovalerate 4.01 vs. 4.94 mmol/mol, valerate 16.4 vs. 16.3 mmol/mol, and NH_3-N concentration 436 vs. 557 mg/L.

Tests for linear effects between the two levels according to Table 1 and the control diet (0 here) revealed no significant effect on in vitro TOMD ($p = 0.148$) for all treatments (data not presented). However, there was a significant linear decrease ($p < 0.01$) in predicted in vivo CH_4 production for propynoic acid (24 and 0 mL/g DM) and p-coumaric acid (27.1 and 19.8 mL/g DM) when the inclusion level was increased.

Table 3. Effect of experimental treatments on predicted in vivo CH_4 production (mL/g DM), in vitro true organic matter digestibility (TOMD, g/kg), total volatile fatty acid production (TVFA, mmol), and molar proportions of acetate, propionate, and butyrate (mmol/mol of TVFA) measured in 48 h gas from the in vitro incubation in experiment 1.

Treatment	CH_4	TOMD	TVFA	Acetate	Propionate	Butyrate	*p*-Value [1]					
							C_{CH4}	C_{TOMD}	C_{TVFA}	$C_{Acetate}$	$C_{Propionate}$	$C_{Butyrate}$
Control	38.7	867	3.62	583	237	125	–	–	–	–	–	–
2-nitroethanol	1.30	858	3.01	440	309	211	<0.01	0.30	0.10	<0.01	<0.01	<0.01
Nitrate [2]	21.3	874	2.96	619	250	87	<0.01	0.82	NA	NA	NA	NA
Propynoic acid	13.9	839	2.57	476	297	209	<0.01	0.25	0.01	<0.01	<0.01	<0.01
Ferulic acid	27.5	859	3.54	597	229	109	0.06	0.71	0.82	0.62	0.68	0.32
p-Coumaric acid	24.2	763	3.01	492	176	121	0.01	<0.01	0.10	<0.01	<0.01	0.82
Bromoform	2.00	822	2.30	436	270	261	<0.01	<0.01	<0.01	<0.01	0.08	<0.01
Fireweed	38.1	858	4.01	583	226	140	0.34	0.69	0.24	0.99	0.55	0.34
Rowan berries	28.9	843	3.71	586	241	117	0.96	0.35	0.80	0.93	0.82	0.64
A. taxiformis	0.20	852	3.61	418	327	184	<0.01	0.97	0.98	<0.01	<0.01	<0.01
Rapeseed oil	38.2	896	4.04	600	217	128	0.82	0.04	0.24	0.56	0.28	0.83
Dried distiller's grain	35.3	877	3.74	549	241	152	0.43	0.48	0.73	0.66	0.91	0.33
Barley: oat	37.4	863	3.66	596	225	124	0.73	0.73	0.89	0.65	0.51	0.98
Maize silage: grass	30.7	846	3.71	577	240	136	0.61	0.56	0.80	0.84	0.87	0.48
Red clover: grass	47.9	882	3.16	599	234	129	0.20	0.53	0.27	0.62	0.88	0.85
Lactic acid	34.1	866	3.65	516	271	158	0.25	0.54	0.93	0.03	0.07	0.05
Lactic acid + acetic acid	35.2	885	3.34	598	224	135	0.34	0.11	0.43	0.59	0.49	0.53
SEM	1.75	4.3	0.120	7.2	4.2	5.5	–	–	–	–	–	–

NA = not analyzed; SEM = standard error mean. [1] Orthogonal contrasts of control diet vs. treatment of the different in vitro traits. [2] Nitrate treatment was compared to the zero nitrate diet made by adding urea and $CaCO_3$ to the control diet according to Table 1; numerical differences of TVFA and molar proportions of volatile fatty acids are given in the text.

Table 4. Effect of experimental treatments on molar proportions of isobutyrate, isovalerate, and valerate (mmol/mol of TVFA), and ammonia concentration (NH_3-N, mg/L) measured in 48 h gas from the in vitro incubation in experiment 1.

Treatments	Isobutyrate	Isovalerate	Valerate	NH_3-N	*p*-Value [1]			
					$C_{Isobutyrate}$	$C_{Isovalerate}$	$C_{Valerate}$	C_{NH3-N}
Control	10.9	0.61	21.4	282	–	–	–	–
2-nitroethanol	5.16	2.77	16.4	436	0.80	<0.01	0.14	0.39
Nitrate [2]	6.63	4.01	16.43	270	NA	NA	NA	NA
Propynoic acid	8.34	2.59	0.77	311	0.91	<0.01	<0.01	0.25
Ferulic acid	14.4	5.23	20.0	320	0.88	0.03	0.67	0.98
p-Coumaric acid	165	4.23	22.6	263	<0.01	<0.01	0.73	0.81
Bromoform	0.00	0.00	28.7	302	0.64	<0.01	0.03	0.17
Fireweed	10.8	5.97	18.5	304	0.99	0.20	0.38	0.78
Rowan berries	10.1	5.85	21.5	289	0.97	0.15	0.98	0.81
A. taxiformis	5.95	5.33	35.4	354	0.83	0.04	0.00	0.52
Rapeseed oil	13.3	8.32	19.1	319	0.92	0.08	0.49	0.25
Dried distiller's grain	11.5	6.87	20.7	301	0.98	0.91	0.83	0.07
Barley: oat	12.5	6.98	18.5	359	0.95	0.98	0.38	0.75
Maize: grass	10.0	6.10	17.8	281	0.97	0.26	0.28	0.20
Red clover: grass	9.04	5.23	14.6	306	0.94	0.06	0.08	0.45
Lactic acid	9.30	5.29	21.3	323	0.94	0.03	0.97	0.86
Lactic acid + acetic acid	8.95	4.61	16.7	282	0.93	<0.01	0.16	0.77
SEM	4.426	0.370	0.87	12.2	–	–	–	–

NA = not analyzed; SEM = standard error mean. [1] Orthogonal contrasts of control diet vs. treatment of the different in vitro traits. [2] Nitrate treatment was compared to the zero nitrate diet made by adding urea and $CaCO_3$ to the control diet according to Table 1; numerical differences molar proportions of branched-chain volatile fatty acids and NH_3-N are given in the text.

3.2. Experiment 2

Predicted in vivo CH_4 production decreased curvilinearly ($p < 0.01$) with increased levels of both 2-NE (Figure 1).

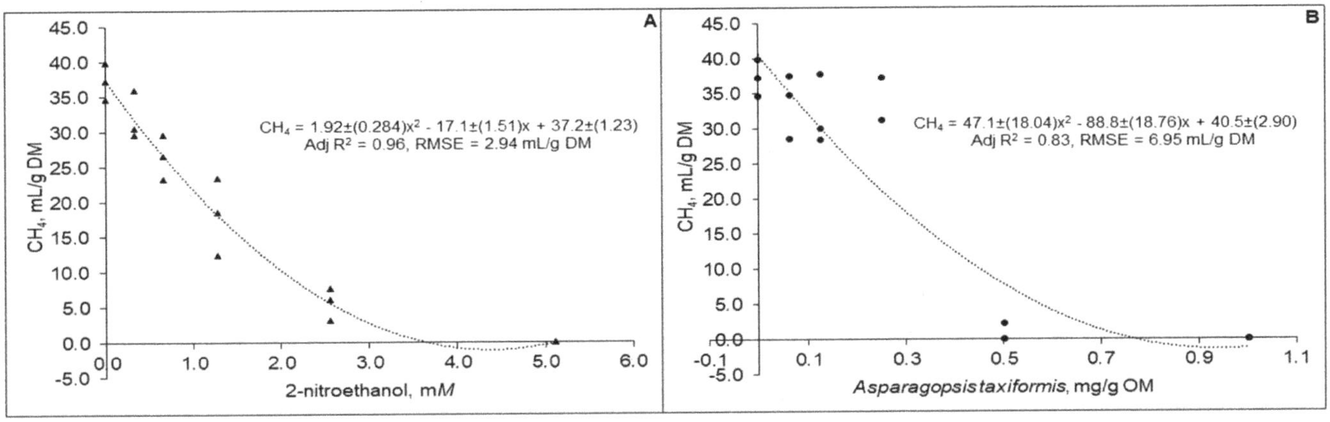

Figure 1. Predicted in vivo methane production based on analysis of 48 h gas from in vitro incubation of a control diet (545:363:92 g/kg of grass silage:barley:rapeseed meal) treated with different levels (three replicates per level) of (**A**) 2-nitroethanol and (**B**) *Asparagopsis taxiformis* in experiment 2.

The TVFA content decreased linearly ($p < 0.01$) from 5.35 to 3.00 mmol at 48 h for 2-NE and from 4.71 to 4.33 mmol at 24 h for AT for the lower to higher level of supplementation (Figure 2). The TVFA content for 2-NE at 8 h ($p < 0.01$; adj $R^2 = 0.38$; RSME = 0.22 mmol) and 24 h ($p < 0.01$; adj $R^2 = 0.55$; RSME = 0.25 mmol) showed curvilinear responses, while for AT the curvilinear pattern was verified at 8 h ($p < 0.01$; $R^2 = 0.55$; RSME = 0.25 mmol) and 48 h ($p = 0.01$; adj $R^2 = 0.33$; RSME = 0.33 mmol).

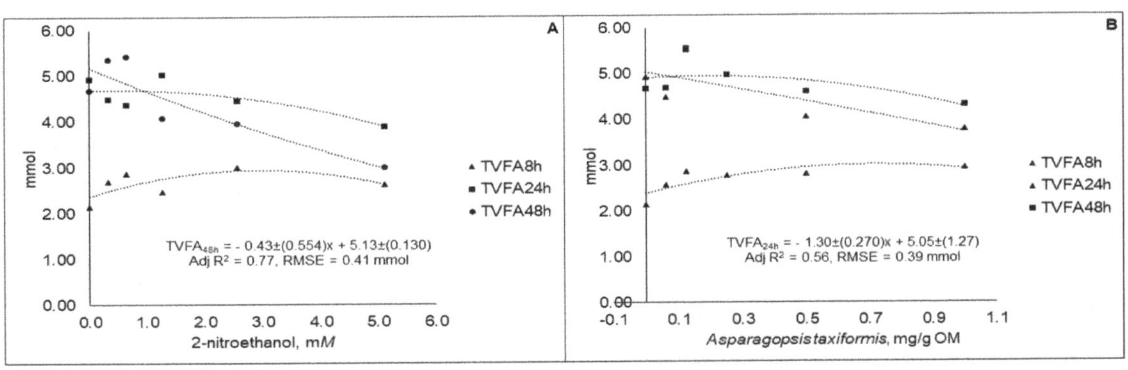

Figure 2. Total volatile fatty acid (TVFA) content in fluid samples taken at different time points during 48 h in vitro incubation of a control diet (545:363:92 g/kg of grass silage:barley:rapeseed meal) treated with different levels (three replicates per level) of (**A**) 2-nitroethanol and (**B**) *Asparagopsis taxiformis* in experiment 2.

Molar proportion of acetate decreased, while propionate and butyrate proportions increased curvilinearly ($p < 0.01$), at all-time points studied for 2-NE and AT (Figure 3). There were no statistical difference ($p > 0.05$) between the coefficients generated for the equations of the different time points. The best fit equations of molar proportions of VFA were generated for both 2-NE and AT from different sampling time points (Figure 3), but the equations generated were not statistically different ($p > 0.05$) from the other sampling time points (results not presented).

There were no linear or curvilinear relationships between TOMD and level of supplementation for 2-NE ($p = 0.152$) or AT ($p = 0.142$) (results not presented).

The equations of the molar proportions of BCVFA (isobutyrate, isovalerate, and valerate) were statistically different ($p < 0.05$) between the different sampling time points. The best fit equations are presented in Figure 4.

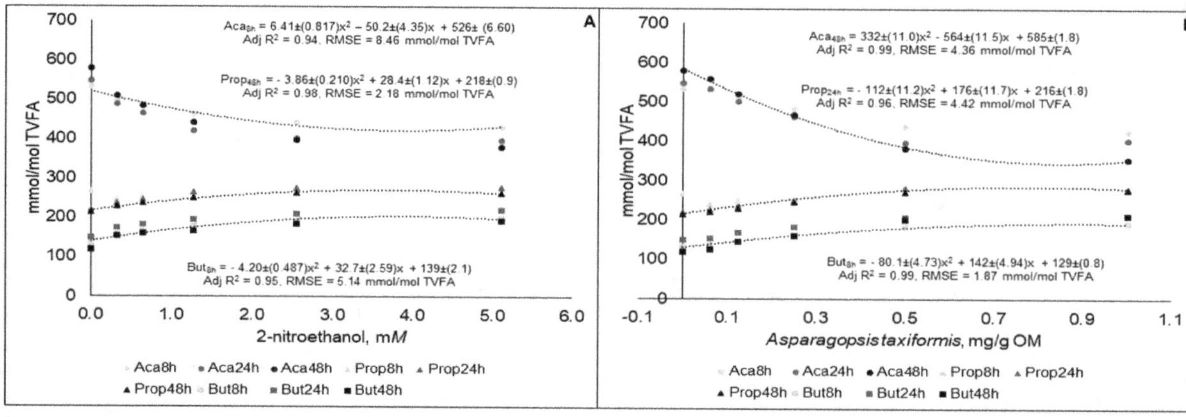

Figure 3. Molar proportions of acetate (Ace), propionate (Prop), and butyrate (But) in fluid samples gas samples taken at different time points during 48 h in vitro incubation of a control diet (545:363:92 g/kg of grass silage:barley:rapeseed meal) treated with different levels (three replicates per level) of (**A**) 2-nitroethanol and (**B**) *Asparagopsis taxiformis* in experiment 2.

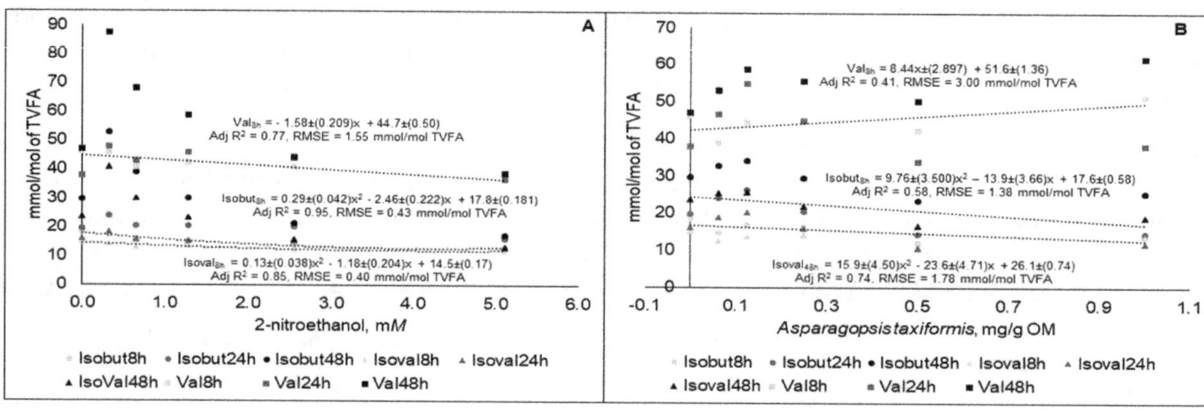

Figure 4. Molar proportions of isobutyrate (Isobut), isovalerate (Isoval), and valerate (Val) in fluid taken at different time points during 48 h in vitro incubation of a control diet (545:363:92 g/kg of grass silage:barley:rapeseed meal) treated with different levels (three replicates per level) of (**A**) 2-nitroethanol and (**B**) *Asparagopsis taxiformis* in experiment 2.

The NH_3-N concentration responses decreased linearly ($p < 0.05$) for both 2-NE and AT at both 8 and 24 h, and the best fit equations are presented in Figure 5.

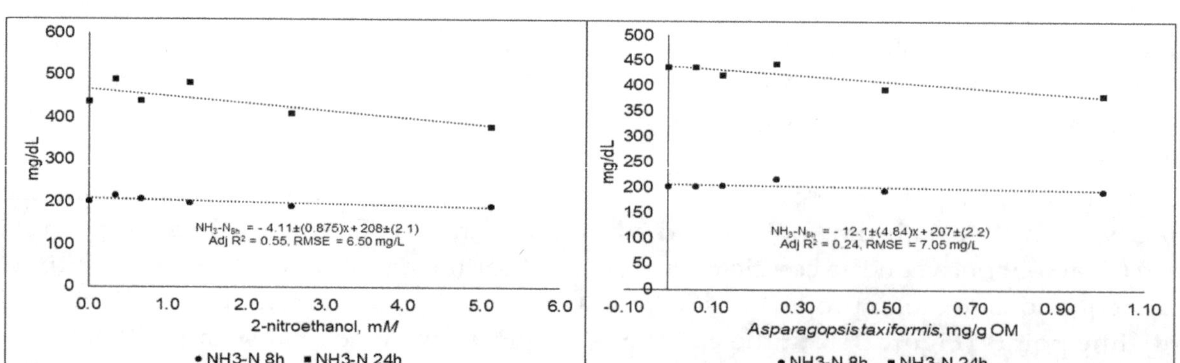

Figure 5. Ammonia concentration (NH_3-N) in fluid samples taken at different time points during 48 h in vitro incubation of a control diet (545:363:92 g/kg of grass silage:barley:rapeseed meal) treated with different levels (three replicates per level) of (**A**) 2-nitroethanol and (**B**) *Asparagopsis taxiformis* in experiment 2.

4. Discussion

The CP concentration of the potential CH_4 reducing diets varied between 140 and 181 g/kg DM, and reflected the characteristics of the dietary ingredient studied in each diets. Regarding that Peptone™ was included in the buffered rumen fluid, none of the diets supplied an insufficient amount of CP in terms of CP available for rumen microbial growth in comparison with in vivo requirements [22].

Ruminants are valuable food producers world-wide, since they are able to utilize fibrous non-human-edible resources (forages and pasture) through microbial fermentation of feed in the rumen. Recent data indicate that domesticated ruminants are not the major contributor to anthropogenic CH_4 emissions. The fermentation of feed and decomposition of manure are the foremost sources of GHG emissions caused by domesticated ruminants [23]. Estimates suggest that livestock are responsible for around 9% and 37% of anthropogenic CO_2 and CH_4 emissions, respectively [24]. Long-term strategies to improve feed efficiency through targeted breeding [25] and improved longevity or lifetime productivity [26] could reduce CH_4 emissions from dairy cows. According to Knapp et al. [9], nutrition and feeding approaches may be able to reduce CH_4 emissions per unit of energy-corrected milk by 2.5–15%, while reductions of 15–30% can be achieved by combined genetic and management approaches.

4.1. In Vitro Measurements of CH_4 Production in Ruminants

This in vitro study evaluated a wide variety of dietary CH_4 inhibitors, which would not have been feasible in an in vivo study. A main advantage of the in vitro gas production system in measuring CH_4 emissions is that it provides a large number of data points, allowing accurate estimates of CH_4 emissions. However, it is a batch culture approach and has some limitations compared with in vivo studies (e.g., no absorption of VFA over time). The in vitro method used here was developed by Ramin and Huhtanen [14] to overcome this problem and involves a modeling approach based on data obtained from the gas in vitro system. They assumed a gross energy concentration of 18.5 MJ/kg DM, while the predicted proportion of CH_4 energy for a sample size of around 1000 mg was calculated to be 0.061. This value is close to observed in vivo values at production levels of intake in dairy cows [27]. Recently, Danielsson et al. [28] evaluated the in vitro technique developed by Ramin and Huhtanen [16] using data (diets) from in vivo studies using a respiration chamber to measure CH_4 emissions. The results showed a high correlation ($R^2 = 0.96$) between observed (chamber) data and predicted in vivo CH_4 values, confirming that the in vitro system is a useful tool for screening diets and evaluating feed additives.

4.2. Dietary Strategies to Decrease CH_4 Production from Ruminants

In this study, we screened many different dietary strategies with known potential to mitigate CH_4 production from ruminants and also a few new potential inhibitors. It is known that improved forage quality, feeding balanced diets to ensure efficient utilization of nutrients, and optimized microbial protein synthesis in the rumen can decrease CH_4 production in relation to animal productivity [29]. With respect to improved forage quality, the effects on CH_4 production reported in the literature are contradictory. Enteric CH_4 production increases with more digestible substrate available for rumen microbes, but overall emissions of CH_4 in lactating dairy cows can decrease per kg increase in digestible OM [8]. The mechanism behind this effect is likely that better forage quality improves intake, and thereby increases passage rate. Increased passage rate (i.e., decreased feed retention) and larger animals (i.e., greater body mass) have been associated with reduced CH_4 emissions in sheep [30–32].

Contrasted to our results, in measurements in vivo, rapeseed oil added at 50g/kg DM to a grass silage-based diet reduced ruminal CH_4 emissions from lactating cows by 22% [33], with the reduction observed being entirely explained by decreases in DM intake and the dilution effect on fermentable OM. Use of dried distiller's grain to replace soybean meal in diets based on grass silage decreased CH_4 production in an in vitro study by Franco et al. [34]. The effect was explained by a shift in the ruminal fermentation pattern to decreased acetate and butyrate production and increased propionate production. A similar shift in ruminal fermentation pattern was observed when rapeseed meal replaced soybean meal in vitro in that study [35]. In the present study, dried distiller's grain replaced rapeseed meal in the control diet and the suggested similarities in ruminal fermentation pattern of these protein supplements would explain the lack of effect on predicted in vivo CH_4 production. In contrast to our results, Fant et al. [35] observed a significant reduction in predicted in vivo CH_4 production of 2.1 mL/g DM when using oats instead of barley as the concentrate carbohydrate source. Inclusion of maize silage is also reported to promote propionate fermentation in the rumen, and thereby decrease CH_4 production in dairy cows [36–38]. However, we did not observe this effect with inclusion of maize silage in the diets. Greater molar proportions of acetate and lower proportions of propionate in VFAs when replacing grass with red clover have been reported both in vivo [39] and in vitro [40], suggesting that CH_4 production potential is greater when ruminants are fed red clover. Maize silage and red clover diets were only numerically lower respectively higher in in vivo predicted CH_4 production compared with the control diet in this study. On the other hand, grasses are generally more likely to accumulate nitrates than legumes, and nitrate inhibits enteric CH_4 production by replacing reduction of CO_2 to CH_4 as a major sink for disposal of H_2 in the rumen [41]. Interactions between ruminant physiological responses and diet quality affecting CH_4 production might explain the lack of impact on in vivo predicted CH_4 production by the potential CH_4-reducing diets screened in vitro in this study.

The chemical inhibitors 2-NE and bromoform, and the plant-derived inhibitor AT, gave a very large reduction in predicted in vivo CH_4 production in this study. The bromoform concentrations of 1.5 and 3.0 mg/kg DM used in this study were representative of concentrations occurring naturally in AT [42]. Vucko et al. [43] analyzed bromoform concentrations in AT biomass subjected to a wide variety of post-harvesting processes and found a maximum concentration of 4.4 mg/g DM for unrinsed, frozen, and freeze-dried AT. Those authors suggest a bromoform threshold of 1.0 mg/g DM in AT for 100% inhibition of CH_4 production in vitro, which corresponds with our results and the levels used in this study.

Machado et al. [44] tested different dosages of AT in vitro and found that production of CH_4 was decreased by 84.7% at an inclusion level of 1% (OM basis), while at AT doses greater than 2% (OM basis), CH_4 production was decreased by more than 99% compared with the control treatment. In the present study, in vivo predicted CH_4 production was inhibited almost completely by AT already at a level of 0.5% on an OM basis. Li et al. [45] added AT to diets fed to sheep and observed a reduction of CH_4 production at inclusion levels exceeding 1% of OM intake, but with altered rumen fermentation at all inclusion rates, i.e., at inclusions $\geq$0.5% of OM intake. On the other hand, in a short-term in vivo experiment by Stefenoni et al. [46], inclusion of AT at 0.5% of DM intake decreased CH_4 emission in lactating dairy cows by 80%, with no negative effects on DM intake and milk yield (rumen fermentation parameters were not measured).

An in vitro study by Zhang and Yang [47] indicated high potential of 2-NE to mitigate CH_4 production, as also found in the present study. However, they observed a negative effect on in vitro digestibility already at their lowest dose of 5 m*M*, which was not observed in this study. Use of 2-NE in an in vivo trial would not be realistic, considering that the concentration we used in vitro would equate to a daily dose of 0.9 L of 2-NE for a dairy cow with a rumen volume of 200 L.

Except for molar proportion of valerate with increased AT supplementation, all of the BCVFA decreased with increased supplementation in the dose response experiment. The BCVFA are mainly a consequence of the degradation of the amino acids valine, isoleucine, leucine and proline and are used for the biosynthesis of those amino acids and higher branched chain volatile fatty acids. The BCVFA are specific nutrients for the ruminal cellulolytic bacteria, and are believed to have a general positive influence on microbial fermentation [48].

Nitrate, propynoic acid, and p-coumaric acid had much lower inhibitory effects on predicted in vivo CH_4 production. Nitrate is reported to be an effective CH_4 production mitigating dietary component [49,50]. For example, a 24.8% reduction in CH_4 production by lactating cows receiving nitrate at 21.1 g NO_3^-/kg of DM was observed by Olijhoek et al. [51]. The dose of nitrate that can be toxic to ruminants' ranges between 198 and 998 mg/kg live weight and is dependent on diet, administration, and consumption [52]. However, the negative effects of nitrate can be reduced through gradual adaptation of animals to consumption of this nitrogen source, which could contribute to reducing CH_4 emissions. The CH_4 inhibitory effect of propynoic acid in this study was lower than that observed by Zhou et al. [53] at a comparable inclusion rate (75.7% reduction compared with a control diet with no inhibitor added). Also the lower amount of TVFA compared with the control diet indicated that propynoic acid can potentially affect digestibility. The inhibitory effect of p-coumaric acid on CH_4 production by ruminants has not been studied previously and there are no in vitro results with which to compare, but the treatment decreased the dietary TOMD. Lactate in the rumen are metabolized to propionate, which could hypothetical induce changes in ruminal fermentation pattern providing an alternative hydrogen sink to reduce methanogenesis. Likely, the lactic acid preservation has to be more extensive than the levels suggested in this study to have an effect on CH_4 production in dairy cows.

5. Conclusions

This study confirmed that natural bioactives produced by the red seaweed *Asparagopsis taxiformis* can act as a strong natural inhibitor of CH_4 production in domesticated ruminants. Use of CH_4 inhibitors with high mitigation potential at a reasonable dietary supplementation level could be an important and effective strategy to mitigate CH_4 emissions by ruminants. However, *Asparagopsis taxiformis* needs to be further evaluated in vivo to ensure it has no negative effects on animal health, productivity, or product quality. It is also important to establish the long-term CH_4 mitigation effect of using this inhibitor.

Author Contributions: S.J.K.: conceived and supervised the study and acquired funding. S.J.K., J.C.C.: inputs to data analysis. S.J.K., M.R.: methodology. J.C.C., S.J.K., M.R.: conducted the experiment and wrote the full paper.

Acknowledgments: The authors would like to extend their sincere appreciation to Ann-Sofi Hahlin, Azam Jafari, and Amanda Poppi for their support in laboratory work, and to Carl Tryggers Foundation for providing a postdoc scholarship for J.C.C.

References

1. FAO. Food and Agriculture Organization of the United Nations. The State of Food Security and Nutrition in the World 2018. Available online: http://www.fao.org/3/i9553en/i9553en.pdf (accessed on 10 August 2019).
2. IPCC. Intergovernmental Panel on Climate Change. Climate Change 2013. Available online: https://www.ipcc.ch/site/assets/uploads/2018/03/WG1AR5_SummaryVolume_FINAL.pdf (accessed on 10 August 2019).
3. Shindell, D. The social cost of atmospheric release. *Clim. Chang.* **2015**, *130*, 313–326. [CrossRef]
4. Howarth, R.W. Ideas and perspectives: Is shale gas a major driver of recent increase in global atmospheric methane? *Biogeosciences* **2019**, *16*, 3033–3046. [CrossRef]
5. Petrenko, V.V.; Smith, A.M.; Schaefer, H.; Riedel, K.; Brook, E.; Baggenstos, D.; Hart, C.; Hua, Q.; Buizert, C.; Schilt, A.; et al. Minimal geological methane emissions during the Younger Dryas-Preboreal abrupt warming event. *Nature* **2017**, *548*, 443–446. [CrossRef] [PubMed]
6. Henderson, G.; Cox, F.; Ganesh, S.; Jonker, A.; Young, W.; Global Rumen Census Collaborators; Janssen, P.H. Rumen microbial community composition varies with diet and host, but a core microbiome is found across a wide geographical range. *Sci. Rep.* **2015**, *5*, 14567. [CrossRef]
7. Johnson, K.A.; Johnson, D.E. Methane emissions from cattle. *J. Anim. Sci.* **1998**, *73*, 2483–2492. [CrossRef]
8. Ramin, M.; Huhtanen, P. Development of equations for predicting methane emissions from ruminants. *J. Dairy Sci.* **2013**, *96*, 2476–2493. [CrossRef]

9. Knapp, J.R.; Laur, G.L.; Vadas, P.A.; Weiss, W.P.; Tricarico, W. Invited review: Enteric methane in dairy cattle production: Quantifying the opportunities and impact of reducing emissions. *J. Dairy Sci.* **2014**, *97*, 3231–3326. [CrossRef]
10. Shindell, D.; Kuylenstierna, J.C.; Vignati, E.; van Dingenen, R.; Amann, M.; Klimont, Z.; Anenberg, S.C.; Muller, N.; Janssens-Maenhout, G.; Raes, F.; et al. Simultaneously mitigating near-term climate change and improving human health and food security. *Science* **2012**, *335*, 183–189. [CrossRef]
11. IPCC. Intergovernmental Panel on Climate Change 2018. Available online: https://report.ipcc.ch/sr15/pdf/sr15_spm_final.pdf (accessed on 10 August 2019).
12. Hristov, A.N.; Oh, J.; Giallongo, F.; Frederick, T.W.; Harper, M.T.; Weeks, H.L.; Brabco, A.F.; Moate, P.J.; Deighton, M.H.; Williams, R.O.; et al. An inhibitor persistently decreased enteric methane emission from dairy cows with no negative effect on milk production. *Proc. Natl. Acad. Sci. USA* **2015**, *112*, 10663–10668. [CrossRef]
13. Machado, L.; Magnusson, M.; Paul, N.A.; de Nys, R.; Tomkins, N. Effects of marine and freshwater macroalgae on in vitro total gas and methane production. *PLoS ONE* **2014**, *9*, e85289. [CrossRef]
14. Ramin, M.; Huhtanen, P. Development of an in vitro method for determination of methane production kinetics using a fully automated in vitro gas system—A modelling approach. *Anim. Feed Sci. Technol.* **2012**, *174*, 190–200. [CrossRef]
15. Menke, K.H.; Steingass, H. Estimation of the energetic feed value obtained from chemical analysis and in vitro gas production using rumen fluid. *Anim. Res. Dev.* **1988**, *28*, 7–25.
16. Cone, J.W.; Van Gelder, A.H.; Visscher, G.J.W.; Oudshoorn, L. Influence of rumen fluid and substrate concentration on fermentation kinetics measured with a fully automated time related gas production apparatus. *Anim. Feed Sci. Technol.* **1996**, *61*, 113–128. [CrossRef]
17. AOAC International. *Official Methods of Analysis of AOAC International*, 18th ed.; AOAC International: Gaithersburg, MD, USA, 2005; pp. 24–56.
18. Mertens, D.R. Gravimetric determination of amylase-treated neutral detergent fiber in feeds using refluxing in beakers or crucibles: Collaborative study. *J. AOAC Int.* **2002**, *85*, 1217–1240.
19. Krizsan, S.J.; Rinne, M.; Nyholm, L.; Huhtanen, P. New recommendations for the ruminal in situ determination of indigestible neutral detergent fibre. *Anim. Feed Sci.* **2015**, *205*, 31–41. [CrossRef]
20. Puhakka, L.; Jaakkola, S.; Simpura, I.; Kokkonen, T.; Vanhatalo, A. Effects of replacing rapeseed meal with fava bean at two concentrate crude protein levels on feed intake, nutrient digestion, and milk production in cows fed grass silage-based diets. *J. Dairy Sci.* **2016**, *99*, 7993–8006. [CrossRef]
21. Roque, B.M.; Salwen, J.K.; Kinley, R.; Kebreab, E. Inclusion of *Asparagopsis armata* in lactating dairy cows' diet reduces enteric methane emission by over 50 percent. *J. Clean. Prod.* **2019**, *234*, 132–138. [CrossRef]
22. Broderick, G.A.; Huhtanen, P.; Ahvenjärvi, S.; Reynal, S.M.; Shingfield, J. Quantifying ruminal nitrogen metabolism using the omasal sampling technique in cattle—A meta-analysis. *J. Dairy Sci.* **2010**, *93*, 3216–3230. [CrossRef]
23. Chang, J.; Peng, S.; Ciais, P.; Saunois, M.; Dangal, S.R.S.; Herrero, M.; Havlík, P.; Tian, H.; Bousquet, P. Revisiting enteric methane emissions from domestic ruminants and their $\delta^{13}C_{CH4}$ source signature. *Nat. Commun.* **2019**, *10*, e3420. [CrossRef]
24. Kingston-Smith, A.H.; Edwards, J.E.; Huws, S.A.; Kim, E.J.; Abberton, M. Plant-based strategies towards minimising 'livestock's long shadow'. *Proc. Nutr. Soc.* **2010**, *69*, 613–620. [CrossRef]
25. Li, B.; Fikse, W.F.; Løvendahl, P.; Lassen, J.; Lidauer, M.H.; Mäntysaari, P.; Berglund, B. Genetic heterogeneity of feed intake, energy-corrected milk and body weight across lactation in Holstein, Nordic Red, and Jersey cows. *J. Dairy Sci.* **2018**, *101*, 10011–10021. [CrossRef] [PubMed]
26. Grandl, F.; Furger, M.; Kreuzer, M.; Zehetmeier, M. Impact of longevity on greenhouse gas emissions and profitability of individual dairy cows analysed with different system boundaries. *Animal* **2019**, *13*, 198–208. [CrossRef] [PubMed]
27. Yan, T.; Agnew, R.E.; Gordon, F.J.; Porter, M.G. Prediction of methane energy output in dairy and beef cattle offered grass silage-based diets. *Livest. Prod. Sci.* **2000**, *64*, 253–263. [CrossRef]
28. Danielsson, R.; Ramin, M.; Bertilsson, J.; Lund, P.; Huhtanen, P. Evaluation of an in vitro system for predicting methane production in vivo. *J. Dairy. Sci.* **2017**, *100*, 8881–8894. [CrossRef] [PubMed]

29. Hristov, N.; Oh, J.; Lee, C.; Meinen, R.; Montes, F.; Ott, T.; Firkins, J.; Rotz, A.; Dell, C.; Adesogan, A.; et al. *Mitigation of Greenhouse Gas Emissions in Livestock Production—A Review of Technical Options for Non-CO_2 emissions*; FAO Animal Production and Health: Rome, Italy, 2013; pp. 10–60.
30. Pinares-Patiño, C.S.; Ulyatt, M.J.; Lassey, K.R.; Barry, T.N.; Holmes, C.W. Rumen function and digestion parameters associated with differences between sheep in methane emissions when fed chaffed lucerne hay. *J. Agric. Sci.* **2003**, *140*, 205–214. [CrossRef]
31. Pinares-Patiño, C.S.; Ebrahimi, S.H.; McEwan, J.C.; Clark, H.; Luo, D. Is rumen retention time implicated in sheep differences in methane emission? In Proceedings of the New Zealand Society of Animal Production, Wellington, New Zeland, 24–26 June 2011; Volume 71, pp. 219–222.
32. Goopy, J.P.; Donaldson, A.; Hegarty, R.; Vercoe, P.E.; Haynes, F.; Barnett, M.; Oddy, V.H. Low-methane yield sheep have smaller rumens and shorter rumen retention time. *Br. J. Nutr.* **2014**, *111*, 578–585. [CrossRef]
33. Bayat, A.R.; Tapio, I.; Vilkki, J.; Shingfield, K.J.; Leskinen, H. Plant oil supplements reduce methane emissions and improve milk fatty acid composition in dairy cows fed grass silage-based diets without affecting milk yield. *J. Dairy Sci.* **2018**, *101*, 1136–1151. [CrossRef]
34. Franco, M.O.; Krizsan, S.J.; Ramin, M.; Spörndly, R.; Huhtanen, P. In vitro evaluation of agro-industrial by-products replacing soybean meal in two different basal diets for ruminants. In Proceedings of the 8th Nordic Feed Science Conference, Uppsala, Sweden, 13–14 June 2017; pp. 170613–170614.
35. Fant, P.; Ramin, M.; Jaakkola, S.; Grimberg, Å.; Carlsson, A.S.; Huhtanen, P. Effects of different barley and oat varieties on methane production, digestibility and fermentation pattern in vitro. *J. Dairy. Sci.* **2019**, in press. [CrossRef]
36. O'Mara, F.P.; Fitzgerald, J.J.; Murphy, J.J.; Rath, M. The effect on milk production of replacing grass silage with maize silage in the diet of dairy cows. *Livest. Prod. Sci.* **1998**, *55*, 79–87. [CrossRef]
37. Beauchemin, K.A.; Kreuzer, M.; O'Mara, F.; McAllister, T.A. Nutritional management for enteric methane abatement: A review. *Aust. J. Exp. Agric.* **2008**, *48*, 21–27. [CrossRef]
38. Haque, M.N. Dietary manipulation: A sustainable way to mitigate methane emissions from ruminants. *J. Anim. Sci. Technol.* **2018**, *60*, 15. [CrossRef] [PubMed]
39. Vanhatalo, A.; Kuoppala, K.; Ahvenjärvi, S.; Rinne, M. Effects of feeding grass or red clover silage cut at two maturity stages in dairy cows. Nitrogen metabolism and supply of amino acids. *J. Dairy Sci.* **2009**, *92*, 5620–5633. [CrossRef] [PubMed]
40. Navarro-Villa, A.; O'Brian, M.; López, S.; Boland, T.M.; O'Kiely, P. In vitro rumen methane output of red clover and perennial ryegrass assayed using the gas production technique (GPT). *Anim. Feed Sci. Tech.* **2011**, *168*, 152–164. [CrossRef]
41. Allison, M.J.; Reddy, C.A.; Cook, H.M. The effects of nitrate and nitrite on VFA and CH_4 production by rumen microbes. *J. Anim.l Sci.* **1981**, *53*, 391–399.
42. Machado, L.; Magnusson, M.; Paul, N.A.; Kinley, R.; Nys, R.; Tomkins, N. Identification of bioactives from the red seaweed *Asparagopsis taxiformis* that promote antimethanogenic activity in vitro. *J. Appl. Phycol.* **2016**, *28*, 3117–3126. [CrossRef]
43. Vucko, M.J.; Magnusson, M.; Kinley, R.D.; Villart, C.; Nys, R. The effects of processing on the in vitro antimethanogenic capacity and concentration of secondary metabolites of *Asparagopsis taxiformis*. *J. Appl. Phycol.* **2017**, *29*, 1577–1586. [CrossRef]
44. Machado, L.; Magnusson, M.; Paul, N.A.; Kinley, R.; de Nys, R.; Tomkins, N. Dose-response effects of *Asparagopsis taxiformis* and *Oedogonium* sp. on in vitro fermentation and methane production. *J. Appl. Phycol.* **2016**, *28*, 1443–1452. [CrossRef]
45. Li, X.; Norman, H.C.; Kinley, R.D.; Laurence, M.; Wilmot, M.; Bender, H.; Nys, R.; Tomkins, N. *Asparagopsis taxiformis* decreases enteric methane production from sheep. *Anim. Prod. Sci.* **2018**, *58*, 681–688. [CrossRef]
46. Stefenoni, H.; Räisänen, S.; Melgar, A.; Lage, C.; Young, M.; Hristov, A. Dose-response effect of the macroalga *Asparagopsis taxiformis* on enteric methane emission in lactating dairy cows. In Proceedings of the American Dairy Science Association Annual Meeting, Cincinnati, OH, USA, 23–26 June 2019; pp. W163, 378.
47. Zhang, D.F.; Yan, H.J. In vitro ruminal methanogenesis of a hay-rich substrate in response to different combination supplements of nitrocompounds; pyromellitic diimide and 2-bromoethanesulphonate. *Anim. Feed Sci. Technol.* **2011**, *163*, 20–23. [CrossRef]

48. Andries, J.I.; Buysse, F.X.; DeBrabander, D.L.; Cottyn, B.G. Isoacids in ruminant nutrition: Their role in ruminal and intermediary metabolism and possible influences on performances—A review. *Anim. Feed Sci. Technol.* **1987**, *18*, 169–180. [CrossRef]
49. Newbold, J.R.; van Zijderveld, S.M.; Hulshof, R.B.A.; Fokkink, W.B.; Leng, R.A.; Terencio, P.; Powers, W.J.; van Adrichem, P.S.J.; Paton, N.D.; Perdok, H.B. The effect of incremental levels of dietary nitrate on methane emissions in Holstein steers and performance in Nelore bulls. *J. Anim. Sci.* **2014**, *92*, 5032–5040. [CrossRef] [PubMed]
50. Klop, G.; Hatew, B.; Bannink, A.; Dijkstra, J. Feeding nitrate and docosahexaenoic acid affects enteric methane production and milk fatty acid composition in lactating dairy cows. *J. Dairy Sci.* **2016**, *99*, 1161–1172. [CrossRef] [PubMed]
51. Olijhoek, D.W.; Hellwing, A.L.F.; Brask, M.; Weisbjerg, M.R.; Højberg, O.; Larsen, M.K.; Dijkstra, E.J.; Erlandsen, E.J.; Lund, P. Effect of dietary nitrate level on enteric methane production, hydrogen emission, rumen fermentation, and nutrient digestibility in dairy cows. *J. Dairy Sci.* **2016**, *99*, 6191–6205. [CrossRef] [PubMed]
52. Department of Climate Change, Commonwealth Government of Australia. The Potential of Feeding Nitrate to Reduce Enteric Methane Production in Ruminants. Available online: http://www.penambulbooks.com/ (accessed on 14 September 2019).
53. Zhou, Z.; Meng, Q.; Yu, Z. Effects of methanogenic inhibitors on methane production and abundances of methanogens and cellulolytic bacteria in in vitro ruminal cultures. *Appl. Environ. Microbiol.* **2011**, *77*, 2634–2639. [CrossRef]

11

Effect of Thymol Addition and Withdrawal on some Blood Parameters, Antioxidative Defence System and Fatty Acid Profile in Rabbit Muscle

Kristina Bacova [1], Karin Zitterl-Eglseer [2], Lubica Chrastinova [3], Andrea Laukova [1], Michaela Madarova [1], Sona Gancarcikova [4], Drahomira Sopkova [5], Zuzana Andrejcakova [5] and Iveta Placha [1,*]

1 Centre of Biosciences, Slovak Academy of Sciences, Institute of Animal Physiology, Soltesovej 4–6, 040 01 Kosice, Slovakia; bacovak@saske.sk (K.B.); laukova@saske.sk (A.L.); madarova@saske.sk (M.M.)

2 Institute of Animal Nutrition and Functional Plant Compounds, University of Veterinary Medicine Vienna, Veterinärplatz 1, A-1210 Wien, Austria; karin.zitterl@vetmeduni.ac.at

3 National Agricultural and Food Centre, Hlohovecka 2, 951 41 Nitra-Luzianky, Slovakia; lubica.chrastinova@nppc.sk

4 Laboratory of Gnotobiology, Department of Microbiology and Immunology, University of Veterinary Medicine and Pharmacy, Komenskeho 73, 041 81 Kosice, Slovakia; sona.gancarcikova@uvlf.sk

5 Department of Anatomy, Histology and Physiology, University of Veterinary Medicine and Pharmacy, Komenskeho 73, 041 81 Kosice, Slovakia; drahomira.sopkova@uvlf.sk (D.S.); zuzana.andrejcakova@uvlf.sk (Z.A.)

* Correspondence: placha@saske sk;

Simple Summary: So far, the study of the bioactivity of thymol, a major constituent of *Thymus vulgaris* L., in the animal organism has received little attention. Our study could give us answers to questions about whether thymol accumulates in the rabbit organism after its sustained administration and if it is also able to exhibit its beneficial properties for a longer period. Thymol in powder form at the concentration 250-mg/kg feed was added to the rabbit diet for 21 days and withdrawn for the next seven days. We confirmed that thymol was sufficiently absorbed from the gastrointestinal tract and was able to express its biological activity not only during application but, also, after withdrawal. Further studies are needed to clarify the biotransformation and bioavailability of thymol in the rabbit organism with respect to the specific features of rabbit digestion.

Abstract: Thymol concentrations in rabbit plasma, intestinal wall (IW) and faeces were detected, and the effects of thymol application and withdrawal on biochemical, antioxidant parameters and fatty acids (FA) in blood (B) and muscle (M) were studied. Forty-eight rabbits were divided into two experimental groups (control, C and with thymol 250-mg/kg feed, T). Thymol was administered for 21 days (TA) and withdrawn for seven days (TW). Thymol in plasma correlated with that in the IW (Spearman's correlation coefficient (r_s) = −1.000, $p = 0.0167$, TA) and was detected in faeces (TA and TW). In TA alkaline phosphatase ($p = 0.0183$), cholesterol ($p = 0.0228$), malondialdehyde ($p = 0.003$), glutathione peroxidase ($p = 0.0177$) in B and lactate dehydrogenase (M, $p = 0.0411$) decreased; monounsaturated FA ($p = 0.0104$) and α-linolenic acid ($p = 0.0227$) in M increased. In TW urea ($p = 0.0079$), docosapentaenoic acid ($p = 0.0069$) in M increased; linoleic acid ($p = 0.0070$), $\sum$ n−6 ($p = 0.0007$) in M and triglycerides decreased (B, $p = 0.0317$). In TA and TW, the total protein ($p = 0.0025$ and 0.0079), creatinine (B; $p = 0.0357$ and 0.0159) and oleic acid (M; $p = 0.0104$ and 0.0006) increased. Thymol was efficiently absorbed from the intestine and demonstrated its biological activity in blood and the muscles.

Keywords: rabbit; thymol; bioavailability; antioxidant

1. Introduction

The supplementation of human and animal diets with *Thymus vulgaris* (thyme) either as dried leaves or its essential oil has often demonstrated its beneficial properties. Since synthetic growth promoters were replaced with alternative herbal products in animal rearing, thyme has started receiving major attention. The most important bioactive compound contained in this plant is thymol, which exhibits antimicrobial, antioxidant, anticarcinogenic and anti-inflammatory activities [1–3]. According to some studies, thyme improves the performance parameters, but some other studies suggest that it has no effect. Some studies report that thyme is able to reduce levels of triglyceride and total cholesterol [4,5]. Yu et al. [6] found that thymol possesses a lipid-reducing function by altering hepatic triglyceride secretion.

The mode of action of herbs and plant extracts and the details about the accumulation of phenolic substances in animal tissues are not known or not completely understood [7,8]. In general, the bioavailability of dietary compounds depends on their digestive stability and the efficiency of their transepithelial passage [9]. Moreover, their influence on each other's intestinal absorption has to be taken into account in studies concerned with the bioavailability of essential oil compounds.

Thymol as one of the major constituents of thyme oil presents a wide range of functional possibilities in pharmacy and the food industry. Besides thymol, thyme contains high concentrations of monoterpene phenols like carvacrol, p-cymene, 1,8-cineole, linalool, borneol, camphor, β-caryophyllene, thymol methyl ether and carvacrol methyl ether, which could have influenced the thymol absorption [2,3].

Oceľová et al. [10] were the first who analysed the thymol concentrations in individual intestinal segments in hens. Sufficient absorption of thymol from the digestive tract and its transport by systemic circulation to tissues in broiler chickens after four weeks of thyme essential oil application were demonstrated by Oceľová et al. [7]. Placha et al. [2] pointed to the sparing effect of thymol against oxidative deterioration of the antioxidant defence system in poultry after sustained thyme oil dietary application at 0.05% (thymol content 248.97 mg/kg dry matter (DM)) and 0.1% (thymol content 460.22 mg/kg DM) concentrations.

Concerning the toxicity of thymol, the data are controversial. According to a recent report of the European Food Safety Authority (EFSA) [11], thymol, when administered by the oral route in a rat, mouse and guinea pig (lethal dose - LD 50, 0.98, 1.80 and 0.88 mg/kg, respectively), suggested a moderately acute toxicity. No significant alteration was observed when a thymol oil-water emulsion was administered at doses (15.39, 30.78 and 61.55 mg/kg, respectively) during 28 days [12]. Based on European Commission (EC) [13], no maximum residual limit for thymol in foodstuffs of animal origin is needed to establish when it is used as a veterinary medicinal product. These data show that further studies are required to establish the thymol appropriate concentration.

Rabbits have a unique digestive system that is represented by an original feature of rabbit feeding behaviour named caecotrophy. Caecotrophy means the excretion and immediate consumption of specific soft faeces termed "caecotrophs". This process is extremely important, because it improves feed utilization by maximizing the digestibility of nutrients [14]. Many scientific studies have tried to find the appropriate dose and form of plant extract application for improving animal health, but insights into the precise mechanism or mode of action of their components are lacking. As far as rabbits are concerned, to our knowledge, only one old study by Takada et al. [15] has been carried out regarding the metabolic outcome of thymol. Probably, the process of caecotrophy could minimize the loss of thymol bioaccessibility during the digestive processes.

The present study provides new insights for understanding the possible processes of absorption and deposition of thymol in the rabbit organism in connection with its protective role against oxidative stress. Based on our previous studies related to the thymol absorption, deposition and beneficial effects against oxidative stress in broiler chickens after a sustained administration of thyme oil, we decided to examine the thymol application into rabbit diets at the concentration 250-mg/kg feed.

2. Materials and Methods

2.1. *Animals and Experimental Design*

After weaning at 35 days of age, 48 rabbits of both sexes (meat line M9) were randomly divided into two experimental groups (control, C and with thymol addition, T), with six replicates in each (one replicate consisting of two cages, one cage/two animals). Initial live weight was 1006 ± 98 g in C and 1035 ± 107 g in T. All experimental wire-net cages (61 cm × 34 cm × 33 cm) were kept in rooms with automatic temperature control (22 ± 4 °C) and photoperiod (16 L:8 D). The rabbits could feed ad libitum and had free access to drinking water. The experiment lasted 28 days. The rabbits were fed with thymol addition for 21 days (TA), and for the next 7 days, the thymol was withdrawn (TW). Eight starved rabbits (6 male and 2 female) in each group were slaughtered in an experimental slaughterhouse at 56 (C and TA) or 63 (C and TW) days of age. Rabbits were stunned with electronarcosis (50 Hz, 0.3 A/rabbit for 5 s), immediately hung by the hind legs on the processing line and quickly bled by cutting the jugular veins and the carotid arteries.

2.2. *Animals Care and Use*

The trial was carried out at the experimental rabbit facility of the National Agricultural and Food Centre, Research Institute for Animal Production, Nitra, Slovakia. The protocol was approved by the Institutional Ethical Committee, and the State Veterinary and Food Office of the Slovak Republic approved the experimental protocol (4047/16-221).

2.3. *Diets and Chemical Analyses*

The basal diet (control, C) was formulated to satisfy growing rabbits' requirements [16] (Table 1) and was tested against the experimental diet (T) containing thymol (≥99.9%, Sigma Aldrich, St. Louis, MO, USA), which was added to the basal diet in white powder form at concentration 250-mg/kg feed. The diets were administered in the form of pellets with an average size of 3.5 mm. The feed was stored in darkness to protect against degradative processes and was analysed to determine the crude protein (CP), ash, ether extract, acid detergent fibre (ADF), starch and dry matter (DM) in the diets, while DM was also determined for the intestinal wall, muscle, liver and faeces according to the Association of Official Analytical Chemists (AOAC) methods [17]. Neutral detergent fibre (NDF) was analysed according to Van Soest et al. [18].

Table 1. Ingredients and chemical composition of rabbit diets.

Ingredients (%)		Chemical Composition (g/kg Feed)	
Dehydrated Lucerne meal	36.0	Dry matter (g/kg)	900.9
Dry malting sprouts	15.0	Organic compounds	831.8
Oats	13.0	Nitrogen free extract	444.3
Wheat bran	9.0	Neutral detergent fibre (NDF)	352.9
Barley	8.0	Acid detergent fibre (ADF)	208.1
Extracted sunflower meal	5.5	Crude fibre	177.8
Extracted rapeseed meal	5.5	Crude protein	176.6
Dried distiller grains with solubles	5.0	Cellulose	163.1
Premix [1]	1.7	Hemicellulose	144.8
Limestone	1.0	Starch	133.1
Sodium chloride	0.3	Ash	69.2
		Fat	33.1
		Metabolic energy, MJ/kg	9.9

[1] The vitamin-mineral premix provided per kg of complete diet: Retinyl acetate 5.16 mg, Cholecalciferol 0.03 mg, Tocopherol 0.03 mg, Thiamine 0.8 mg, Riboflavin 3.0 mg, Pyridoxin 2.0 mg, Cyanocobalamin 0.02 mg, Niacin 38 mg, Folic acid 0.6 mg, Calcium 1.8 mg, Iron 70 mg, Zinc 66 mg, Copper 15 and Selenium 0.25 mg.

2.4. *Growth Performance and Health Status*

Body weight (BW) and feed intake (FI) were recorded individually once a week. The average daily FI, average daily weight gain (WG) and feed conversion ratio (FCR) were calculated at the end of the trial (on 56 and 63 d of age). Data pertaining to any animal that died during the experiment were excluded from the calculations. Mortality was recorded daily throughout the experimental periods.

2.5. *Thymol Antioxidant Capacity and Stability in Feed*

The Trolox equivalent capacity (TEAC) was determined in thymol and in the experimental feed according with the method described by Karamać et al. [19] using the 2,2′-Azinobis-(3-Ethylbenzthiazolin-6-Sulfonic Acid (ABTS •+) decolorization assay. The results were expressed as mmol Trolox equivalents (TE) per g. Thymol evaporation in the feed was analysed every week during thymol application using High-Performance Liquid Chromatography HPLC according to the modified method of Pisarčíková et al. [20]. Samples were analysed in triplicate and were relatively stable (0 d—151.89, 7 d—134.75 and 14 d—128.30 μg/g DM, respectively).

2.6. *Sampling*

Blood samples for biochemical analyses were collected from the marginal ear vein (Vena auricularis) into dry nonheparinized Eppendorf tubes at experimental days 21 and 28 and were left to clot in a standing position for approximately 2 h to obtain the serum, and then, the serum was separated by centrifugation at 700× *g* for 15 min. Blood for analyses of antioxidant parameters was collected into heparinized tubes, and plasma was obtained after centrifugation at 1180× *g* for 15 min. Samples of serum, plasma, muscle (*Longissimus thoracis et lumborum*, LTL), small intestinal wall, liver and hard faeces (freshly voided, collected using nets mounted under the cages) were immediately frozen in liquid nitrogen and stored at −70 °C until analysis.

2.7. *Thymol Analyses in Plasma, Small Intestinal Wall and Faeces*

Detection of thymol in samples of plasma, intestinal walls and faeces was performed using headspace solid-phase microextraction followed by gas chromatography coupled with the mass spectrometry method, as described by Placha et al. [2]. Briefly, detection and quantification were carried out using a gas chromatography/mass spectrometry (GC/MS) (type HP 6890 GC) coupled with a 5972 quadrupole-mass selective detector (Agilent Technologies GmbH, Wilmington, DE, USA). Detection of thymol was confirmed by comparing its specific mass spectrum and retention time with those of the reference compound. Additionally, the Kovats index was calculated. Calibration curves were generated by plotting the peak area ratios of thymol to o-cresol used as the internal standard (Sigma-Aldrich, St. Louis, MO, USA) against the known thymol concentrations. The selective ion mode was used for the quantitative analysis of thymol. The mass fragments *m/z* 135 and *m/z* 150, as well as *m/z* 107 and *m/z* 108, were monitored as characteristic for thymol and o-cresol, respectively. Calibration curves were prepared from blank samples spiked directly with thymol (AppliChem, Darmstadt, Germany) in standard solutions with known concentrations. Each point on the calibration curve was analysed as a duplicate. The peak of thymol was detected around 19 min, and the o-cresol peak occurred around 10 min in all samples. Samples for the detection of thymol were prepared using the method described by Ocel'ová et al. [10]. Enzyme β–Glucuronidase Helix pomatia Type HP-2 (aqueous solution, ≥100,000 units/mL, Sigma-Aldrich, St Louis, MO, USA) was added to samples to cleave thymol from its glucuronide and sulphate to obtain the total amount of thymol in the plasma.

2.8. *Biochemical and Antioxidant Parameters and Activity of Lactate Dehydrogenase in Blood and Tissues*

Total proteins (TP; g/L), creatinine (μmol/L), urea (mmol/L), triglycerides (mmol/L), total cholesterol (mmol/L), alanine aminotransferase (ALT; μkat/L), aspartate aminotransferase (AST; μkat/L) and alkaline phosphatase (ALP; µkat/L) were analysed using a DIALAB commercial kit (Prague, Czech Republic) and an ELLIPSE analyser (AMS, Guidonia, Rome, Italy).

Activity of glutathione peroxidase (GPx, EC 1.11.1.9) in blood was measured by monitoring the oxidation of Nicotinamide Adenine Dinucleotide Phosphate (NADPH) at 340 nm in-line with Paglia and Valentine [21] using a commercial kit (Ransel, Randox, London, UK). Haemoglobin (Hb) content in blood was analysed using a commercial kit from Randox, UK. The samples of LTL muscle and liver for malondialdehyde (MDA) measurement and activity of lactate dehydrogenase (LDH, EC 1.1.1.27) were washed in buffered saline to remove excess blood and connective tissue. Samples for MDA analyses were homogenised with deionized distilled water and 50 µL of 7.2% butylated hydroxytoluene and for LDH activity in cold buffer (0.05-mol/L Tris-HCl buffer, pH 7.3). The homogenates were subsequently centrifuged at 13,680× *g* at 4 °C for 20 min. MDA concentrations in these tissues and plasma were measured using the modified fluorometric method of Jo and Ahn [22]. The enzyme activity of LDH was measured using a commercial diagnostic kit (Crumlin, Randox, UK) with an Alizé automatic biochemical analyser (Lisabio, Pouilly-en-Auxois, France) at 340 nm, as described by Andrejčáková et al. [23]. The protein concentration in the muscle and liver was quantified using the spectrophotometric method published by Bradford [24].

2.9. Fatty Acids in Muscle Tissue

The fatty acid (FA) composition in the muscle tissue was determined using the method of Ouhayoun et al. [25]. Fatty acid methyl esters (FAME) were prepared by means of alcoholises in an essential nonalcoholic solution and analysed using gas chromatography on GC 6890N (Agilent Technologies, Basel, Switzerland). Results were expressed as percentages of total fatty acids.

2.10. Statistical Analyses

Values of thymol, GPx, LDH, MDA and FA concentrations were tested for normal distribution with the Kolmogorov-Smirnov test. The Mann-Whitney U test was used for statistical analysis. Results were presented as the mean value ± standard deviation. Significant differences were considered at $p < 0.05$. Correlations of thymol concentrations between plasma and feed and plasma and intestinal wall were analysed using nonparametric Spearman's rank correlation and expressed as Spearman's correlation coefficient (r_S). Statistical analyses were performed using Graph Pad Prism 5.0. (GraphPad Software, San Diego, CA, USA).

3. Results

3.1. Growth Performance

All broiler rabbits in the present trial were in good health, and the growth performance was normal and was not affected by the addition of thymol. Five animals (three/control and two/experimental group) died during the whole experiment. If there were no differences in the given parameters, we do not include them in the table.

3.2. Thymol Antioxidative Capacity

Approximately equal TEAC value was found in thymol and in the feed of the experimental group (0.76 vs. 0.74 mmol TE/g).

3.3. Thymol in Feed, Plasma, Small Intestinal Wall and Faeces

Thymol content in feed amounted to 148.9 ± 16.7 µg/g DM. Thymol concentration in plasma was 0.05 ± 0.02 µg/L and, in intestinal wall, 0.04 ± 0.03 µg/g DM in TA, but, in TW, it was not detected. Thymol concentration in faeces in TA was 0.89 ± 0.45 µg/g DM, and, in TW, it was 0.08 ± 0.04 µg/g DM (Table 2). Thymol concentration in plasma significantly correlated with the thymol concentration in the intestinal wall ($r_S = -1.000$, $p = 0.0167$).

Table 2. Thymol content in plasma (μg/L), feed, intestinal wall and faeces (μg/g dry matter (DM)).

Substance	TA	TW
Feed	148.90 ± 16.65	-
Plasma	0.05 ± 0.02	ND
Intestinal wall	0.04 ± 0.03	ND
Faeces	0.89 ± 0.45	0.08 ± 0.04

ND—not detected, TA—thymol addition and TW—thymol withdrawal.

3.4. Biochemical Parameters in Blood

Thymol addition significantly decreased the levels of ALP ($p = 0.0183$) and cholesterol ($p = 0.0228$). Urea significantly increased ($p = 0.0079$), and triglycerides decreased ($p = 0.0317$) in TW. TA and TW had significantly increased TP ($p = 0.0025$ vs. $p = 0.0079$) and creatinine ($p = 0.0357$ vs. $p = 0.0159$) (Table 3).

Table 3. Effects of thymol on aspartate aminotransferase (AST, μkat/L), alanine aminotransferase (ALT, μkat/L), alkaline phosphatase (ALP, μkat/L), total proteins (TP, g/L), urea (mmol/L), creatinine (μmol/L), triglycerides (mmol/L) and cholesterol (mmol/L) in rabbit blood.

Parameter	TA			*p*-Value	TW			*p*-Value
	C	T	SD		C	T	SD	
AST (μkat/L)	0.45	0.34	0.12	0.1199	0.49	0.34	0.20	0.2948
ALT (μkat/L)	0.77	0.62	0.14	0.2677	0.76	0.69	0.32	1.0000
ALP (μkat/L)	2.00 ^{a}	1.52 ^{b}	0.42	0.0183	2.42	2.19	0.83	0.5476
TP (g/L)	36.20 ^{a}	61.86 ^{b}	13.88	0.0025	36.84 ^{a}	55.90 ^{b}	11.79	0.0079
Urea	5.18	5.31	0.50	1.0000	5.03 ^{a}	6.54 ^{b}	0.95	0.0079
Creatinine	79.67 ^{a}	115.40 ^{b}	19.55	0.0357	64.25 ^{a}	115.60 ^{b}	28.52	0.0159
Triglycerides	1.15	1.45	0.29	0.1508	1.53 ^{a}	0.84 ^{b}	0.49	0.0317
Cholesterol	2.20 ^{a}	1.17 ^{b}	0.80	0.0228	1.76	0.95	0.60	0.0556

^{a,b} Values within a row with different superscript letters differed significantly ($p < 0.05$). Data are presented as mean ± standard deviation (SD). C—control diet, TA—thymol addition and TW—thymol withdrawal.

3.5. Antioxidant Parameters in Blood and Tissues

MDA and GPx in blood ($p = 0.003$ and $p = 0.0177$) and LDH in muscle ($p = 0.0411$) significantly decreased in TA (Table 4).

Table 4. Effects of thymol on the antioxidant parameters and activity of lactate dehydrogenase (LDH) in rabbit blood, liver and muscle.

Parameter	TA			*p*-Value	TW			*p*-Value
	C	T	SD		C	T	SD	
Blood								
MDA (nmol/mL)	0.35 ^{a}	0.254 ^{b}	0.06	0.0030	0.32	0.30	0.04	0.1049
GPx (μkat/g Hb)	3.61 ^{a}	2.52 ^{b}	0.78	0.0177	2.61	2.32	0.71	0.8763
Liver								
MDA (nmol/g protein)	83.11	83.40	18.31	0.7984	83.81	68.69	21.51	0.1605
LDH (μkat/g protein)	41.56	41.62	6.50	0.6454	48.88	42.75	7.83	0.1304
Muscle								
MDA (nmol/g protein)	23.63	28.45	11.20	0.5737	18.34	23.45	5.82	0.0650
LDH (μkat/g protein)	78.74 ^{a}	35.89 ^{b}	36.47	0.0411	78.50	50.62	43.60	0.2026

^{a,b} Values within a row with different superscript letters differed significantly ($p < 0.05$). Data are presented as mean ± standard deviation (SD). C—control diet, TA—thymol addition, TW—thymol withdrawal, MDA—malondialdehyde and GPx—glutathione peroxidase.

3.6. Fatty Acids in Muscle

Concentrations of oleic acid (C 18:1 n-9) significantly increased in TA and TW ($p = 0.0104$ vs. 0.0006). MUFA and α-linolenic acid (C 18:3 n-3) significantly increased in TA ($p = 0.0104$ and $p = 0.0227$).

Docosapentaenoic acid (C 22:5 n-3) significantly increased ($p = 0.0069$), and linoleic acid (C 18:2 n-6) and ∑ n-6 significantly decreased in TW ($p = 0.0070$ and $p = 0.0007$; Table 5).

Table 5. Effects of thymol on fatty acids profile (% of total FA) in rabbit muscle.

Parameter	TA			*p*-Value	TW			*p*-Value
	C	T	SD		C	T	SD	
C 12:0	0.107	0.109	0.006	0.3428	0.105	0.108	0.004	0.1015
C 14:0	1.383	1.392	0.030	0.5950	1.357	1.386	0.039	0.1375
C 16:0	24.400	24.320	0.159	0.3823	24.390	24.580	0.222	0.1049
C 18:0	10.460	10.550	0.251	0.5049	10.750	10.620	0.210	0.2392
C 17:0	0.302	0.308	0.029	0.3713	0.312	0.286	0.037	0.4001
∑ SFA	36.650	36.680	0.344	1.0000	36.910	36.980	0.280	1.0000
C 18:1 n-7	4.945	4.975	0.109	0.6454	4.854	4.899	0.105	0.3439
C 18:1 n-9	30.030	34.470	3.754	0.0104	31.960	37.270	3.660	0.0006
C 20:1	0.597	0.589	0.091	0.8785	0.545	0.561	0.072	0.5054
∑ MUFA	35.600	40.020	3.730	0.0104	37.420	42.670	4.640	0.2345
C 18:2 n-6	52.59	50.73	6.435	0.5054	51.36	48.83	8.458	0.0070
C 20:4 n-6	1.765	1.978	0.244	0.1049	1.842	1.708	0.279	0.5054
∑ n-6	54.35	52.70	6.341	0.5054	53.20	50.53	4.756	0.0007
C 18:3 n-3	1.794	1.958	0.149	0.0227	1.793	1.885	0.191	0.2786
C 20:5 n-3	1.088	1.108	0.103	0.6350	1.096	0.935	0.173	0.1409
C 22:5 n-3	1.270	1.393	0.127	0.0513	1.276	1.381	0.081	0.0069
C 22:6 n-3	0.328	0.298	0.049	0.3431	0.341	0.310	0.046	0.2677
∑ n-3	4.479	4.755	0.236	0.0519	4.506	4.511	0.199	1.000
∑ n-6/∑ n-3	12.14	11.10	1.444	0.1949	11.86	11.19	1.223	0.3282

Data are presented as mean ± standard deviation (SD). C—control diet, TA—thymol addition, TW—thymol withdrawal, SFA—saturated fatty acids and MUFA—monounsaturated fatty acids.

4. Discussion

4.1. Growth Performance

The inclusion of thymol at the concentration used in our experiment did not show any significant effect on the animals' weight, weight gain and conversion ratio. Although thyme has been shown to improve the palatability and feed intake in growing rabbits, its beneficial effects on the live growth performance in rabbits has not yet been confirmed [26]. According to Erdelyi et al. [27], due to the specific digestive physiology of rabbits, essential oils or plant extracts have much lesser positive effects than in broilers or piglets. Windisch et al. [28] reported that the use of thyme in animal diets could be limited when applied in certain amounts, because it is highly aromatic. According to Gerencsér et al. [29], thyme leaves did not demonstrate any substantial effects on the growth performance or health status. This statement is in agreement with our study, in that the thymol did not affect the feed intake, and most animals remained in optimal condition.

4.2. Thymol in Plasma, Intestinal Wall and Faeces

We found a significant correlation between thymol levels in the small intestinal wall and plasma during the period of thymol addition to the feed. This result is in agreement with Placha et al. [2], who confirmed the efficient absorption of thymol from the digestive tract into the systemic circulation in broiler chickens. According to Ocel'ová [30], after the absorption of plant compounds from the intestine, they are metabolised and eliminated from the organism.

One of the most original features of rabbit-feeding behaviour is caecotrophy. The result of this process, in which soft faeces are swallowed and then stored intact in the fundus of the stomach, is that they undergo the same digestive processes as normal feed. Some parts of the initial food intake may be recycled even up to four times in this way [31]. The detection of thymol in faeces points to its excretion from the organism in an unmetabolised form (Table 2). This unmetabolised thymol could be recycled and again absorbed in the intestinal wall, where it is finally metabolised by the processes of biotransformation. The part of the metabolites in the intestinal wall can be transported back into the intestinal lumen by efflux transporters, converted to thymol (parental compound) and again reabsorbed by the enterocytes. Another molecule can be transported through the basolateral membrane of enterocytes directly to the blood circulation. All these processes play a crucial role in affecting the thymol concentration in blood [7].

Thymol sulphate and thymol glucuronide are the main metabolites of thymol biotransformation, and so far, little is known about their bioactivity or whether these compounds are only inactive forms [3]. Pisarčíková et al. [20] first detected thymol sulphate and thymol glucuronide in the liver and duodenal wall of broiler chickens after a sustained four-week administration of thyme essential oil, and according to Kohlert et al. [32] and Rubió et al. [9], thymol metabolites could be deconjugated to the parental compounds and, in this way, express their pharmacological properties. We have to bear in mind that, during biotransformation, plant compounds change their pharmacological properties, which usually differ from the properties of the parental compounds.

There are only a few studies concerning thymol distributions in animal tissues. Thymol conjugates have been detected in the plasma of humans and animals, and thymol after enzymatic cleavage was detected in the plasma of horses, chickens and pigs [1,10,32–35]. According to Placha et al. [2], erythrocytes can act as depots of polyphenols due to their ability to bind to the surface of red blood cells, and in this way, they can circulate in the organism. There is the question whether bound thymol is able to unbind from tissue depots if the amount of circulated thymol decreases. Since we detected thymol in rabbit faeces also after its withdrawal, this mechanism could be indicated (Table 2).

4.3. *Biochemical Parameters in Blood*

Despite the fact that plant bioactive compounds undergo fast biotransformation and elimination, Ocel'ová et al. [10] and Placha et al. [2] confirmed the accumulation of thymol in the breast muscles, kidneys and livers of broiler chickens after four weeks of thyme essential oil diet supplementation. Moreover, they found the highest concentration of thymol in the kidneys and the lowest in liver tissues, which could point to intensive metabolism in the liver and accumulation in the kidneys. Bardal et al. [36] showed that the distribution of plant compounds from systemic circulation to the tissues is restricted, since only free drugs in plasma are able to diffuse into the target tissue.

Gumus et al. [37] reported that specific polyphenols such as thymol may reduce plasma lipids by altering the hepatic triglyceride secretion, as well as inhibiting the activity of cholesterol-synthesizing enzyme 3-hydroxy-3-methylglutaryl-coenzyme A reductase. If some imbalance between the cellular free radical formation and the antioxidant defence system exists, excessive amounts of free radicals are produced, and cellular components such as the lipids are attacked. The administration of antioxidants such as thymol could combat oxidative stress by the scavenging of free radicals and, in this way, effectively reduce the serum levels of lipid parameters such as triglycerides and cholesterol [6]. Our study suggests that thymol in the rabbit organism possesses a lipid-reducing function by this mechanism (Table 3).

AST, ALT and ALP are enzymes that significantly reflect the liver function. These parameters in our study were in the range of the reference interval. We assumed that thymol in this concentration was able to express its antioxidant properties and positively affected these parameters, although only ALP was affected significantly. The values of urea and creatinine also remained within the normal ranges. This suggests that no hepatic or renal injuries occurred in this experiment. The slight increase in the urea and creatinine amounts could imply the effect of thymol and/or its metabolites on the kidney function, particularly on glomerular filtration. Ocel'ová [30] and Kohlert et al. [32] suggested the role of the kidneys in the metabolism and elimination of phenolic compounds. They assumed that thymol metabolites could be cleaved at the brush border of the renal tubule and reabsorbed as thymol back into the peritubular capillaries. As mentioned above, thymol can accumulate in the kidneys, and based on this finding, we can hypothesise that thymol concentrations in the kidneys in our experiment could lead during the metabolic processes to the formation of some toxic substances that could be responsible for altering the renal function (Table 3).

The nutritive value of the protein is determined not only by its amino acid composition but, also, by its digestibility [38]. The antioxidative and antimicrobial properties of thymol may beneficially affect the gastrointestinal microbiota ecosystem and, in this way, improve the nutrient utilisation. Moreover, phytogenics can stimulate the production of digestive enzymes and, also, beneficially affect the nutrient digestibility [39]. Ocel'ová [30] found that the duodenal wall in chickens was repeatedly exposed to thymol molecules after the four-week addition of thyme essential oil to their diet. Based on the significant correlation between thymol in the plasma and intestinal wall in our experiment, we assumed that the same processes were also present in the rabbit intestinal wall. Many studies have suggested that thymol possesses useful antioxidant properties that are responsible, among other things, for the renewal rate of mature enterocytes at the surface of the intestinal villi, which causes an increase in their absorption capacity [40]. Gasco et al. [41] reported that serum TP synthesis depends on the content of available protein in the diet. We assume that the increased TP in our groups with the thymol addition, as well as thymol withdrawal, was caused by extending the absorption surface of the intestinal wall, which probably evoked a better absorption of the proteins from the diet, and/or also by a higher production of the digestive enzymes (Table 3).

4.4. Antioxidant Parameters and LDH in Blood and Tissues

In the present study, the positive effect of thymol on the activity of GPx, which protects intracellular lipids against peroxidation, as well as on MDA levels as an indicator of lipid peroxidation, was expressed in plasma. This finding reflects the fact that thymol was efficiently absorbed from the digestive tract to the blood, where it manifested its antioxidant properties. The antioxidant properties of thymol were also demonstrated in muscles during the thymol addition, as well as after its withdrawal. Oxidative stress is associated with damage in a wide range of macromolecular species and derangements of the cellular metabolism, which permeabilise the cellular or organelle membranes and cause apoptosis [42]. LDH is an intracellular enzyme, and its increased level is an indicator of cellular damage. We detected decreased LDH levels, which points to the sparing effect of thymol on cellular metabolism (Table 4).

4.5. Fatty Acids in Muscle

Rabbits need adequate amounts of essential FA, which are primarily represented by linoleic (C18:2, n-6) and α-linolenic acids (C18:3, n-3) in their diet. The stability of these essential FA is very low, and for example, linolenic acid is oxidized ten times more rapidly than linoleic acid [43]. Polyunsaturated FA are easily oxidized because of the unstable double bonds between their carbon atoms and are rapidly broken down into short-chain compounds, which is undesirable from the nutritional aspect. The lipoperoxidation of the cell structure is a consequence of oxidative stress. These biological reactions are initiated by the production of ROS, which remove protons from FA [44]. Dabbou et al. [45] demonstrated that bilberry pomace, which is characterised by high antioxidant activity due to high contents of phenols, effectively prevented the oxidation of unsaturated lipids in the muscles of rabbits. An increase in unsaturated FA and, mainly, linolenic acid were observed in our study after the addition of thymol, which probably blocked the oxidation of the lipids with its strong antioxidative properties. After thymol withdrawal, linoleic acid, as well as $\sum$n-6, decreased, which could be the reason for the insufficient effect of thymol in connection with its low deposition in the muscles (Table 5). Moreover, as was mentioned above, linoleic acid is more rapidly oxidized than other polyunsaturated FA.

5. Conclusions

In conclusion, after its sustained application, thymol is able to accumulate and circulate in the rabbit organism, particularly due to the rabbits' specific digestive characteristic, i.e., caecotrophy. The investigated parameters were affected not only during the thymol addition to the feed but, also, after its withdrawal, which points to thymol accumulation in the organism. The thymol concentration used in this experiment clearly demonstrated its antioxidant properties and inhibited the lipid peroxidation and formation of oxidative deterioration compounds and may become an important antioxidant food supplement. However, further studies should be conducted to confirm the thymol distribution and deposition within the rabbit organism. Since there is a lack of literature describing the absorption and distribution of thymol in rabbit tissues in connection with its beneficial effects on animal health, to our knowledge, this is the first study attempting to explain these processes in the rabbit organism.

Author Contributions: Conceptualisation: I.P.; methodology: I.P. and K.B.; validation: I.P.; formal analysis: I.P., K.Z.-E., K.B., M.M., S.G., D.S. and Z.A.; investigation: I.P. and K.B.; resources: I.P., L.C., K.Z.-E., S.G. and D.S.; data curation: I.P., L.C. and A.L.; writing—original draft preparation: I.P. and K.B.; writing—review and editing: K.Z.-E.; visualisation: I.P. and K.B.; project administration: I.P. and funding acquisition: I.P. All authors have read and agreed to the published version of the manuscript.

Acknowledgments: The authors gratefully acknowledge the technical support provided by L. Ondruska, V. Parkanyi, R. Jurcik and J. Pecho from the National Agricultural and Food Centre, Research Institute for Animal Production, Nitra. The authors also thank Andrew Billingham for improving the written English of the manuscript.

References

1. Haselmeyer, A.; Zentek, J.; Chizzola, R. Effects of thyme as a feed additive in broiler chickens on thymol in gut contents, blood plasma, liver and muscle. *J. Sci. Food Agric.* **2015**, *95*, 504–508. [CrossRef] [PubMed]
2. Placha, I.; Ocelova, V.; Chizzola, R.; Battelli, G.; Gai, F.; Bacova, K.; Faix, S. Effect of thymol on the broiler chicken antioxidative defence system after sustained dietary thyme oil application. *Br. Poult. Sci.* **2019**, *60*, 589–596. [CrossRef] [PubMed]
3. Salehi, B.; Mishra, A.P.; Shukla, I.; Sharifi-Rad, M.; Contreras, M.D.M.; Segura-Carretero, A.; Fathi, H.; Nasrabadi, N.N.; Kobarfard, F.; Sharifi-Rad, J. Thymol, thyme, and other plant sources: Health and potential uses. *Phytother. Res.* **2018**, *32*, 1688–1706. [CrossRef] [PubMed]
4. Mehdipour, Z.; Afsharmanesh, M.; Sami, M. Effects of supplemental thyme extract (Thymus vulgaris L.) on growth performance, intestinal microbial populations, and meat quality in Japanese quails. *Comp. Clin. Pathol.* **2014**, *23*, 1503–1508. [CrossRef]
5. Popović, S.; Puvača, N.; Kostadinović, L.; Džinić, N.; Bošnjak, J.; Vasiljević, M.; Djuragic, O. Effects of dietary essential oils on productive performance, blood lipid profile, enzyme activity and immunological response of broiler chickens. *Eur. Poult. Sci.* **2016**, 80. [CrossRef]
6. Yu, Y.M.; Chao, T.Y.; Chang, W.C.; Chang, M.J.; Lee, M.F. Thymol reduces oxidative stress, aortic intimal thickening, and inflammation-related gene expression in hyperlipidemic rabbits. *J. Food Drug Anal.* **2016**, *24*, 556–563. [CrossRef]
7. Ocel'ová, V.; Chizzola, R.; Battelli, G.; Pisarcikova, J.; Faix, S.; Gai, F.; Placha, I. Thymol in the intestinal tract of broiler chickens after sustained administration of thyme essential oil in feed. *J. Anim. Physiol. Anim. Nutr.* **2018**, *103*, 204–209. [CrossRef]
8. Botsoglou, N.A.; Christaki, E.; Fletouris, D.J.; Florou-Paneri, P.; Spais, A.B. The effect of dietary oregano essential oil on lipid oxidation in raw and cooked chicken during refrigerated storage. *Meat Sci.* **2002**, *62*, 259–265. [CrossRef]
9. Rubió, L.; Macià, A.; Castell-Auví, A.; Pinent, M.; Blay, M.T.; Ardévol, A.; Romero, M.P.; Motilva, M.J. Effect of the co-occurring olive oil and thyme extracts on the phenolic bioaccessibility and bioavailability assessed by in vitro digestion and cell models. *Food Chem.* **2014**, *149*, 277–284. [CrossRef]

10. Ocel'ová, V.; Chizzola, R.; Pisarčíková, J.; Novak, J.; Ivanišinová, O.; Faix, Š. Effect of thyme essential oil supplementation on thymol content in blood plasma, liver, kidney and muscle in broiler chickens. *Nat. Prod. Commun.* **2016**, *11*, 1545–1550. [CrossRef]
11. EFSA (European Food Safety Authority). Conclusion on the peer review of the pesticide risk assessment of the active substance thymol. *EFSA J.* **2012**, *10*, 1–43.
12. Gad, M.M.E. Acute and Repeated-Doses (28 Days) Toxicity of Thymol Formulation in Male Albino Rats. *Aust. J. Basic Appl. Sci.* **2012**, *7*, 915–922.
13. EC (European Commission). Commission Regulation (EU) No 37/2010 of 22 December 2009 on pharmacologically active substances and their classification regarding maximum residue limits in foodstuffs of animal origin. *Off. J. Eur. Union* **2010**, *L15*, 1–72.
14. Irlbeck, N.A. How to feed the rabbit (Oryctolagus cuniculus) gastrointestinal tract. *J. Anim. Sci.* **2001**, *79*, E343–E346. [CrossRef]
15. Takada, M.; Agata, I.; Sakamoto, M.; Yagi, N.; Hayashi, N. On the metabolic detoxication of thymol in rabbit and man. *J. Toxicol. Sci.* **1979**, *4*, 341–350. [CrossRef]
16. de Blas, C.; Mateos, G.G. Feed formulation. In *Nutrition of the Rabbit*, 2nd ed.; de Blas, C., Wiseman, J., Eds.; CAB International: Wallingford, UK, 2010; pp. 222–232.
17. AOAC (Association of official analytical chemists). *International Official Methods of Analysis*, 17th ed.; AOAC: Arlington, VA, USA, 2000.
18. Van Soest, P.J.; Robertson, J.B.; Lewis, B. A Methods for dietary fiber, neutral detergent fiber, and nonstarch polysaccharides in relation to animal nutrition. *J. Dairy Sci.* **1991**, *74*, 3583–3597. [CrossRef]
19. Karamać, M.; Gai, F.; Peiretti, P.G. Effect of the growth stage of false flax (Camelina sativa L.) on the phenolic compound content and antioxidant potential of the aerial part of the plant. *Polish J. Food Nutr. Sci.* **2020**, *70*, 189–198. [CrossRef]
20. Pisarčíková, J.; Ocel'ová, V.; Faix, Š.; Plachá, I.; Calderón, A.I. Identification and quantification of thymol metabolites in plasma, liver and duodenal wall of broiler chickens using UHPLC-ESI-QTOF-MS. *Biomed. Chromatogr.* **2017**, *31*, 1–12. [CrossRef]
21. Paglia, D.E.; Valentine, W.N. Studies on the quantitative and qualitative characterization of erythrocyte glutathione peroxidase. *J. Lab. Clin. Med.* **1967**, *70*, 158–169.
22. Jo, C.; Ahn, D.U. Fluorometric analysis of 2-thiobarbituric acid reactive substances in turkey. *Poult. Sci.* **1998**, *77*, 475–480. [CrossRef]
23. Andrejčáková, Z.; Sopková, D.; Vlčková, R.; Kulichová, L.; Gancarčíková, S.; Almášiová, V.; Holovská, K.; Petrilla, V.; Krešáková, L. Synbiotics suppress the release of lactate dehydrogenase, promote non-specific immunity and integrity of jejunum mucosa in piglets. *Anim. Sci. J.* **2016**, *87*, 1157–1166. [CrossRef] [PubMed]
24. Bradford, M. A rapid and sensitive method for the quantitation of microgram quantities of protein utilising the principle of protein-dye binding. *Anal. Biochem.* **1976**, *72*, 248–254. [CrossRef]
25. Ouhayoun, J. Rabbit meat characteristics and qualitative variability. *Cuni-Sciences* **1992**, *7*, 1–15.
26. Fekete, S.; Lebas, F. Effect of a natural flavour (Thyme extract) on the spontaneous feed ingestion, digestion coefficients and fattening parameters. *Magy. Allatorv. Lapja* **1983**, *38*, 121–125.
27. Erdelyi, M.; Matics, Z.; Gerencsér, Z.; Princz, Z.; Szendrő, Z.; Mézes, M. Study of the effect of rosemary (Rosmarinus officinalis) and garlic (Allium sativum) essential oils on the performance of rabbit. In *Nutrition and Digestive Physiology, Proceedings of the 9th World Rabbit Congress, Verona, Italy, 10–13 June 2008*; Xicato, G., Trocino, A., Lukefahr, S.D., Eds.; World Rabbit Science Association: Castanet-Tolosan, France, 2008.
28. Windisch, W.; Schedle, K.; Plitzner, C.; Kroismayr, A. Use of phytogenic products as feed additives for swine and poultry. *J. Anim. Sci.* **2008**, *86*, E140–E148. [CrossRef] [PubMed]
29. Gerencsér, Z.; Szendro, Z.; Matics, Z.; Radnai, I.; Kovács, M.; Nagy, I.; Cullere, M.; Dal Bosco, A.; Dalle Zotte, A. Effect of dietary supplementation of spirulina (Arthrospira platensis) and thyme (Thymus vulgaris) on apparent digestibility and productive performance of growing rabbits. *World Rabbit Sci.* **2014**, *22*, 1–9. [CrossRef]
30. Ocel'ová, V. Plant Additives in Relation to the Animal Gastrointestinal Tract and Metabolism of Their Main Compounds. Ph.D Thesis, Institute of Animal Physiology, Slovak Academy of Sciences, University of Veterinary Medicine and Pharmacy in Kosice, Košice, Slovakia, 2017.
31. Lebas, F.; Gidene, T. Feeding Behaviour in Rabbits. In Proceedings of the III International Rabbit Production Symposium, Villareal, Portugal, 2 November 2005.

32. Kohlert, C.; Schindler, G.; März, R.W.; Abel, G.; Brinkhaus, B.; Derendorf, H.; Gräfe, E.U.; Veit, M. Systemic availability and pharmacokinetics of thymol in humans. *J. Clin. Pharmacol.* **2002**, *42*, 731–737. [CrossRef]
33. Rubió, L.; Serra, A.; Macià, A.; Borràs, X.; Romero, M.P.; Motilva, M.J. Validation of determination of plasma metabolites derived from thyme bioactive compounds by improved liquid chromatography coupled to tandem mass spectrometry. *J. Chromatogr. B* **2012**, *905*, 75–84. [CrossRef]
34. Van den Hoven, R.; Zappe, H.; Zitterl-Eglseer, K.; Jugl, M.; Franz, C. Study of the effect of Bronchipret on the lung function of five Austrian saddle horses suffering recurrent airway obstruction (heaves). *Vet. Rec.* **2003**, *152*, 555–557. [CrossRef]
35. Zitterl-Eglseer, K.; Wetscherek, W.; Stoni, A.; Kroismayr, A.; Windisch, W. Bioavailability of essential oils of a phytobiotic feed additive and impact of performance and nutrient digestibility in weaned piglets. *Bodenkult.-J. Land Manag. Food Environ.* **2008**, *59*, 121–129.
36. Bardal, S.K.; Waechter, J.E.; Martin, D.S. *Applied Pharmacology*; Elsevier Saunders: Louis, MO, USA, 2011; pp. 17–34.
37. Gumus, R.; Ercan, N.; Imik, H. The effect of thyme essential oil (Thymus Vulgaris) added to quail diets on performance, some blood parameters, and the antioxidative metabolism of the serum and liver tissues. *Rev. Bras. Cienc. Avic.* **2017**, *19*, 297–304. [CrossRef]
38. Villamide, M.J.; Nicodemus, N.; Fraga, M.J.; Carabaño, R. Protein Digestion. In *Nutrition of the Rabbit*, 2nd ed.; de Blas, C., Wiseman, J., Eds.; CAB International: Wallingford, UK, 2010; pp. 222–232.
39. Dalle Zotte, A.; Sartori, A.; Bohatir, P.; Rémignon, H.; Ricci, R. Effect of dietary supplementation of Spirulina (Arthrospira platensis) and Thyme (Thymus vulgaris) on growth performance, apparent digestibility and health status of companion dwarf rabbits. *Livest. Sci.* **2013**, *152*, 182–191. [CrossRef]
40. Bozkurt, M.; Küçükyilmaz, K.; Çatli, A.U.; Özyildiz, Z.; Çinar, M.; Çabuk, M.; Çoven, F. Influences of an essential oil mixture supplementation to corn versus wheat-based practical diets on growth, organ size, intestinal morphology and immune response of male and female broilers. *Ital. J. Anim. Sci.* **2012**, *11*, e54. [CrossRef]
41. Gasco, L.; Dabbou, S.; Trocino, A.; Xiccato, G.; Capucchio, M.T.; Biasato, I.; Dezzutto, D.; Birolo, M.; Meneguz, M.; Schiavone, A.; et al. Effect of dietary supplementation with insect fats on growth performance, digestive efficiency and health of rabbits. *J. Anim. Sci. Biotechnol.* **2019**, *10*, 4. [CrossRef]
42. Jayakumar, S.; Madankumar, A.; Asokkumar, S.; Raghunandhakumar, S.; Gokula Dhas, K.; Kamaraj, S.; Divya, M.G.; Devaki, T. Potential preventive effect of carvacrol against diethylnitrosamine-induced hepatocellular carcinoma in rats. *Mol. Cell. Biochem.* **2012**, *360*, 51–60. [CrossRef]
43. Xiccato, G. Fat digestion. In *Nutrition of the Rabbit*, 2nd ed.; de Blas, C., Wiseman, J., Eds.; CAB International: Wallingford, UK, 2010; pp. 222–232.
44. Niki, E.; Yoshida, Y.; Saito, Y.; Noguchi, N. Lipid peroxidation: Mechanisms, inhibition, and biological effects. *Biochem. Biophys. Res. Commun.* **2005**, *338*, 668–676. [CrossRef]
45. Dabbou, S.; Renna, M.; Lussiana, C.; Gai, F.; Rotolo, L.; Kovitvadhi, A.; Brugiapaglia, A.; Helal, A.N.; Schiavone, A.; Zoccarato, I.; et al. Bilberry pomace in growing rabbit diets: Effects on quality traits of hind leg meat. *Ital. J. Anim. Sci.* **2017**, *16*, 371–379. [CrossRef]

12

Application of Meta-Analysis and Machine Learning Methods to the Prediction of Methane Production from In Vitro Mixed Ruminal Micro-Organism Fermentation

Jennifer L. Ellis [1,*], Héctor Alaiz-Moretón [2], Alberto Navarro-Villa [3,4], Emma J. McGeough [3,5], Peter Purcell [3], Christopher D. Powell [1], Padraig O'Kiely [3], James France [1] and Secundino López [6,*]

1 Centre for Nutrition Modelling, Department of Animal Biosciences, University of Guelph, 50 Stone Road East, Guelph, ON N1G 2W1, Canada; cpowell@uoguelph.ca (C.D.P.); jfrance@uoguelph.ca (J.F.)
2 Departamento de Ingeniería Eléctrica de Sistemas y Automática, Escuela de Ingeniería Industrial e Informática, Universidad de León, Campus Universitario de Vegazana, 24071 León, Spain; hector.moreton@unileon.es
3 Animal & Grassland Research and Innovation Centre, Teagasc, Grange, Dunsany, Co. Meath C15 PW93, Ireland; alberto.navarro.villa@trouwnutrition.com (A.N.-V.); emma.mcgeough@umanitoba.ca (E.J.M.); peter.purcell86@gmail.com (P.P.); Padraig.OKiely@teagasc.ie (P.O.)
4 Trouw Nutrition R&D, Ctra. CM-4004 km 10.5, 45950 El Viso de San Juan, Spain
5 Department of Animal Science, University of Manitoba, Winnipeg, MB R3T 2N2, Canada
6 Instituto de Ganadería de Montaña (IGM), CSIC-Universidad de León, Departamento de Producción Animal, Universidad de León, 24007 León, Spain
* Correspondence: jellis@uoguelph.ca (J.L.E.); s.lopez@unileon.es (S.L.);

Simple Summary: In vitro gas production systems are regularly utilized to screen feed ingredients for inclusion in ruminant diets. However, not all in vitro systems are set up to measure methane (CH_4) production, nor do all papers report in vitro CH_4. Therefore, the objective of this study was to develop models to predict in vitro production of CH_4, a greenhouse gas produced by ruminants, from in vitro gas and volatile fatty acid (VFA) production data, and to identify the major drivers of CH_4 production in these systems. Meta-analysis and machine learning (ML) methodologies were applied to predict CH_4 production from in vitro gas parameters. Meta-analysis results indicate that equations containing apparent dry matter (DM) digestibility, total VFA production, propionate, valerate and feed type (forage vs. concentrate) resulted in best prediction of CH_4. The ML models far exceeded the predictability achieved using meta-analysis, but further evaluation on an external database would be required to assess their generalization capacity. The models developed can be utilized to estimate CH_4 emissions in vitro.

Abstract: In vitro gas production systems are utilized to screen feed ingredients for inclusion in ruminant diets. However, not all in vitro systems are set up to measure methane (CH_4) production, nor do all publications report in vitro CH_4. Therefore, the objective of this study was to develop models to predict in vitro CH_4 production from total gas and volatile fatty acid (VFA) production data and to identify the major drivers of CH_4 production in these systems. Meta-analysis and machine learning (ML) methodologies were applied to a database of 354 data points from 11 studies to predict CH_4 production from total gas production, apparent DM digestibility (DMD), final pH, feed type (forage or concentrate), and acetate, propionate, butyrate and valerate production. Model evaluation was performed on an internal dataset of 107 data points. Meta-analysis results indicate that equations containing DMD, total VFA production, propionate, feed type and valerate resulted in best predictability of CH_4 on the internal evaluation dataset. The ML models far exceeded the predictability achieved using meta-analysis, but further evaluation on an external database would be required to assess generalization ability on unrelated data. Between the ML methodologies assessed, artificial neural networks and support vector regression resulted in very similar predictability, but differed in fitting, as assessed by behaviour analysis. The models developed can be utilized to estimate CH_4 emissions in vitro.

Keywords: in vitro gas production; methane; rumen; feed; meta-analysis; machine learning; neural network

---

1. Introduction

Globally, greenhouse gas (GHG) emissions from the agriculture, forestry and other land use (AFOL) sector account for ~23% of the global anthropogenic GHG total emissions [1], with enteric methane (CH_4) from fermentation in the forestomach of ruminants representing 32%–40% of that total [1] (thereby 7.4%–9.2% of the global anthropogenic total). From the farmer's perspective, CH_4 also represents an energy loss and an inefficiency of production, ranging from approximately 3.0 (feedlot cattle) to 7.0 (forage fed cattle) percent of gross energy intake, with a ±20% uncertainty [2]. As a result, and to meet public expectation for sustainably produced food products, the agriculture sector has mobilized to examine a large array of potential CH_4 (as well as N and P excretion) mitigation strategies [3–5], to reduce the environmental impact of livestock and food production.

At the animal level, CH_4 is produced as a byproduct of anaerobic fermentation in the rumen and hindgut of ruminants, whereby methanogens utilize H_2 to obtain ATP by reducing CO_2 to CH_4 [6]. The removal of H_2 through methanogenesis, the main H-sink in the rumen [6], prevents the inhibitory effect of H_2 on ruminal fermentation and allows for the degradation and fermentation of feed to proceed. When methanogenesis is reduced, other pathways must be promoted to utilize H_2 or otherwise fermentation, digestibility and intake may be negatively affected [6].

As animal experiments to evaluate feedstuffs and feed additives are costly, time consuming and do not guarantee conclusive outcomes, the in vitro gas production technique represents a viable option for prescreening or screening of feedstuffs/additives for potential inclusion in the ration of modern dairy cows, beef cattle and other ruminants. However, CH_4 is often, but not always, included in the gases measured during in vitro incubation (particularly in developing countries where equipment may be unaffordable, unavailable or limited, for example). A reliable measure of CH_4 from in vitro cultures of mixed ruminal micro-organisms would be a useful tool to assess the potential dietary effects on methanogenesis. Estimation of CH_4 from the output of other fermentation end-products commonly measured in vitro could be a suitable alternative, and Jayanegara [7] proposed the use of the stoichiometric equations of Hegarty and Nolan [8] and of Moss et al. [9] to predict in vitro CH_4. However, using this approach CH_4 was generally overpredicted, presumably because in vitro H_2 recovery observed in practice was substantially less than that assumed by the stoichiometric models. The objectives of this study were therefore to: (1) to develop empirical models to predict in vitro CH_4 production from in vitro gas production measures—via meta-analysis (multiple linear regression) and machine learning (ML) methods (artificial neural networks, ANN, and support vector regression, SVR), and (2) to identify the fermentation parameters most closely related to CH_4 production in vitro.

2. Materials and Methods

2.1. *Database*

The database compiled for this study consisted of 397 in vitro rumen fermentation bottle means (each the average of 3–5 replicate measurements), taken after 24 h of incubation, from 13 experiments reported in 10 publications [10–19] (experiments 1–3 were from publication [10]), plus 1 unpublished study [20]. As a result, experimental animals were not directly employed in this study. In accordance with the National Centre for the Replacement Refinement and Reduction of Animals in Research (NC3Rs), per Directive 2010/63/EU, all study data used were publicly available (with the exception of the one unpublished study) as reported in the aforementioned articles. Studies evaluated the in vitro gas and CH_4 production from oven-dried feedstuffs, including ryegrass, forbs, grass silages, clover, maize silage and other whole-crop cereal silages and concentrate feeds (no feed additives or rumen modifiers were included in the database). Feed type (FT) was categorized as either forage (FT = 1) or concentrate (FT = 2). The database included in vitro measurements of CH_4 gas production (CH_4i, mL/g DM incubated, and CH_4d, mL/g DM apparently digested), total gas production (TGP, mL/g DM incubated), apparent DM digestibility (DMD, g/g), volatile fatty acid production (VFA, mmol/g DM incubated), molar proportions of acetic acid (AC, mmol/mol VFA), propionic acid (PR, mmol/mol VFA), butyric acid (BT, mmol/mol VFA) and valeric acid (VL, mmol/mol VFA), the acetate to propionate ratio (C2C3) and the incubation medium final pH (pH). Daily production of each volatile fatty acid (ACp, PRp, BTp or VLp for mmol AC, PR, BT or VL produced per g DM incubated, respectively) was calculated from total VFA and the corresponding molar proportions. Variable abbreviations, units and descriptions are also summarized in Appendix A (Table A1). When digestibility is measured during in vitro batch cultures of mixed ruminal micro-organisms, it is assumed that DM disappearance after the incubation time (in this particular case 24 h) is an acceptable metric of apparent DM digestibility. Due to missing data, two studies [11,12], were removed from the database, leaving 354 observations from 11 experiments.

For model development and evaluation purposes, the dataset (n = 354) was divided into two subsets, the first one for training and model development purposes (70% of data, n = 247, with 4 outlier data points removed for meta-analysis), and the second one for model testing and evaluation (internal evaluation) purposes (30% of data, n = 107). Aside from 4 data points which were removed for the meta-analysis (statistical outliers), the 'two' developmental datasets were identical. Division of data points into the training or evaluation datasets was via random assignment, but each contained a proportional number of observations relative to the FT variable. Descriptive statistics for the training and evaluation datasets are provided in Table 1.

Table 1. Summary of the training (n = 247, 243) and internal evaluation (n = 107) datasets.

Variable [1]	pH	DMD	TGP	CH_4i	CH_4d	VFA	AC	PR	BT	VL	C2C3
Training Dataset (Machine learning, n = 247)											
Mean	6.59	0.67	163	26.6	40.7	5.48	632.3	238.5	94.8	31.8	2.7
Median	6.64	0.68	160	25.3	38.0	5.33	632.2	237.7	94.4	31.8	2.7
Minimum	5.45	0.20	51	6.9	19.5	2.12	477.6	117.5	43.4	5.0	1.4
Maximum	6.78	0.91	276	50.8	71.5	9.79	812.1	346.9	181.3	79.3	7.0
Training Dataset (Meta-analysis, n = 243)											
Mean	6.59	0.67	164	26.7	40.6	5.50	631.7	239.2	94.9	31.9	2.7
Median	6.64	0.68	161	25.3	38.0	5.35	632.1	237.9	94.4	31.8	2.7
Minimum	5.45	0.22	71	11.2	19.5	2.12	477.6	117.5	60.0	10.3	1.4
Maximum	6.78	0.91	276	50.8	71.5	9.79	812.1	346.9	181.3	79.3	7.0
Evaluation Dataset (Machine learning/Meta-Analysis, n = 107)											
Mean	6.60	0.66	162	26.0	40.2	5.51	630.6	241.5	94.5	31.8	2.7
Median	6.64	0.68	162	25.4	37.9	5.25	629.0	238.5	96.4	32.5	2.7
Minimum	5.49	0.21	61	8.7	22.0	4.02	503.7	152.1	53.8	6.9	1.7
Maximum	6.84	0.87	246	43.0	70.9	9.81	787.5	333.1	178.9	60.8	5.1

[1] Variables (units): pH = final pH in the incubation medium. DMD = apparent dry matter (DM) digestibility (g DM disappeared/g DM incubated). TGP = total gas production (mL gas/g DM incubated). CH_4i = methane production (mL CH_4/g DM incubated). CH_4d = methane production (mL CH_4/g DM apparently digested). VFA production (mmol total VFA/g DM incubated). AC = acetic acid (mmol AC/mol VFA). PR = propionic acid (mmol PR/mol VFA). BT = butyric acid (mmol BT/mol VFA). VL = valeric acid (mmol VL/mol VFA). C2C3 = acetate to propionate ratio.

Independent testing of the model gives a measure of the model's 'generalization ability' ('test error'), or the ability to make predictions on unseen data. This is particularly important for some ML approaches, which may achieve very accurate predictions, but essentially model the noise in the data.

2.2. Model Fitting—Meta-Analysis

The main effects of in vitro fermentation variables (TGP, DMD, VFA, AC, ACp, PR, PRp, BT, BTp, VL, VLp, C2C3 and final pH) were analyzed for inclusion in predictive models using the PROC MIXED procedure of SAS [21] to predict CH_4i (mL/g incubated DM), or CH_4d (mL/g DM apparently digested). Equations were fitted to the training dataset (Table 1—meta-analysis).

The mixed model analysis was chosen because the data were compiled from multiple studies, and thus the experiment was considered as a random effect [22]. If, when running the model, the random covariance or the random slope was not significant, they were removed from the model or simplified [22], though the random intercept term was always retained. The dual quasi-Newton technique was used for optimization with an adaptive Gaussian quadrature as the integration method. Normal distribution of the random study effect was assessed via Q–Q distribution plot, and normality of residuals via examination of the residual plots (PROC MIXED).

Three approaches were taken to fitting mixed models to this dataset: (1) univariate analysis of each dependent–independent variable combination (explanatory variable in linear, quadratic or cubic form); (2) multivariate analysis, preceded by examination in PROC REG (MaxR) and assessment for collinearity between driving variables in PROC CORR/visual plotting; and (3) multivariate analysis based on known biological principles. Approaches (2) and (3) are not distinguished/presented separately in the results, as both are considered 'multivariate'. A fourth approach was also included for comparison with the ML models (described below): (4) where all driving variables were included, irrespective of significance or collinearity (linear equation form). With the exception of approach (4), only equations with significant slope parameters ($p < 0.05$) and normally distributed residuals/random effects were retained and evaluated.

2.3. Model Fitting—Machine Learning

The in vitro fermentation variables (TGP, DMD, VFA, AC, PR, BT, VL, C2C3 and final pH) were retained as potential driving variables for development of ML-based predictive models for CH_4i and CH_4d. Predictive models were fitted on the training dataset (Table 1—Machine learning). The raw dataset was subjected to a preprocessing normalization process (standard scalar) [23] according to:

$$Z = (X - u)/S \tag{1}$$

where Z is the normalized value, X is the raw value, u is the mean of the training samples and S is the standard deviation of the training samples. The objective of this normalization step was to improve the convergence of the training process in the regression methods utilized [24]. Subsequently, two ML techniques (support vector regression and artificial neural network) were implemented using the Scikit-learn software library [25] for the Python programming language [26]. For both ML approaches, a 10-fold cross-validation procedure was used to fit the predictive models to the training dataset ($n = 247$; the evaluation database, $n = 107$, was therefore not included in this analysis). The training dataset was subsequently randomly split into 10 equal subgroups, and the model was trained using nine of the subsets and validated on the remaining one part of the data to compute a performance measure. This holdout process was repeated for each of the 10-folds, such that each subset was utilized for validation, whereas the other nine subsets were pooled for the training, in turn. The error estimation was averaged over the 10 iterations to assess the fit performance.

2.3.1. Support Vector Regression

Support Vector Machine (SVM) is a ML technique based on supervised learning with a modality oriented for regression problems, namely Support Vector Regression (SVR), able to forecast continuous variables [27] (in this case, CH_4 from in vitro cultures of mixed ruminal micro-organisms). The SVR method transforms the input data (previously normalized) into a multidimensional space by using nonlinear mapping, and a linear regression procedure is applied to each hyperplane obtained to calculate the desired output. The SVR method is developed by changing the kernel function and tuning the parameters C (the regularization parameter), γ (the kernel coefficient), Tol (tolerance for stopping criterion) and degree of the polynomial. Three 'kernel' functions were considered—linear, radial basis function and polynomial. The ranges of values used for the parameter optimization were $C \in \{1, 10, 100, 1000\}$, Tol $\in \{0.1, 0.01\}$, $\gamma \in \{1, 0.1, 0.01, 0.001\}$ and degree (only for the polynomial function) $\in \{2, 3, 4, 5, 6\}$. Grid search combined with cross-validation [27] were used to achieve the best combination of parameters resulting in the optimal and most robust SVR model solution, on the basis of the ε-insensitive loss function. The best SVR models for both variables to be predicted (CH_4i or CH_4d) were obtained using the radial basis function as kernel, with the parameter values $C = 1000$, Tol $= 0.1$ and $\gamma = 1$.

2.3.2. Artificial Neural Network—Multilayer Perceptron

A multilayer perceptron (MLP) is a machine learning method based on supervised learning, and is a specific topology of a feedforward artificial neural network (ANN) [28]. The MLP network used for the current study was composed of three layers of nodes: the input layer, one hidden layer and an output layer. Achieving the optimal MLP architecture can require tuning a number of hyperparameters such as the number of hidden layers, neurons or iterations. For the current study, one hidden layer was applied and the rectified linear unit (ReLU) nonlinear activation function was implemented in each node (neuron) of this hidden layer (except the input nodes). On the other hand, the single neuron of the output layer utilized the linear activation function [28]. The training procedure was based on the backpropagation technique, using grid search combined with cross-validation [28] to derive the best combination of parameters resulting in the optimal and most robust MLP model solution. The square-error loss function and the limited-memory Broyden–Fletcher–Goldfarb–Shanno (L-BFGS) numerical method were used for optimization. The number of hidden neurons in the best MLP models was 26 and 27 for CH_4i and CH_4d predictions, respectively, (the range tested was from 12–30 neurons in the hidden layer). Other hyperparameters were included in the grid for tuning aiming to optimize the training of the ANN. The best results were observed when early stopping was activated (to prevent overfitting), any prior attributes stored on the estimator were cleared ("warm start" disabled), initial learning rate was set at 10^{-7} and kept constant, and the batch size for each iteration was equal to 1.

2.4. Model Evaluation

Model predictions developed in the current study (via meta-analysis, SVR or ANN) were evaluated using an independent data subset (internal evaluation, as the data are independent but related to the training dataset), described in Section 2.1 and in Table 1. Models were evaluated for their predictability using mean square prediction error (MSPE), calculated as:

$$\text{MSPE} = \sum_{i=1}^{n} \frac{(O_i - P_i)^2}{n} \tag{2}$$

where O_i is the observed value, P_i is the predicted value and n is the number of observations. Square root of the MSPE (RMSPE), expressed as a proportion of the observed mean (RMSPE, %), gives an estimate of the overall prediction error. The RMSPE was decomposed into random (disturbance) error

(ED), error due to deviation of the regression slope from unity (ER), and error in central tendency due to overall bias (EB) [29]. The EB, ER and ED fractions of MSPE were calculated as:

$$\mathrm{EB} = \left(\overline{P} - \overline{O}\right)^2 \tag{3}$$

$$\mathrm{ER} = \left(S_p - R \times S_o\right)^2 \tag{4}$$

$$\mathrm{ED} = \left(1 - R^2\right) \times {S_o}^2 \tag{5}$$

where $\overline{P}$ and $\overline{O}$ are the predicted and observed means, S_p is the standard deviation of predicted values, S_o is the standard deviation of observed values and R is the Pearson correlation coefficient.

Correspondence between predicted and observed values was also assessed by the concordance correlation coefficient (CCC) [30], which was calculated as:

$$\mathrm{CCC} = R \times C_b \tag{6}$$

where C_b is a bias correction factor (a measure of accuracy), and R is the Pearson correlation coefficient (a measure of precision). The C_b variable is calculated as:

$$C_b = \frac{2}{\left(\nu + \frac{1}{\nu} + \mu^2\right)} \tag{7}$$

where

$$\nu = \frac{S_o}{S_p} \tag{8}$$

$$\mu = \frac{\overline{O} - \overline{P}}{\sqrt{\left(S_o - S_p\right)}} \tag{9}$$

so that ν provides a measure of scale shift, while μ provides a measure of location shift. The ν value indicates the change in standard deviation, if any, between predicted and observed values. A positive μ value indicates underprediction, while a negative μ indicates overprediction. Predictions were further evaluated visually against observations (via predicted vs. observed plots) as well as against residuals (residual vs. predicted, not shown).

As one criticism of many ML methodologies remains their lack of transparency (i.e., no predictive equation is produced), models developed via ANN and SVR were further evaluated using behaviour analysis, where model inputs were systematically altered ±10% (in isolation) and the model's 'behavioural' response (% change in output prediction) was assessed (direction and magnitude).

3. Results

3.1. Correlation Matrix Analysis

Potential X variables were evaluated against each other via correlation matrix analysis, to determine the extent of collinearity between X variables (Table 2). X variables that were highly collinear with each other (correlation >0.500) are highlighted in grey (Table 2). X variables that were highly collinear included TGP and DMD, TGP and VFA/ACp/PRp, VFA and ACp/PRp/BTp, AC and PR/PRp, ACp and PRp/BTp, PR and PRp, PRp and BTp, BT and BTp, pH and BTp. These combinations were therefore avoided in multivariate meta-analysis equation development.

Correlation analysis was also used to examine potential correlations between X and Y variables (Table 2). The X variables moderately correlated (>0.300) with CH_4d included DMD (−0.408), AC (0.405), ACp (0.350) and PR (−0.396), while X variables moderately correlated (>0.300) with CH_4i included DMD (0.399), TGP (0.755), VFA (0.472), ACp (0.472), PRp (0.326) and BTp (0.315).

Table 2. Correlation matrix (Pearson correlation coefficients, *R*) of dependent and independent variables, meta-analysis development dataset (n = 243) [1,2].

Variable	CH_4d	CH_4i	pH	DMD	TGP	VFA	AC	ACp	PR	PRp	BT	BTp	VL	VLp
pH	0.027	0.007												
DMD	−0.408	0.399	−0.060											
TGP	0.146	0.755	−0.206	0.707										
VFA	0.223	0.472	−0.437	0.282	0.648									
AC	0.405	−0.014	0.208	−0.474	−0.341	−0.159								
ACp	0.350	0.472	−0.340	0.136	0.544	0.944	0.168							
PR	−0.396	−0.045	−0.125	0.426	0.316	0.094	−0.863	−0.184						
PRp	−0.050	0.326	−0.431	0.439	0.663	0.817	−0.622	0.605	0.640					
BT	−0.035	0.045	−0.360	0.065	0.115	0.250	−0.279	0.136	−0.150	0.143				
BTp	0.103	0.315	−0.566	0.234	0.482	0.798	−0.294	0.679	−0.001	0.633	0.762			
VL	−0.113	0.223	0.273	0.369	0.096	0.007	−0.246	−0.071	0.023	0.016	0.145	0.071		
VLp	0.059	0.458	−0.021	0.439	0.435	0.555	−0.279	0.458	0.069	0.465	0.236	0.491	0.820	
C2C3	0.420	−0.038	0.070	−0.516	−0.373	−0.103	0.921	0.194	−0.924	−0.606	−0.078	−0.133	−0.186	−0.198

[1] Statistical significance of *R* values (n = 243): $p < 0.05$ if $|R| > 0.126$, $p < 0.01$ if $|R| > 0.165$, $p < 0.001$ if $|R| > 0.210$. Light grey boxes (for pairwise correlations between *X* variables) show $|R| > 0.500$, light blue boxes (for correlations between *X* and *Y* variables) show $|R| > 0.300$, [2] Variables (units): CH_4i = methane production (mL CH_4/g DM incubated). CH_4d = methane production (mL CH_4/g DM apparently digested). pH = final pH in the incubation medium. DMD = apparent dry matter (DM) digestibility (g DM disappeared/g DM incubated). TGP = total gas production (mL gas/g DM incubated). VFA production (mmol total VFA/g DM incubated). AC = acetic acid (mmol AC/mol VFA). ACp = AC production (mmol AC/g DM incubated). PR = propionic acid (mmol PR/mol VFA). PRp = PR production (mmol PR/g DM incubated). BT = butyric acid (mmol BT/mol VFA). BTp = BT production (mmol BT/g DM incubated). VL = valeric acid (mmol VL/mol VFA). VLp = VL production (mmol VL/g DM incubated). C2C3 = acetate to propionate ratio.

3.2. Univariate Meta-Analysis Models

Seventy-eight univariate equations to predict CH_4d or CH_4i were developed and evaluated with the variables presented in Table 2, in linear, quadratic or cubic form. Those with nonsignificant slope parameters or model fitting (fixed or random) problems were discarded, and the remaining equations (n = 22) were assessed on the evaluation dataset. On average, the CH_4i outcome was predicted with higher CCC and lower RMSPE values compared to the CH_4d outcome (Table 3). The best six performing equations have their model evaluation results presented in Table 3. The best performing univariate equations included the X variables ACp, PRp, DMD, VLp, VFA and TGP. The best performing univariate equations were those predicting CH_4i with TGP as a driving X variable, with a CCC on the evaluation database of 0.644 (quadratic) and 0.650 (linear) (Table 3).

Table 3. Univariate equations—meta-analysis model evaluation via root mean square prediction error (RMSPE) and concordance correlation coefficient (CCC) analysis on the internal evaluation (n = 107) database.

Equation [1]	Y	X	Form	Mean [2]	SEM	RMSPE, %	EB, %	ER, %	ED, %	CCC	R	C_b
U1	CH_4d	C2C3	Linear	44.7	0.37	25.7	19	2	79	0.113	0.194	0.579
U2	CH_4d	PR	Quad	44.7	0.35	25.9	18	3	78	0.116	0.185	0.623
U3	CH_4d	AC	Linear	44.8	0.29	25.6	20	1	79	0.124	0.228	0.541
U4	CH_4d	PR	Linear	44.8	0.31	26.1	19	4	77	0.128	0.198	0.649
U5	CH_4d	DMD	Quad	44.3	0.23	24.0	18	2	81	0.182	0.391	0.466
U6	CH_4d	DMD	Linear	44.4	0.24	23.9	19	3	78	0.196	0.432	0.454
U7	CH_4i	PRp	Quad	27.7	0.29	21.0	11	3	87	0.303	0.377	0.803
U8	CH_4i	DMD	Cubic	28.2	0.19	21.8	15	4	80	0.305	0.375	0.813
U9	CH_4i	VLp	Quad	28.0	0.26	20.7	14	0	85	0.314	0.420	0.747
U10	CH_4i	VFA	Linear	27.4	0.16	20.9	7	7	86	0.346	0.390	0.889
U11	CH_4i	TGP	Quad	27.2	0.33	15.5	10	1	90	0.644	0.717	0.898
U12	CH_4i	TGP	Linear	27.3	0.31	15.5	10	0	89	0.650	0.717	0.906

[1] Equation ID corresponds to equations presented in subsequent tables, with Y as the response (predicted) variable [either CH_4i (observed mean 26.0 ± 0.53 mL CH_4/g DM incubated) or CH_4d (observed mean 40.2 ± 0.91 mL CH_4/g DM apparently digested) and X as the explanatory variable (see Tables 1 and 2 for abbreviations and units of each variable), [2] Mean = mean of predicted values; SEM = standard error of the mean of predicted values; RMSPE = root mean square prediction error expressed as a percentage of the observed mean; EB, ER and ED = error due to bias, regression and disturbance, respectively (all as % of total MSPE); CCC = concordance correlation coefficient; R = Pearson correlation coefficient (measure of precision); C_b = bias correction factor (measure of accuracy).

The results of univariate equation development (Table 3) agreed roughly with the correlation analysis (Table 2), where the variables most highly correlated with CH_4i (TGP, VFA, ACp, VLp, DMD) and CH_4d (DMD, AC, C2C3, PR) (Table 2) appeared in the best performing univariate equations (Table 3). Some differences were evident, for example in the R-values, which may be explained by the difference in approach (correlation across all data points vs. correlation within study). The best performing univariate equations (U6, U12) were as follows:

$$CH_4d \text{ (mL } CH_4\text{/g DM digested)} = 58.52\ (\pm\ 3.210) - 21.24\ (\pm\ 3.045) \times \text{DMD (DM digestibility) (U6)} \quad (10)$$

$$CH_4i \text{ (mL } CH_4\text{/g DM incubated)} = 3.00\ (\pm\ 1.546) + 0.149\ (\pm\ 0.005) \times \text{TGP (mL/g DM incubated) (U12)} \quad (11)$$

3.3. Multivariate Meta-Analysis Models

Seventy-two multivariate equations, to predict CH_4d or CH_4i, were developed and evaluated with the variables presented in Table 2, in linear combinations. Those with nonsignificant slope parameters, model fitting problems (fixed or random) or had multiple X variables which were previously deemed to be collinear (Table 2) were discarded, and the remaining equations (n = 29) were evaluated on the evaluation dataset. Evaluation of the top six performing multivariate equations (for each of CH_4d and CH_4i) is reported in Table 4.

Best performing equations for CH_4d included (1) equation M5 (CCC = 0.419) with DMD, VFA, PR, FT and VL as *X* variables, as well as (2) M6 (CCC = 0.425) with DMD and VFA as *X* variables (Table 4). Best performing equations for CH_4i included (1) equation M11 (CCC = 0.438) with VFA and FT as *X* variables, and (2) equation M12 (CCC = 0.703) with PR, VL and TGP as *X* variables (Table 4).

The overall best performing equations (from univariate or multivariate origin, CH_4d, CH_4i) are presented in Table 5, and their predicted vs. observed plots are illustrated in Figure 1.

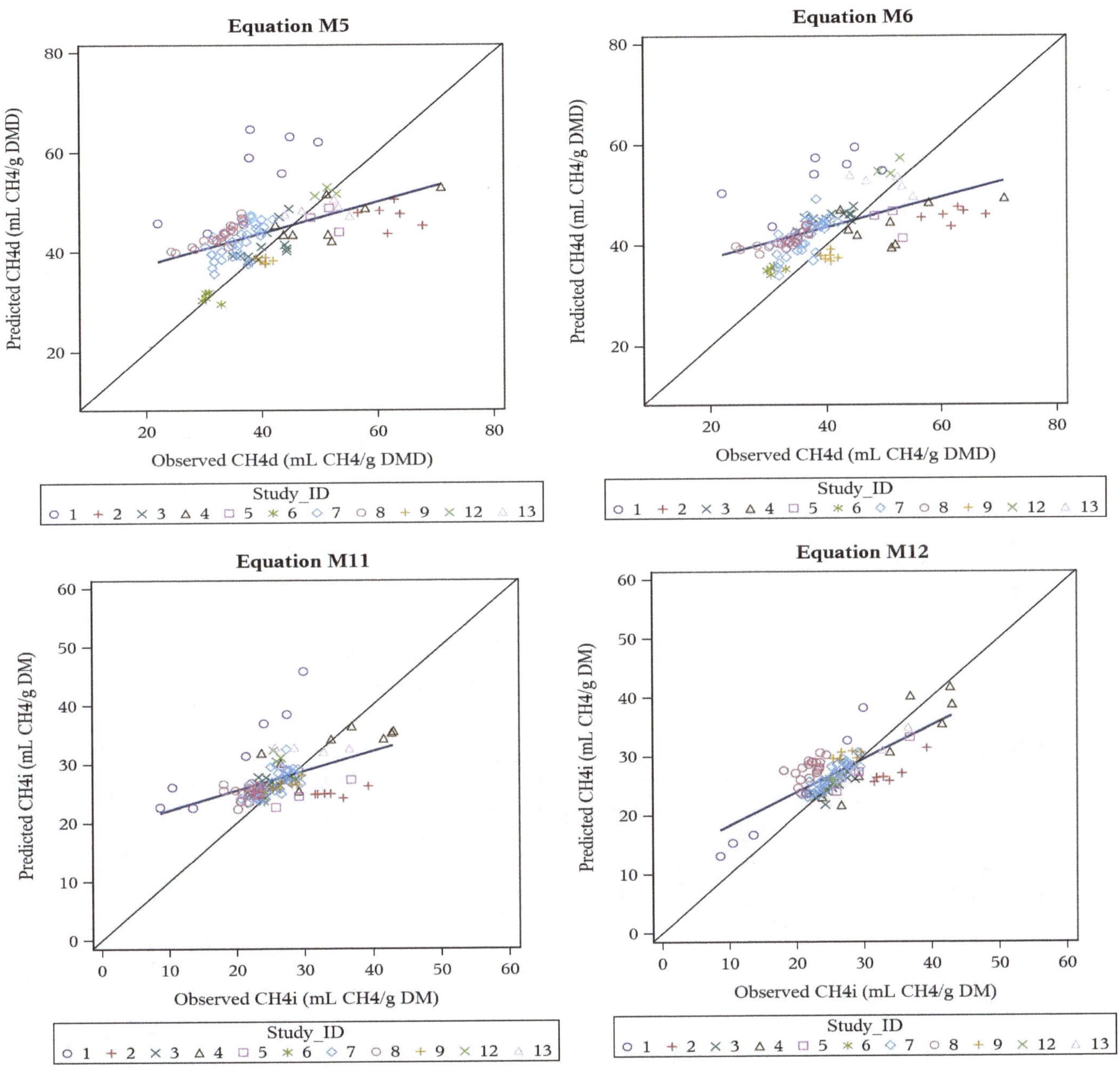

Figure 1. Predicted vs. observed plots for the top four performing meta-analysis equations M5 (CH_4d), M6 (CH_4d), M11 (CH_4i) and M12 (CH_4i), as evaluated on the evaluation dataset (n = 107).

Table 4. Linear multivariate equations—meta-analysis model evaluation via root mean square prediction error (RMSPE) and concordance correlation coefficient (CCC) analysis on the internal evaluation (n = 107) database.

Equation [1]	*Y*	*X*	Mean [2]	SEM	RMSPE, %	EB, %	ER, %	ED, %	CCC	*R*	C_b
M1	CH_4d	BTp, DMD	44.0	0.36	22.7	17	1	82	0.306	0.476	0.643
M2	CH_4d	PRp, VLp, DMD	43.9	0.51	22.6	16	1	83	0.379	0.473	0.800
M3	CH_4d	PR, VL, VFA, DMD	43.8	0.55	22.9	16	2	82	0.383	0.461	0.830
M4	CH_4d	DMD, VFA, pH, PR	43.5	0.65	23.3	12	7	81	0.401	0.448	0.896
M5	CH_4d	DMD, VFA, PR, FT, VL	43.8	0.58	22.5	16	2	82	0.419	0.492	0.853
M6	CH_4d	DMD, VFA	43.5	0.53	21.7	14	1	85	0.425	0.516	0.823
M7	CH_4i	pH, DMD, VLp, FT, BTp	28.5	0.34	21.1	21	3	76	0.407	0.496	0.833
M8	CH_4i	pH, DMD, PRp, VLp, FT	28.4	0.33	20.8	21	2	78	0.410	0.520	0.826
M9	CH_4i	pH, DMD, BTp, FT	28.3	0.32	20.2	20	1	79	0.428	0.520	0.823
M10	CH_4i	DMD, VFA, FT	27.7	0.32	19.4	12	1	87	0.434	0.514	0.844
M11	CH_4i	VFA, FT	27.4	0.37	19.8	8	5	87	0.438	0.484	0.905
M12	CH_4i	PR, VL, TGP	27.2	0.40	14.8	11	0	89	0.703	0.752	0.936

[1] Equation ID corresponds to equations presented in subsequent tables, with *Y* as the response (predicted) variable [either CH_4i (observed mean 26.0 ± 0.53 mL CH_4/g DM incubated) or CH_4d (observed mean 40.2 ± 0.91 mL CH_4/g DM apparently digested)] and *X* as the explanatory variables (see Tables 1 and 2 for abbreviations and units of each variable; FT is feed type either forage or concentrate), [2] Mean = mean of predicted values; SEM = standard error of the mean of predicted values; RMSPE = root mean square prediction error expressed as a percentage of the observed mean; EB, ER and ED = error due to bias, regression and disturbance, respectively (all as % of total MSPE); CCC = concordance correlation coefficient; *R* = Pearson correlation coefficient (measure of precision); C_b = bias correction factor (measure of accuracy).

Table 5. Best performing equations from the meta-analysis [1].

Equation ID	*Y*	Intercept	*X1*	*X2*	*X3*	*X4*	*X5*
M5	CH_4d	76.35 (± 4.511)	−31.03 (± 3.922) × DMD	3.21 (± 0.352) × VFA	−0.094 (± 0.01202) × PR	−3.017 (± 1.460) (if FT = 1)	−0.133 (± 0.0380) × VL
M6	CH_4d	51.35 (± 3.086)	−41.98 (± 3.429) × DMD	3.65 (± 0.390) × VFA			
M11	CH_4i	15.8 (± 2.614)	3.06 (± 0.241) × VFA	−5.70 (± 1.028) (if FT = 1)			
M12	CH_4i	11.58 (± 1.627)	−0.0633 (± 0.0057) × PR	0.0947 (± 0.01728) × VL	0.172 (± 0.0046) × TGP		

[1] Variables (units): CH_4i = methane production (mL CH_4/g DM incubated). CH_4d = methane production (mL CH_4/g DM apparently digested). DMD = apparent dry matter (DM) digestibility (g DM disappeared/g DM incubated). TGP = total gas production (mL gas/g DM incubated). VFA production (mmol total VFA/g DM incubated). PR = propionic acid (mmol PR/mol VFA). VL = valeric acid (mmol VL/mol VFA). FT = feed type either forage or concentrate.

3.4. Support Vector Regression and Artificial Neural Network Models

Evaluation of SVR and ANN models developed are presented in Table 6. Both SVR and ANN models demonstrated high predictability on the test dataset, with CCC values >0.90 for both CH_4d and CH_4i. For comparison purposes, meta-analysis equations METd and METi were also developed, via meta-analysis, but included all *X* variables (in linear form, regardless of significance). The CCC values for these equations were 0.645 and 0.734, respectively (Table 6), indicating that the SVR and MLP models must consider a complex multiple-nonlinear response surface between the *X* variables and Y variables, in order to achieve substantially higher CCC values. The predicted vs. observed plots for these models are illustrated in Figure 2.

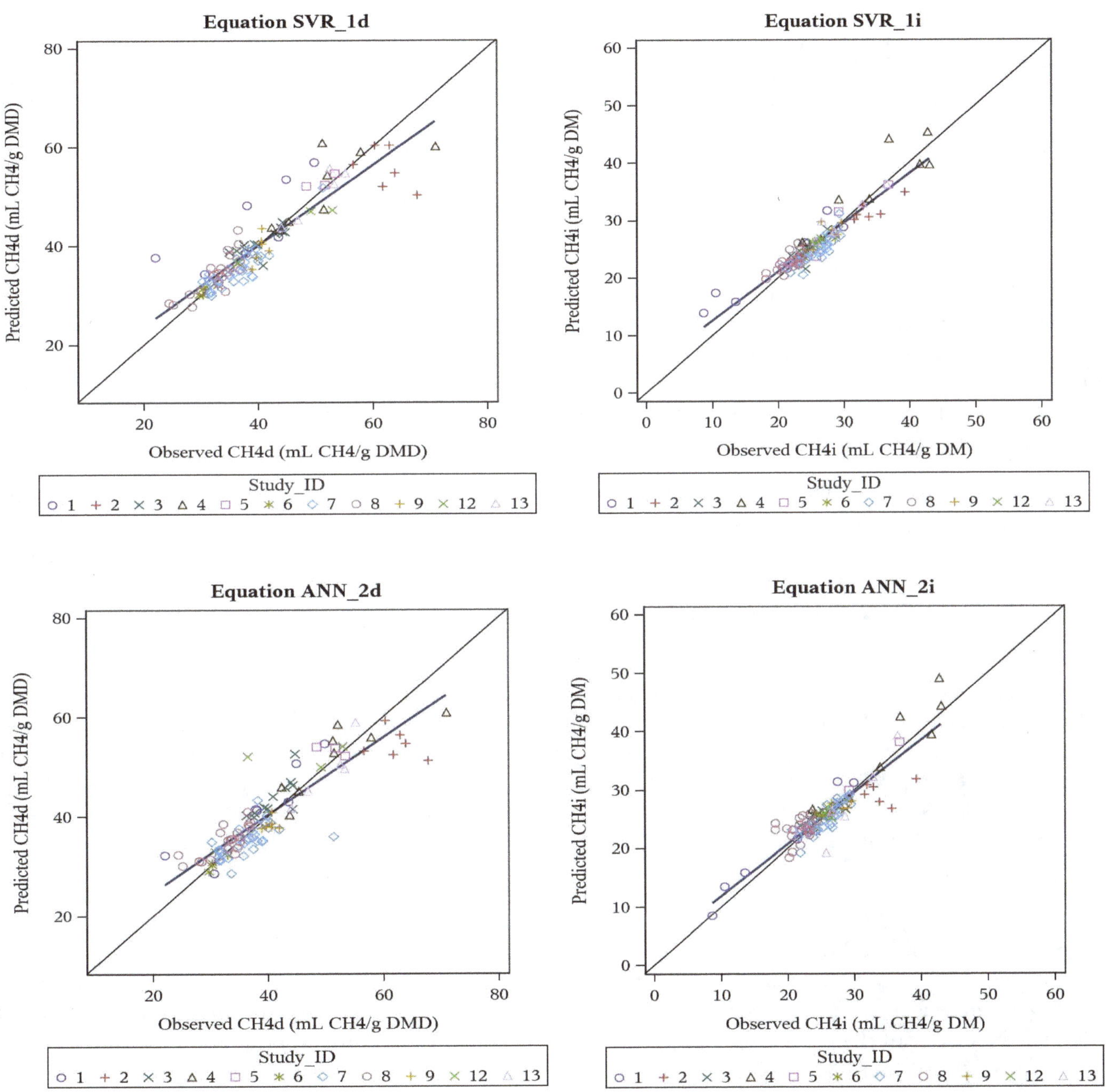

Figure 2. Predicted vs. observed plots for the machine learning equations SVR_1d (CH_4d), SVR_1i (CH_4i), ANN_2d (CH_4d) and ANN_2i (CH_4i), as evaluated on the evaluation dataset (n = 107).

Table 6. Machine learning and meta-analysis (including all the *X* variables) model evaluation via root mean square prediction error (RMSPE) and concordance correlation coefficient (CCC) analysis on the internal evaluation (n = 107) database.

Equation [1]	*Y*	*X*	Mean [2]	SEM	RMSPE, %	EB, %	ER, %	ED, %	CCC	*R*	C_b
SVR_1d	CH_4d	all, nonlinear	40.2	0.82	9.9	0.5	0	99.5	0.899	0.905	0.994
SVR_1i	CH_4i	all, nonlinear	26.1	0.49	8.3	0.6	0.1	99.3	0.917	0.920	0.997
ANN_2d	CH_4d	all, nonlinear	40.5	0.80	9.5	0.5	0.8	98.7	0.907	0.915	0.991
ANN_2i	CH_4i	all, nonlinear	26.0	0.52	9.1	0	2.9	97.1	0.906	0.906	1.000
METd	CH_4d	all, linear	42.9	0.54	17.1	16	6	79	0.643	0.762	0.844
METi	CH_4i	all, linear	27.2	0.40	14.0	12	0	88	0.734	0.782	0.939

[1] Equation IDs with 'd' refer to CH_4d, and 'i' refer to CH_4i equations. METd and METi are meta-analysis equations. *Y* is the response (predicted) variable [either CH_4i (observed mean 26.0 ± 0.53 mL CH_4/g DM incubated) or CH_4d (observed mean 40.2 ± 0.91 mL CH_4/g DM apparently digested) and *X* are the explanatory variables (all variables included in this analysis for comparison purposes), [2] Mean = mean of predicted values; SEM = standard error of the mean of predicted values; RMSPE = root mean square prediction error expressed as a percentage of the observed mean; EB, ER and ED = error due to bias, regression and disturbance, respectively (all as % of total MSPE); CCC = concordance correlation coefficient; *R* = Pearson correlation coefficient (measure of precision); C_b = bias correction factor (measure of accuracy).

3.5. Behaviour Analysis—Machine Learning Models

Unlike the meta-analysis method that results in a predictive equation, the ML methods SVR and ANN do not have the same degree of transparency. To understand the causal pathways to obtain the predictive result, behaviour analysis was performed (Table 7) by systematically changing the inputs in isolation and determining the degree of change in the output prediction. This was performed at +10% and −10% to determine direction of change in the response variable.

Results show (Table 7) that the models ANN_2i and SVR_1i (predicting CH_4i) were highly sensitive to the *X* variables pH and TGP, to varying extents (dependent on the model and FT). Secondary to these variables, the CH_4i predictions were sensitive to AC, PR, BT and DMD. Each model (ANN, SVR) demonstrated different sensitivity to these driving variables, and the sensitivity differed between the FT 1 (forage) and FT 2 (concentrate) substrates (Table 7).

For the models ANN_2d and SVR_1d (predicting CH_4d), these were shown to be highly sensitive to the *X* variables pH, DMD, TGP, AC and BT (Table 7), again dependent on the method (ANN, SVR) and the FT (forage vs. concentrate). Driving variables that differed greatly in sensitivity between models (ANN vs. SVR) included pH (14% and 36% vs. −6% and 5% change with ±10%, FT = 1, CH_4i), DMD (0% vs. −5% and 9% change with ±10%, FT = 1, CH_4i) and AC (11% and −11% vs. 2% and 0% change with ±10%, FT = 1, CH_4d) (Table 7), indicating that each approach fit the data slightly differently.

Some responses had different directional effects in the different models. For example, increasing pH increased CH_4i by 14% in ANN_2i, but decreased it by 6% in SVR_1i (FT = 1) (and similarly for FT = 2, pH increased CH_4i in ANN_2i by 4% and by 30% in SVR_1i); increasing BT did not change CH_4i in ANN_2i, but decreased CH_4i by 6% in SVR1i (FT = 1); and increasing AC reduced CH_4i by 3% in ANN_2i but increased CH_4i by 7% in SVR_1i (FT = 2). Similar results were found for CH_4d predictions, where, for example, increasing pH decreased CH_4d with ANN_2d by 14%, but increased it by 9% with SVR_1d (FT = 1), and for FT = 2, raising pH increased CH_4d by 11% with ANN_2d, and by 37% with SVR_1d.

Table 7. Behaviour analysis [1] of the artificial neural network (ANN) and support vector regression (SVR) models.

X-Variable	CH_4i (on Average 25.5 and 36.3 mL CH_4/g Dry Matter Incubated, for Forage and Concentrate, Respectively)				CH_4d (on Average 40.0 and 46.4 mL CH_4/g Dry Matter Apparently Digested for Forage and Concentrate, Respectively)			
	ANN (ANN_2i)		SVR (SVR_1i)		ANN (ANN_2d)		SVR (SVR_1d)	
Change in X-variable	+10% [2]	−10% [3]	+10% [2]	−10% [3]	+10% [2]	−10% [3]	+10% [2]	−10% [3]
Feed type = forage (FT = 1)								
pH	14%	36%	−6%	5%	−14%	7%	9%	35%
DMD	0%	0%	−5%	9%	−7%	18%	−13%	18%
TGP	12%	−10%	20%	−16%	15%	−8%	20%	−16%
Total VFA	−1%	1%	−1%	0%	−1%	1%	−1%	0%
Acetate (AC)	5%	11%	4%	3%	11%	−11%	2%	0%
Propionate (PR)	0%	0%	−2%	2%	−1%	1%	−1%	2%
Butyrate (BT)	0%	0%	−6%	6%	−1%	1%	−6%	8%
Valerate (VL)	−2%	2%	0%	−1%	−1%	1%	1%	−1%
C2C3	−1%	1%	1%	0%	−4%	4%	1%	0%
Feed type = concentrate (FT = 2)								
pH	4%	−24%	30%	−39%	11%	−23%	37%	−37%
DMD	−2%	−1%	5%	−2%	−8%	8%	−3%	5%
TGP	11%	−11%	12%	−10%	10%	−10%	9%	−8%
Total VFA	−4%	2%	−3%	2%	−2%	2%	−2%	1%
Acetate (AC)	−3%	3%	7%	−6%	3%	−3%	6%	−6%
Propionate (PR)	−3%	3%	−4%	4%	−3%	0%	−3%	3%
Butyrate (BT)	2%	−2%	−1%	1%	2%	−3%	−2%	2%
Valerate (VL)	1%	−1%	0%	0%	1%	−1%	0%	0%
C2C3	0%	−2%	1%	−1%	0%	0%	1%	−1%

[1] Expected change (in %) in the predicted variable (either CH_4i or CH_4d), with a ±10% change in the driving variable. Changes in the predicted variable exceeding |10%| are highlighted in grey, [2] Increase of 10% in the explanatory variable, [3] Decrease of 10% in the explanatory variable.

Some behaviour responses within the ML methods were also directionally different between FT. For example, CH_4i (ANN_2i) increased as pH was increased (+14%, FT = 1), but also increased when pH was decreased (+36%), indicating a nonlinear/polynomial response surface. This is in contrast to when FT = 2, where increasing pH increased CH_4i by 4%, and decreasing pH decreased CH_4i by 24% (Table 7). For CH_4i (SVR_1i, FT = 1), increasing pH decreased CH_4i by 6%, while increasing pH increased CH_4i by 5%. This is in contrast to when FT = 2, where increasing pH increased CH_4i by 30%, and decreasing it decreased CH_4i by 39%. Similar directional differences were observed for CH_4d (FT = 1 vs. 2).

4. Discussion

To predict CH_4 (in vivo or in vitro) based on stoichiometry principles alone and considering a H recovery of 100%, Hegarty and Nolan [8] proposed the equation: CH_4 (mmol/L) = 0.5AC + 0.5BT − 0.25PR − 0.25VL (where all VFAs are expressed in mmol/L). Similarly, Moss et al. [9], considering a H recovery of 90%, proposed the equation: CH_4 (mmol/L) = 0.45AC − 0.275PR + 0.40BT (where all VFAs are expressed in mmol/L). These equations reflect the net production of H as a result of AC and BT synthesis by rumen microbes and the net utilization of H as a result of PR and VL synthesis by rumen microbes during fermentation of feed. The resulting H is utilized by methanogens to reduce CO_2 to H_2O ($CO_2 + 8H \rightarrow CH_4 + 2H_2O$). However, predicting CH_4 from the above stoichiometric equations is only valid if (1) these VFAs are the only end-products of fermentation, (2) no free H_2 accumulates or escapes, (3) the microbial digestion process is strictly anaerobic, and (4) H_2 is not used in other reactions (e.g., reduction of sulphates to sulphides, or saturation of double bonds in fatty acids) [8]. In practice, the production of CH_4 will be less than the stoichiometry prediction given by the above equations, because these assumptions are generally not held. Jayanegara et al. [7] used both stoichiometric equations to predict CH_4 from VFA concentrations in vitro, and found that indeed, the equations overpredicted CH_4, likely due to a much lower observed H recovery (observed range of 28.9% to 56.2%) compared to the recoveries assumed by the models (100% and 90%). In agreement with Jayanegara et al. [7], when the stoichiometric equations [8,9] were applied to the current test dataset, CH_4i was overpredicted (observed CH_4i (mmol/L) = 11.6 ± 2.44, using [8], predicted CH_4i = 16.2 ± 3.07; using [9], predicted CH_4i = 14.0 ± 2.68) and had poor CCC evaluation statistics (0.135 and 0.227 for [8,9], respectively). For the test dataset, the average H_2 recovery, calculated according to [31], was 80%, a value that is substantially lower than the theoretical recovery rates [8,9], and also different from those observed by Jayanegara et al. [7], indicating the potential value of an empirical approach, such as those developed in our work.

The objective of the current study was to utilize meta-analysis and ML methodologies to predict CH_4 emissions from in vitro gas and VFA production data. Results of this work found that via meta-analysis, the best predictive equations of in vitro CH_4d included the variables − DMD + VFA, or − DMD + VFA − PR − FT − VL (Equations M6 and M5 respectively, Table 5), while the best predictive equations of in vitro CH_4i included the variables + VFA − FT, or − PR + VL + TGP (Equations M11 and M12, respectively, Table 5). The significant positive sign on VL in Equation M12 is concerning, as stoichiometrically, the production of VL utilizes H and therefore is associated with a lower CH_4 emission. This illustrates a limitation of empirical modelling (whether it be meta-analysis or a ML), that the resulting equations strive to find the best statistical relationship to the data, regardless of biological principles. It is possible this could be related to the relatively small contribution to total VFA made by VL, or correlation with specific feed ingredient properties.

The best performing univariate equations (U6 (CH_4d), U12 (CH_4i)) were based on DMD and TGP, respectively. The correlation between CH_4d and TGP was low (Table 2), indicating that the DMD correction to CH_4i (CH_4d) accounted for much of the strong relationship between CH_4i and TGP. Such simple regressions may be used when VFA data are not reported, but would miss considerable variance explained by defining the type of VFA being produced (see multivariate equations).

The models produced via ML methodologies, ANN and SVR, have much higher predictability (CCC, RMSPE analysis) of the CH_4i and CH_4d outcomes compared to the meta-analysis models. This was a result of the meta-analysis models being limited to including only significant X variables ($p < 0.05$), while the ML methodologies have no such limitation. As well, the ML methodologies mapped more complex response surfaces between multiple X variables and the Y variable, based on linear, radial or polynomial shapes. While this resulted in a greatly improved prediction on related (internal evaluation) data (Table 6), it may end up fitting noise or other unrelated data characteristics in the training dataset, resulting in a diminished predictability on unrelated data (external evaluation). Such an external evaluation would be a required next step to test the globalization ability of such ML models—in particular considering the relatively small size of the training dataset and the data hungry nature of ML models.

Unsurprisingly, in both meta-analysis and ML models TGP was a significant driving variable, as an indicator of the overall extent of fermentation occurring in vitro. Directionally, the meta-analysis and ML methods agree, whereby increasing TGP increases CH_4i and CH_4d. The variable DMD was particularly relevant with the CH_4d models, where increasing DMD resulted in a lower CH_4d (Tables 5 and 7).

Interestingly, while pH did not appear in many highly significant meta-analysis equations (Tables 3 and 4), it did appear to have a strong presence in the ML models, as illustrated by the behaviour analysis (Table 7). According to the pH dependent VFA stoichiometry of [32], an increase in ruminal pH causes a shift in soluble carbohydrate fermentation towards AC and away from PR and BT, and a shift in starch fermentation towards AC and BT and away from PR. For FT = 2 (concentrates), the ANN_2d and SVR_1d equations to predict CH_4d show a tendency for CH_4 to increase as pH increases (by 11% and 37%, respectively) (Table 7). In line, when pH is decreased, CH_4d also decreased (Table 7). However, when FT = 1 (forages), and pH increases the ANN_2d prediction decreases (−14%), while the SVR_1d prediction increases (9%). For the ANN_2d equation, it is difficult to conceptualize where the −14% in CH_4d comes from, aside from a nuance in the database.

5. Conclusions

The current study successfully delivered models (using both meta-analysis and ML methodologies) which can be used to estimate CH_4 production from in vitro fermentation systems. Meta-analysis results indicate that equations containing DMD, VFA, PR, FT and VL resulted in the best prediction of CH_4 on an internal evaluation dataset of in vitro data. The ML models by far exceed the predictability achieved using meta-analysis methods, but should be evaluated on an external database to assess predictability and generalization potential on unrelated data, in particular given the limited database size and the data hungry nature of such ML methodologies. Between the ML methodologies assessed, ANN and SVR resulted in very similar predictive performance, but differences in fitting, as assessed by behaviour analysis, were evident. The models developed may be utilized to estimate CH_4 emissions in vitro, in instances where total gas and VFA production, but not CH_4, are measured.

Author Contributions: Conceptualization, S.L. and J.F.; methodology, J.L.E. and H.A.-M.; software, J.L.E. and H.A.-M.; validation, J.L.E., H.A.-M. and S.L.; formal analysis, J.L.E., H.A.-M.; investigation, J.L.E.; resources, S.L.; data curation, S.L., A.N.-V., E.J.M., P.P. and P.O.; writing—original draft preparation, J.L.E.; writing—review and editing, J.L.E., S.L., H.A.-M., J.F., C.D.P., A.N.-V., E.J.M., P.P. and P.O.; visualization, J.L.E.; supervision, S.L. and J.F.; project administration, S.L.; funding acquisition, S.L. All authors have read and agreed to the published version of the manuscript.

Appendix A

Table A1. Summary of variable abbreviations and units/descriptions.

Variable Abbreviation	Unit	Description
CH_4i	mL CH_4/g DM incubated	In vitro methane production
CH_4d	mL CH_4/g DM apparently digested	In vitro methane production
pH	-	Final pH in the incubation medium
DMD	g DM disappeared/g DM incubated	Apparent dry matter (DM) digestibility
TGP	mL gas/g DM incubated	Total gas production
VFA	mmol total VFA/g DM incubated	Total VFA production
AC	mmol AC/mol VFA	Acetic acid, proportion of total VFA
PR	mmol PR/mol VFA	Propionic acid, proportion of total VFA
BT	mmol BT/mol VFA	Butyric acid, proportion of total VFA
VL	mmol VL/mol VFA	Valeric acid, proportion of total VFA
ACp	mmol AC/g DM incubated	Acetic acid production
PRp	mmol PR/g DM incubated	Propionic acid production
BTp	mmol BT/g DM incubated	Butyric acid production
VLp	mmol VL/g DM incubated	Valeric acid production
C2C3	AC/PR	Acetate to propionate ratio

References

1. IPCC. *Climate Change 2014 Synthesis Report. Contribution of Working Groups I, II and III to the Fifth Assessment Report of the Intergovernmental Panel on Climate Change*; Core Writing Team, Pachauri, R.K., Meyer, L.A., Eds.; IPCC: Geneva, Switzerland, 2014; pp. 1–151.
2. IPCC. *2019 Refinement to the 2006 IPCC Guidelines for National Greenhouse Gas Inventories*; Volume 4: Agriculture, Forestry and Other Land Use - Chapter 10: Emissions from livestock and manure management; Calvo Buendia, E., Tanabe, K., Kranjc, A., Baasansuren, J., Fukuda, M., Ngarize, S., Osako, A., Pyrozhenko, Y., Shermanau, P., Federici, S., Eds.; IPCC: Geneva, Switzerland, 2019; pp. 10.1–10.209.
3. Martin, C.; Morgavi, D.P.; Doreau, M. Methane mitigation in ruminants: From microbe to the farm scale. *Animal* **2010**, *4*, 351–365. [CrossRef] [PubMed]
4. Kumar, S.; Choudhury, P.K.; Carro, M.D.; Griffith, G.W.; Dagar, S.S.; Puniya, M.; Calabro, S.; Ravella, S.R.; Dhewa, T.; Upadhyay, R.C.; et al. New aspects and strategies for methane mitigation from ruminants. *Appl. Microbiol. Biotechnol.* **2014**, *98*, 31–44. [CrossRef] [PubMed]
5. Hristov, A.N.; Oh, J.; Firkins, J.L.; Dijkstra, J.; Kebreab, E.; Waghorn, G.; Makkar, H.P.S.; Adesogan, A.T.; Yang, W.; Lee, C.; et al. Special Topics—Mitigation of methane and nitrous oxide emissions from animal operations: I. A review of enteric methane mitigation options. *J. Anim. Sci.* **2013**, *91*, 5045–5069. [CrossRef] [PubMed]
6. Ellis, L.L.; Dijkstra, J.; Kebreab, E.; Bannink, A.; Odongo, N.E.; McBride, B.W.; France, J. Aspects of rumen microbiology central to mechanistic modelling of methane production in cattle. *J. Agric. Sci.* **2008**, *146*, 213–233. [CrossRef]
7. Jayanegara, A.; Ikhsan, I.; Toharmat, T. Assessment of methane estimation from volatile fatty acid stoichiometry in the rumen in vitro. *J. Ind. Trop. Anim. Agric.* **2013**, *38*, 103–108. [CrossRef]
8. Hegarty, R.S.; Nolan, J.V. Estimation of ruminal methane production from measurement of volatile fatty acid production. In *Measuring Methane Production from Ruminants*; Makkar, H.P.S., Vercoe, P.E., Eds.; Springer: Dordrecht, The Netherlands, 2007; pp. 69–92.
9. Moss, A.; Jouany, J.-P.; Newbold, J. Methane production by ruminants: Its contribution to global warming. *Ann. Zootech.* **2000**, *49*, 231–253. [CrossRef]
10. McGeough, E.J.; O'Kiely, P.; O'Brien, M.; Kenny, D.A. An evaluation of the methane output associated with high-moisture grains and silages using the in vitro total gas production technique. *Anim. Prod. Sci.* **2011**, *51*, 627–634. [CrossRef]
11. Navarro-Villa, A.; O'Brien, M.; López, S.; Boland, T.M.; O'Kiely, P. Modifications of a gas production technique for assessing in vitro rumen methane production from feedstuffs. *Anim. Feed Sci. Technol.* **2011**, *166*, 163–174. [CrossRef]

12. Navarro-Villa, A.; O'Brien, M.; López, S.; Boland, T.M.; O'Kiely, P. In vitro rumen methane output of red clover and perennial ryegrass assayed using the gas production technique (GPT). *Anim. Feed Sci. Technol.* **2011**, *168*, 152–164. [CrossRef]
13. Navarro-Villa, A.; O'Brien, M.; López, S.; Boland, T.M.; O'Kiely, P. In vitro rumen methane output of grasses and grass silages differing in fermentation characteristics using the gas-production technique (GPT). *Grass Forage Sci.* **2013**, *68*, 228–244. [CrossRef]
14. Purcell, P.J.; O'Brien, M.; Boland, T.M.; O'Kiely, P. In vitro rumen methane output of perennial ryegrass samples prepared by freeze drying or thermal drying (40 °C). *Anim. Feed Sci. Technol.* **2011**, *166*, 175–182. [CrossRef]
15. Purcell, P.J.; O'Brien, M.; Boland, T.M.; O'Donovan, M.; O'Kiely, P. Impacts of herbage mass and sward allowance of perennial ryegrass sampled throughout the growing season on in vitro rumen methane production. *Anim. Feed Sci. Technol.* **2011**, *166*, 405–411. [CrossRef]
16. Purcell, P.J.; Boland, T.M.; O'Brien, M.; O'Kiely, P. In vitro rumen methane output of forb species sampled in spring and summer. *Agric. Food Sci.* **2012**, *21*, 83–90. [CrossRef]
17. Purcell, P.J.; O'Brien, M.; Navarro-Villa, A.; Boland, T.M.; McEvoy, M.; Grogan, D.; O'Kiely, P. In vitro rumen methane output of perennial ryegrass varieties and perennial grass species harvested throughout the growing season: In vitro rumen methane output of perennial grasses. *Grass Forage Sci.* **2012**, *67*, 280–298. [CrossRef]
18. Purcell, P.J.; Grant, J.; Boland, T.M.; Grogan, D.; O'Kiely, P. The in vitro rumen methane output of perennial grass species and white clover varieties, and associative effects for their binary mixtures, evaluated using a batch-culture technique. *Anim. Prod. Sci.* **2012**, *52*, 1077. [CrossRef]
19. Purcell, P.J.; Boland, T.M.; O'Kiely, P. The effect of water-soluble carbohydrate concentration and type on in vitro rumen methane output of perennial ryegrass determined using a 24-hour batch-culture gas production technique. *Irish J. Food Agric. Res.* **2014**, *53*, 21–36.
20. Navarro-Villa, A.; O'Brien, M.; López, S.; Boland, T.M.; O'Kiely, P. Determination of the in vitro rumen methane output of contrasting feeds using the gas production technique (GPT). (unpublished).
21. SAS Institute Inc. *SAS/STAT® 14.1 User's Guide*; SAS Institute Inc.: Cary, NC, USA, 2015.
22. St-Pierre, N.R. Invited review: Integrating quantitative findings from multiple studies using mixed model methodology. *J. Dairy Sci.* **2001**, *84*, 741–755. [CrossRef]
23. Coelho, L.P.; Richert, W.; Brucher, M. *Building Machine Learning Systems with Python: Explore Machine Learning and Deep Learning Techniques for Building Intelligent Systems Using Scikit-Learn and TensorFlow*; Packt Publishing: Birmingham, UK, 2018.
24. Buitinck, L.; Louppe, G.; Blondel, M.; Pedregosa, F.; Mueller, A.; Grisel, O.; Niculae, V.; Prettenhofer, P.; Gramfort, A.; Grobler, J.; et al. API design for machine learning software: Experiences from the scikit-learn project. In Proceedings of the European Conference on Machine Learning and Principles and Practice of Knowledge Discovery in Databases, Workshop on Languages for Data Mining and Machine Learning, Prague, Czech Republic, 23–27 September 2013; pp. 108–122. Available online: https://arxiv.org/pdf/1309.0238.pdf (accessed on 15 April 2020).
25. Pedregosa, F.; Varoquaux, G.; Gramfort, A.; Michel, V.; Thirion, B.; Grisel, O.; Blondel, M.; Prettenhofer, P.; Weiss, R.; Dubourg, V.; et al. Scikit-learn: Machine learning in Python. *J. Mach. Learn. Res.* **2011**, *12*, 2825–2830.
26. Rossum, G.V. *Python Tutorial Release 3.6.4*; Python Software Foundation: Wilmington, DE, USA, 2018.
27. Drucker, H.; Burges, C.J.C.; Kaufman, L.; Smola, A.J.; Vapnik, V. Support vector regression machines. In *Advances in Neural Information Processing Systems 9*; Mozer, M.C., Jordan, M.I., Petsche, T., Eds.; MIT Press: Cambridge, MA, USA, 1997; pp. 155–161.
28. Paliwal, M.; Kumar, U.A. Neural networks and statistical techniques: A review of applications. *Expert Syst. Appl.* **2009**, *36*, 2–17. [CrossRef]
29. Bibby, J.; Toutenburg, T. *Prediction and Improved Estimation in Linear Models*; John Wiley & Sons: Chichester, UK, 1977.
30. Lin, L.I. A concordance correlation coefficient to evaluate reproducibility. *Biometrics* **1989**, *45*, 255–268. [CrossRef]
31. Demeyer, D.; Van Nevel, C. Protein fermentation and growth by rumen microbes. *Ann. Rech. Vet.* **1979**, *10*, 277–279.
32. Bannink, A.; France, J.; López, S.; Gerrits, W.J.J.; Kebreab, E.; Tamminga, S.; Dijkstra, J. Modelling the implications of feeding strategy on rumen fermentation and functioning of the rumen wall. *Anim. Feed Sci. Technol.* **2008**, *143*, 3–26. [CrossRef]

13

Variability and Potential of Seaweeds as Ingredients of Ruminant Diets

Ana de la Moneda [1], Maria Dolores Carro [2], Martin R. Weisbjerg [3], Michael Y. Roleda [4,5], Vibeke Lind [4], Margarita Novoa-Garrido [4,6] and Eduarda Molina-Alcaide [1,*]

1 Estación Experimental del Zaidin (Consejo Superior de Investigaciones Cientificas), Profesor Albareda, 1, 18008 Granada, Spain; anamonedarod@gmail.com

2 Departamento de Producción Agraria. Escuela Técnica Superior de Ingeniería Agronómica, Alimentaria y de Biosistemas, Universidad Politécnica de Madrid, Ciudad Universitaria, 28040 Madrid, Spain; mariadolores.carro@upm.es

3 Aarhus University, AU Foulum, Blichers Allé 20, 8830 Tjele, Denmark; martin.weisbjerg@anis.au.dk

4 Norwegian Institute of Bioeconomy Research (NIBIO), PB 115, 1431 Ås, Norway; Michael.Roleda@nibio.no (M.Y.R.); vibeke.lind@nibio.no (V.L.); margarita.novoa-garrido@nord.no (M.N.-G.)

5 The Marine Science Institute, College of Science, University of the Philippines, Diliman, Quezon City 1101, Philippines

6 Faculty of Biosciences and Aquaculture, Nord University, 8049 Bodø, Nordland, Norway

* Correspondence: molina@eez.csic.es;

Simple Summary: The use of seaweeds as ingredients of ruminant diets can be an alternative to conventional feedstuffs, but it is necessary to assess their nutritive value. The aim of this study was to analyze the chemical composition and in vitro rumen fermentation of eight brown, red and green seaweed species collected in Norway during both spring and autumn. The in vitro ruminal fermentation characteristics of 17 diets composed of oat hay:concentrate in a 1:1 ratio, with the concentrate containing no seaweed or including one of the 16 seaweed samples, was also studied. Species and season determined differences in chemical composition and in vitro fermentation of seaweeds. Most of the tested seaweeds can be included in the diet (up to 200 g/kg concentrate) without negative effects on in vitro ruminal fermentation.

Abstract: This study was designed to analyze the chemical composition and in vitro rumen fermentation of eight seaweed species (Brown: *Alaria esculenta*, *Laminaria digitata*, *Pelvetia canaliculata*, *Saccharina latissima*; Red: *Mastocarpus stellatus*, *Palmaria palmata* and *Porphyra* sp.; Green: *Cladophora rupestris*) collected in Norway during spring and autumn. Moreover, the in vitro ruminal fermentation of seventeen diets composed of 1:1 oat hay:concentrate, without (control diet) or including seaweeds was studied. The ash and N contents were greater ($p < 0.001$) in seaweeds collected during spring than in autumn, but autumn-seaweeds had greater total extractable polyphenols. Nitrogen in red and green seaweeds was greater than 2.20 and in brown seaweeds, it was lower than 1.92 g/kg DM. Degradability after 24 h of fermentation was greater in spring seaweeds than in autumn, with *Palmaria palmata* showing the greatest value and *Pelvetia canaliculata* the lowest. Seaweeds differed in their fermentation pattern, and autumn *Alaria esculenta*, *Laminaria digitata*, *Saccharina latissima* and *Palmaria palmata* were similar to high-starch feeds. The inclusion of seaweeds in the concentrate of a diet up to 200 g/kg concentrate produced only subtle effects on in vitro ruminal fermentation.

Keywords: seaweeds; chemical composition; in vitro rumen fermentation; goats; methane

---

1. Introduction

The expected growth in the human population and the demand for animal products in the forthcoming years have increased the need for searching for alternative sources of nutrients for livestock feeding [1]. Seaweeds had been proposed as alternative feeds that might also have potential benefits on the health of the animals and the consumers of animal products due to their content in bioactive compounds [2,3]. Moreover, seaweeds offer additional advantages, as their cultivation does not compete with terrestrial agriculture, do not need fresh water, and the aquatic photosynthesis contribute to reduce CO_2 levels. The use of seaweeds in animal feeding could also help to alleviate the environmental pollution caused by management of seaweeds in coastal zones. On the other hand, seaweed farming is known to render environmental benefits by recycling nutrients and preventing eutrophication [4].

Although there are studies [5] reporting the traditional use of seaweeds for feeding sheep in the Artic coastal areas and deers in Scotland and Alaska, their widespread use in ruminants is still limited, partly due to the lack of information on the species-specific variability in their the nutritional value and consistency in their chemical composition that may exhibit spatial (site-specific or regional) and temporal (i.e., seasonal and interannual) variations [6–9]. A characteristic common to all seaweeds is their high water content, which may be an important limitation to their direct use in livestock feeding. Another possible limitation is their high salt content [10]. In addition, the presence of compounds that can be a challenge for the digestive system of terrestrial animals may also limit the use of seaweeds in animal feeding [2]. Some recent studies have shown that seaweeds can contain bioactive compounds with antimetanogenic activity, and therefore, they could contribute to reducing the enteric CH_4 emission from ruminants [11–14].

The use of seaweeds as ingredients of ruminant diets requires the assessment of their nutritive value. The first objective of this study was to investigate the chemical composition and in vitro ruminal fermentation of eight different species of seaweeds (three brown, four red and one green) harvested in Norway during spring and autumn. The second objective was to compare the in vitro ruminal fermentation of diets containing these seaweeds with a control diet not including seaweed that was formulated for goat feeding. The gas production technique was used for this study, as it is a relatively cheap and rapid technique that has being widely used in recent years for nutritive evaluation of different ruminant feeds, including seaweeds [8].

2. Materials and Methods

2.1. *Seaweeds*

The seaweeds used in the present study were chosen based on their biomass availability, potential for cultivation and traditional use for feeding livestock in the Artic areas where they were collected [15]. Eight different seaweed species were collected manually both in spring (March–April) and autumn (October–November) of 2015 in Bodø (northern Norway, 67°19′00″ N, 14°28′60″ E) during low tide. The tested seaweed species corresponded to three groups (Phyla) of seaweeds: the brown (Ochrophyta: *Alaria esculenta, Laminaria digitata, Pelvetia canaliculata* and *Saccharina latissima*), the red (Rhodophyta*: Mastocarpus stellatus, Palmaria palmata and Porphyra* sp.), and the green (Chlorophyta: *Cladophora rupestris*). The collected biomass was cleaned in a seawater bath to remove the remains of sand and associated fauna. Then, they were washed with a 30:70 mixture of seawater:freshwater, and finally, in fresh water to reduce the surface salt. The excess of surface water was manually drained and the seaweeds were frozen at −20 °C until their subsequent lyophilization. Once lyophilized, they were ground through a 1 mm sieve in a ZM 200 mill (Retsch GmbH, Haan, Germany).

2.2. *Experimental Diets*

Seventeen diets based on oat hay and concentrate in a 1:1 ratio were studied. The concentrate in the control diet was high in cereals (633 g/kg fresh matter) to be representative of those fed to goats in the practice and did not include any seaweed. Concentrates in the other 16 experimental diets included seaweeds (Table 1) replacing different amounts of feed ingredients (corn, wheat, soyabean meal, sunflower meal, palm soap and salts) present in the control concentrate.

Table 1. Ingredient composition (g/kg fresh matter) of the experimental concentrates [1].

Ingredient	Control	AS	AA	LS	LA	PS	PA	SS	SA	MS	MA	PPS	PPA	POS	POA	CS	CA
Wheat bran	250	165	165	166	166	186	186	250	250	250	250	195	195	250	250	250	250
Corn	250	250	250	245	245	164	164	100	100	159	159	206	206	250	250	250	250
Wheat	133	18	18	22	22	133	133	133	133	133	133	133	133	133	133	133	133
Soybean meal	124	124	124	124	124	124	124	124	124	104	104	93	93	100	100	96	96
Sunflower meal	84	84	84	84	84	84	84	84	84	45	45	84	84	24	24	28	28
Soybean husk	100	100	100	100	100	100	100	100	100	100	100	100	100	100	100	100	100
Others [2]	59	59	59	59	59	59	59	59	59	59	59	59	59	59	59	59	59
Seaweed	-	200	200	200	200	150	150	150	150	150	150	130	130	84	84	84	84

[1] Each experimental concentrate contained one seaweed (A: Alaria esculenta; L: Laminaria digitata; P: Pelvetia canaliculata; S: Saccharina latissima; M: Mastocarpus stellatus; PP: Palmaria palmata; PO: Porphyra sp.; C: Cladophora rupestris harvested either in spring (AS, LS, PS, SS, MS, PPS, POS and CS concentrates) or in autumn (AA, LA, PA, SA, MA, PPA, POA and CA concentrates). [2] All the concentrates included: 10 g of calcium carbonate, 10 g of sugarbeet molasses, 10 g of sepiolite, 14 g of palm soap, 5 g of NaCl, 5 g of Na_2CO_3, and 5 g of mineral-vitamin mixture per kg.

2.3. Donor Animals and Feeding

Four rumen-cannulated Murciano-Granadina goats with an average body weight of 43.8 ± 3.95 kg were used as donors of ruminal content for the in vitro incubations. The animals were fed a diet composed of oat hay and a commercial concentrate in a 50:50 ratio and were housed in pens in pairs with free access to drinking water. The level of intake was that of energy maintenance requirements [16] and the diet was supplied twice a day in equal amounts. The care and handling of the goats were carried out by trained personnel in accordance with the Spanish guidelines for the protection of animals used for experimentation or other purposes, and the experimental procedures were approved by the Animal Welfare Committee at the Zaidín Experimental Station of the Spanish National Research Council (Approval number: 05/24/2016/091).

2.4. In Vitro Trials

In vitro incubations were conducted using the seaweed samples alone and the 17 experimental diets (oat hay and concentrate 1:1) as substrates. Incubations were carried out in batch cultures of ruminal microorganisms using 120-mL glass bottles and ruminal fluid from goats as inoculum. The ruminal content was obtained from each of the four goats before the morning feeding, mixed, and immediately transported to the laboratory in thermal flasks pre-warmed at 39 °C. The ruminal content was filtered through four layers of surgical gauze and mixed with a buffer solution in a 1:4 ratio [17]; no trypticase added and under a continuous CO_2 flow. A total of six incubation runs were carried out. In the first three incubation runs, seaweeds were used as substrate and three feeds commonly used in goat feeding (oat hay, barley straw and a commercial concentrate) were also included for comparative purposes. In the last three incubation runs, the substrates were the 17 experimental diets. In all the incubation runs, four bottles per substrate were used, and four blanks (bottles without substrate) were included.

Five hundred mg of each substrate were carefully weighed in each bottle and 60 mL of the mixture of ruminal fluid and buffer solution were added under a continuous flow of CO_2. Bottles were sealed with butyl rubber stoppers and aluminum caps and placed in a water bath at 39 °C. The pressure inside the bottles and the volume of gas produced in two bottles per substrate and two blanks were measured at 2, 4, 6, 8, 12, 24, 48, 72, 96, 120 and 144 h of incubation using a pressure gauge scope (Sper Scientific LTD, Scottsdale, AZ, USA) and a calibrated glass syringe (Ruthe®, Normax Marinha Grande, Portugal). Additionally, in the incubations using seaweeds as substrate, the content of each bottle at the end of the 144 h of incubation was weighed, frozen at −20 °C and analyzed for neutral detergent fibre (NDF) content to estimate the true dry matter (DM) digestibility ($TDMD_{144}$) as described by Van Soest et al. [18].

In each incubation run, the other two bottles for each substrate (either seaweeds or the experimental diets) and blanks were incubated for 24 h. Gas production measurement was done as described above and a gas sample (5 mL) was stored in a vacuum tube (Terumo Europe NV, Leuven, Belgium) for analysis of CH_4. The fermentation was stopped by chilling on ice. The content of the bottles was homogenized and the following samples were taken: 2 mL were added to 2 mL of a deproteinizing solution (20 g of metaphosphoric acid and 0.6 g of crotonic acid per liter) for the analysis of volatile fatty acids (VFA), and 1 mL was mixed with 1 mL of 0.5 M HCl for the analysis of NH_3-N. Additionally, in the incubations using seaweeds as substrate, the content of each bottle was weighed (before sampling), frozen at −20 °C, and analyzed for neutral detergent fibre (NDF) content to estimate the true dry matter (DM) digestibility ($TDMD_{24}$) as described by Van Soest et al. [18].

2.5. Chemical Analyses

The DM content of the seaweeds and experimental concentrates was determined by lyophilization and subsequent drying of the lyophilized material in an oven at 103 °C for 24 h [19]. Ash content in seaweed (ID 048.13) and ether extract (ID 945.16) were determined according the AOAC procedures [19].

The total N content was analyzed by the Kjeldahl method. The NDF content in the in vitro incubation residues was determined following the procedure of Goering and van Soest [17] using a FibertecTMM6 system (Foss Analytical, Hillerød, Denmark). In the NDF analysis of concentrates, heat-stable amylase was added [20], and all results are expressed as ash-free. The content in total extractable polyphenols (TEP) was analyzed following the procedure of Julkunen-Tiito [21]. The concentrations of individual VFA in the content of the bottles and CH_4 in the gas produced were analyzed by gas chromatography using a HP Hewlett 5890 Packard Series II gas chromatograph (Waldbronn, Germany) equipped with a flame ionization detector (FID) and an HPINNOWAX cross linked polyethylene glycol column (25 m× 0.2 mm× 0.2 μm; Teknokroma, Madrid, Spain) as described by Molina-Alcaide et al. [8]. The concentration of $N\text{-}NH_3$ was determined following the colorimetric method of Weatherburn [22] using a spectrophotometer (Thermo Scientific, Genesys 10 uV Scanning, Madison, WI 53711 USA).

2.6. Calculations and Statistical Analysis

The gas production data were adjusted to the exponential model: gas = $A\,(1 - e^{(-c\,(t\text{-}lag))})$, where A is the asymptotic gas production, c is the gas production rate, lag is the delay at the start of gas production, and t is the time of gas measurement. Parameters A, c and lag were estimated using an iterative least-square procedure following the NLIN procedure of SAS (version 9.4, SAS Inst. Inc., Cary, NC, USA). The average gas production rate (AGPR, ml/h) is defined as the average rate of gas production between the start of incubation and the time at which half of A is reached, and was calculated as $AGPR = A\,c/[2\,(\ln 2 + c\,lag)]$. The amount of VFA in each bottle after 24 h of incubation was corrected by the amount of VFA added with the ruminal fluid used as inoculum.

Data on the chemical composition of seaweed were analyzed by ANOVA using the PROC GLM of SAS (version 9.4, SAS Inst. Inc., Cary, NC, USA) in which the seaweed species and harvest season were the main effects. Fermentation data of seaweeds were analyzed using the PROC MIXED of SAS as a mixed model (version 9.4, SAS Inst. Inc., Cary, NC, USA), in which the seaweed species, harvest season and seaweed species x season interaction were considered as fixed effects, and the incubation run was considered random. The model for the analysis of data of experimental diets included the fixed effect of diet and the random effect of the incubation run. When a significant effect was detected ($p \leq 0.05$), the differences between the means were tested using Tukey's multiple comparison test.

3. Results

3.1. Chemical Composition and In Vitro Fermentation of Seaweeds

Both seaweed species and harvest season affected ($p < 0.001$) all chemical fractions analyzed (Table 2). Ash and N content was greater ($p < 0.001$) in seaweeds collected in spring than in those harvested in autumn (224 vs. 121 g/kg DM and 3.08 vs. 1.92 g/kg DM, respectively). Ash content ranged from 88.2 g/kg DM in *Porphyra* sp. to 225 g/kg DM in *Laminaria digitata* and *Saccharina latissima* (values averaged across seasons). There were also wide variations in total N content, with red and green seaweeds having values greater than 2.20 g/kg DM (values averaged for both collection seasons) and brown species showing values lower than 1.90 g/kg DM. The TEP content was greater ($p < 0.001$) in autumn than in spring (12.1 vs. 6.82 g/kg DM), and the greatest values corresponded to *Alaria esculenta* and *Pelvetia canaliculata*.

Table 2. Chemical composition (g/kg dry matter unless otherwise stated) of different seaweed species harvested in spring and autumn in northern Norway and of feeds commonly used in ruminant diets.

Species	Season	Dry Matter (g/100 g Fresh Matter)	Ash	Nitrogen	Total Extractable Polyphenols
Brown seaweeds					
Alaria esculenta	Spring	110	288	23.4	4.51
	Autumn	277	73.6	13.1	28.1
	Average	193 [f]	181 [d]	18.2 [d]	16.3 [e]
Laminaria digitata	Spring	115	311	23.0	1.44
	Autumn	189	138	6.77	6.08
	Average	152 [b]	225 [e]	14.9 [c]	3.76 [d]
Pelvetia canaliculata	Spring	237	199	16.3	26.9
	Autumn	237	174	6.88	40.4
	Average	237 [g]	187 [d]	11.6 [a]	33.7 [f]
Saccharina latissima	Spring	87.0	350	17.6	3.87
	Autumn	220	100	6.03	5.21
	Average	154 [c]	225 [e]	11.8 [a]	4.54 [d]
Red seaweeds					
Mastocarpus stellatus	Spring	261	183	26.4	4.36
	Autumn	245	194	18.1	3.57
	Average	253 [h]	189 [d]	22.2 [e]	3.97 [ab]
Palmaria palmata	Spring	121	213	43.0	3.86
	Autumn	191	103	14.6	1.93
	Average	156 [d]	158 [c]	28.8 [f]	2.89 [b]
Porphyra sp.	Spring	90.0	97.9	59.8	4.75
	Autumn	116	78.4	50.9	5.85
	Average	103 [a]	88.2 [a]	55.4 [h]	5.30 [c]
Green seaweeds					
Cladophora rupestris	Spring	191	149	37.1	4.88
	Autumn	181	105	37.0	5.39
	Average	186 [e]	127 [b]	37.1 [g]	5.14 [bc]
p value					
Species		<0.001	<0.001	<0.001	<0.001
Season		<0.001	<0.001	<0.001	<0.001
SEM		0.004	0.670	0.128	0.035
Feeds					
Oat hay		896	62.7	12.7	6.82
Barley straw		941	43.7	3.07	NA [1]
Commercial concentrate		933	77.4	23.0	NA [1]

[a–e] For each parameter, average values for each seaweed not sharing the same superscript differ ($p < 0.001$); [1] NA: not analysed.

As shown in Table 3, seaweed species x season interactions ($p < 0.001$) were detected for all the parameters of gas production (A, *c*, *lag* and AGPR) and $TDMD_{244}$. There were differences ($p < 0.001$) among seaweed species in all the parameters of gas production and in vitro digestibility values. *Palmaria palmata* had the greatest ($p < 0.05$) A and AGPR values (143 mL and 4.95 mL/g DM, respectively) with A values being similar to those in the three feedstuffs used as reference and AGPR values higher than those for feedstuffs. The lowest ($p < 0.05$) values were shown by *Pelvetia canaliculata* (8.2 mL and 1.38 mL/g DM, respectively, for A and AGPR) and were much lower than A for any of the feedstuffs and AGPR similar to this value in barley straw. The *lag* values were 0.00 for most seaweed samples, with the exception of *Alaria esculenta* in autumn, *Saccharina latissima* in spring and *Palmaria palmata*, but all the values were lower than 1 h except those for *Alaria esculenta* in autumn (2.58 h). The collecting season affected ($p < 0.001$) the values of A, *lag*, AGPR and $TDMD_{24}$. Compared with spring seaweeds, those collected in autumn had greater A (65.5 vs. 87.5 mL), lag (0.01 vs. 0.42 mL) and AGPR (2.14 vs. 2.93), but lower $TDMD_{144}$ values (87.9 vs. 83.0%).

Table 3. Parameters of gas production kinetics (A, c, *lag* and AGPR) and true dry matter (DM) digestibility ($TDMD_{144}$) after 144 h of in vitro incubation of different seaweed species harvested in spring and autumn in northern Norway and of feeds commonly used in ruminant diets [1].

Seaweed Species	Season	A (ml)	c (h^{-1})	*lag* (h)	AGPR (ml/h)	$TDMD_{144}$ (%)
Brown seaweeds						
Alaria esculenta	Spring	85.9	0.034	0.00	2.11	93.2
	Autumn	104.9	0.033	2.58	2.20	75.4
	Average	95.4 ^{e}	0.034 ^{a}	1.29 ^{c}	2.16 ^{b}	84.3 ^{c}
Laminaria digitata	Spring	85.2	0.027	0.00	1.68	98.3
	Autumn	107.4	0.034	0.00	2.59	79.1
	Average	96.3 ^{e}	0.031 ^{a}	0.00 ^{a}	2.14 ^{b}	88.7 ^{d}
Pelvetia canaliculata	Spring	6.3	0.351	0.00	1.58	67.8
	Autumn	10.0	0.162	0.00	1.17	68.4
	Average	8.15 ^{a}	0.257 ^{b}	0.00 ^{a}	1.38 ^{a}	68.1 ^{a}
Saccharina latissima	Spring	84.0	0.030	0.07	1.82	97.6
	Autumn	147.1	0.043	0.00	4.58	94.6
	Average	116 ^{f}	0.037 ^{a}	0.04 ^{a}	3.20 ^{c}	96.1 ^{e}
Red seaweeds						
Mastocarpus stellatus	Spring	31.0	0.068	0.00	1.52	89.3
	Autumn	20.6	0.078	0.00	1.16	91.0
	Average	25.8 ^{b}	0.073 ^{a}	0.00 ^{a}	1.34 ^{a}	90.2 ^{d}
Palmaria palmata	Spring	114.6	0.060	0.03	4.93	95.8
	Autumn	171.9	0.042	0.74	4.97	96.4
	Average	143 ^{g}	0.051 ^{a}	0.39 ^{b}	4.95 ^{d}	96.1 ^{e}
Porphyra sp.	Spring	54.8	0.063	0.00	2.51	87.3
	Autumn	64.7	0.071	0.00	3.31	90.0
	Average	59.8 ^{c}	0.067 ^{a}	0.00 ^{a}	2.91 ^{c}	88.7 ^{d}
Green seaweeds						
Cladophora rupestris	Spring	62.4	0.020	0.00	0.99	73.5
	Autumn	73.1	0.066	0.00	3.47	74.3
	Average	67.8 ^{d}	0.043 ^{a}	0.00 ^{a}	2.19 ^{b}	73.9 ^{b}
p value						
Species		<0.001	<0.001	<0.001	<0.001	<0.001
Season		<0.001	0.301	<0.001	<0.001	<0.001
Species x season		<0.001	<0.001	<0.001	<0.001	<0.001
SEM		0.296	0.008	0.029	0.057	0.347
Feeds						
Oat hay		129.1	0.037	0.00	3.43	79.7
Barley straw		124.6	0.017	0.41	1.53	56.9
Commercial concentrate		146.3	0.064	0.00	6.75	91.4

$^{a-e}$ For each parameter, the average values for each seaweed not sharing the same superscript differ ($P < 0.05$). [1] A: asymptotic gas production; *c:* rate of gas production; lag: lag time before fermentation starts; AGPR: average gas production rate; $DMED_{24}$: dry matter effective degradability calculated for a rumen passage rate of 0.041 per h. Data are expressed per 0.5 g DM fermented.

There were differences ($p < 0.001$ to 0.003) among seaweed species in total VFA production, VFA profile and acetate/propionate ratio (Table 4). *Pelvetia canaliculata* had the lowest ($p < 0.05$) VFA production, whereas *Alaria esculenta* and *Saccharina latissima* had the greatest production ($p < 0.05$). The VFA production was not affected ($p = 0.821$) by the harvesting season, and no seaweed species x season interaction ($p = 0.609$) was detected. In contrast, seaweed species x season interactions ($p < 0.001$) were detected for molar proportions of acetate, propionate, isobutyrate and isovalerate. *Palmaria palmata* had the lowest proportion of acetate and the greatest propionate proportion (58.5% and 30.1%, respectively), whereas *Mastocarpus stellatus* had the lowest proportion of propionate and the greatest of butyrate (15.1% and 9.50%). The production of minor VFA (isobutyrate, isovalerate and valerate) also differed among seaweed species, with *Porphyra* sp. having the greatest ($p < 0.05$) proportions of isobutyrate and isovalerate and *Pelvetia canaliculata* the greatest valerate proportions. Compared to seaweeds harvested in spring, autumn seaweeds had lower ($p < 0.001$ to 0.020) proportions of acetate (69.0% vs. 59.3%) and minor VFA, as well as greater propionate (18.4% vs. 27.1%) and butyrate (6.37% vs. 8.87%) proportions. The acetate/propionate ratio was highly variable, with values ranging from 1.47 mol/mol in *Alaria esculenta* to 4.76 mol/mol for *Mastocarpus stellatus* both collected in autumn. Spring seaweeds had greater ($p < 0.001$) acetate/propionate ratios than those collected in autumn (3.91 vs. 2.52).

Table 4. Fermentation parameters and true dry matter (DM) digestibility ($TDMD_{24}$) after 24 h of in vitro incubation of different seaweed species harvested in spring and autumn in northern Norway and of feeds commonly used in ruminant diets.

Seaweed Species	Season	VFA (mmol/g DM)	Molar Proportions (mol/100 mol)						Acetate/Propionate (mol/mol)	NH_3-N (mg/100 mL)	CH_4 (mL/g DM)	CH_4/VFA (mL/mmol)	$TDMD_{24}$ (%)
			Acetate	Propionate	Butyrate	Isobutyrate	Isovalerate	Valerate					
Brown Seaweeds													
Alaria esculenta	Spring	3.44	71.1	19.3	5.80	0.78	1.09	1.86	3.69	7.52	28.6	8.31	90.8
	Autumn	3.42	53.6	37.1	8.18	0.02	0.03	1.15	1.47	1.58	28.7	8.39	75.5
	Average	3.43 ^{cd}	62.4 ^{b}	28.2 ^{de}	7.0 ^{ab}	0.40 ^{a}	0.56 ^{a}	1.51 ^{a}	2.58 ^{b}	4.55 ^{a}	28.7 ^{b}	8.37 ^{b}	83.2 ^{c}
Laminaria digitata	Spring	3.16	76.1	16.2	4.00	0.80	1.17	1.75	4.73	10.7	17.0	5.38	92.6
	Autumn	3.24	54.8	32.7	10.0	0.66	0.50	1.32	1.70	0.80	38.5	11.9	75.7
	Average	3.20 ^{bc}	65.5 ^{cd}	24.4 ^{c}	7.00 ^{ab}	0.73 ^{b}	0.84 ^{ab}	1.54 ^{a}	3.22 ^{cd}	5.75 ^{bc}	27.8 ^{b}	8.69 ^{bc}	84.2 ^{c}
Pelvetia canaliculata	Spring	1.08	65.9	21.1	7.05	0.46	1.06	4.37	3.13	7.28	4.91	4.54	70.1
	Autumn	1.10	64.6	19.4	10.1	0.53	1.01	4.34	3.40	5.58	6.10	5.55	69.3
	Average	1.09 ^{a}	65.3 ^{cd}	20.3 ^{b}	8.60 ^{bc}	0.50 ^{a}	1.04 ^{b}	4.36 ^{c}	3.27 ^{cde}	6.43 ^{c}	5.50 ^{a}	5.05 ^{a}	69.7 ^{b}
Saccharina latissima	Spring	3.56	74.9	16.3	4.85	0.93	1.30	1.74	4.61	9.86	24.6	6.91	92.8
	Autumn	4.62	52.7	35.4	9.22	0.78	0.65	1.29	1.49	0.51	47.1	10.2	85.2
	Average	4.09 ^{d}	63.8 ^{bc}	25.8 ^{cd}	7.00 ^{ab}	0.86 ^{c}	0.98 ^{b}	1.52 ^{a}	3.05 ^{c}	5.18 ^{ab}	35.9 ^{b}	8.78 ^{b}	89.0 ^{de}
Red seaweeds													
Mastocarpus stellatus	Spring	2.64	67.6	15.8	9.30	1.38	2.81	3.05	4.28	15.1	18.0	6.82	88.3
	Autumn	1.44	68.3	14.3	9.66	1.49	2.99	3.22	4.76	13.5	11.1	7.71	88.8
	Average	2.04 ^{ab}	67.9 ^{e}	15.1 ^{a}	9.50 ^{c}	1.44 ^{e}	2.90 ^{c}	3.14 ^{b}	4.52 ^{f}	14.3 ^{e}	14.6 ^{a}	7.16 ^{a}	88.6 ^{d}
Palmaria palmata	Spring	7.18	61.1	26.4	6.65	1.33	1.71	2.76	2.31	22.1	64.6	9.00	94.9
	Autumn	6.56	55.8	33.7	7.89	0.70	0.55	1.35	1.67	0.78	60.2	9.18	88.7
	Average	6.87 ^{e}	58.5 ^{a}	30.1 ^{e}	7.30 ^{ab}	1.02 ^{d}	1.13 ^{b}	2.06 ^{a}	1.99 ^{a}	11.4 ^{d}	62.4 ^{c}	9.08 ^{bc}	91.8 ^{e}
Porphyra sp.	Spring	3.28	66.4	16.1	8.33	2.42	3.66	3.17	4.18	39.9	34.3	10.5	82.7
	Autumn	3.00	61.0	19.9	9.28	2.62	4.00	3.23	3.09	37.7	39.3	13.1	79.9
	Average	3.14 ^{bc}	63.7 ^{bc}	17.9 ^{b}	8.80 ^{bc}	2.52 ^{g}	3.83 ^{d}	3.20 ^{b}	3.64 ^{e}	38.8 ^{g}	36.8 ^{b}	11.7 ^{cd}	81.3 ^{c}
Green seaweeds													
Cladophorarupestris	Spring	2.38	69.0	16.0	4.97	2.07	3.66	4.25	4.32	29.8	31.9	13.4	63.9
	Autumn	3.00	63.3	24.4	6.59	1.40	1.85	2.43	2.60	19.5	35.9	12.0	62.0
	Average	2.70 ^{bc}	66.2 ^{de}	20.2 ^{b}	5.80 ^{a}	1.74 ^{f}	2.76 ^{c}	3.34 ^{b}	3.46 ^{de}	24.7 ^{f}	33.9 ^{b}	12.6 ^{e}	63.0 ^{a}
p value													
Species		<0.001	<0.001	<0.001	0.013	<0.001	<0.001	<0.001	<0.001	<0.001	<0.001	0.001	<0.001
Season		0.821	<0.001	<0.001	<0.001	<0.001	<0.001	0.021	<0.001	<0.001	<0.001	0.083	<0.001
Species x season		0.609	<0.001	<0.001	0.104	<0.001	<0.001	0.066	<0.001	<0.001	0.267	0.312	<0.001
SEM		0.154	0.310	0.330	0.249	0.015	0.041	0.086	0.047	0.164	1.588	0.394	0.355
Feeds													
Oat hay		3.84	68.6	19.5	8.10	0.75	0.86	2.13	3.52	6.09	28.2	7.34	64.0
Barley straw		2.68	69.3	20.4	7.03	0.70	0.91	1.66	3.40	2.97	21.7	8.10	38.0
Commercial concentrate		5.48	57.5	25.7	1.10	1.44	1.87	2.53	2.24	12.2	56.4	10.3	86.8

$^{a–g}$ For each parameter, the average values for each seaweed not sharing the same superscript differ ($p < 0.05$).

Both species and season affected ($p < 0.001$) NH_3-N concentrations, CH_4 production and $TDMD_{24}$, (Table 4), and seaweed species × season interactions were detected for NH_3-N concentrations and $TDMD_{24}$. *Alaria esculenta* and *Porphyra* sp. had the lowest and greatest NH_3-N concentrations, respectively, whereas *Pelvetia canaliculata* and *Palmaria palmata* had the lowest and greatest CH_4 productions, respectively. The values of $TDMD_{24}$ ranged from 63.0% to 91.8%, the lowest and greatest values corresponding to *Cladophora rupestris* and *Palmaria palmata*, respectively. Greater ($p < 0.001$) NH_3-N concentrations and $TDMD_{24}$ values and lower ($p < 0.001$) CH_4 production were observed for the samples collected in spring (17.8 mg/100 mL, 84.5% and 26.7 mL, respectively) compared to those collected in autumn (9.99 mg/100 mL, 78.1% and 33.4 mL).

3.2. Chemical Composition and In Vitro Fermentation of Experimental Diets

The chemical composition of the experimental diets is shown in Table 5. In general, ash content was greater in the diets containing seaweeds than in the control diet, whereas the opposite was observed for ether extract content. There were only small differences among diets in N content, which ranged from 17.6 to 19.8 g/kg DM, whereas TEP content varied from 4.75 to 7.98 g/kg DM.

Table 5. Chemical composition (g/kg dry matter (DM) unless otherwise stated) of diets containing 50% of oat hay and 50% of concentrate either including no seaweeds (control) or different seaweed species harvested in spring and autumn in northern Norway.

Seaweed Species	Harvesting Season	Concentrate	Dry Matter (g/kg Fresh Matter)	Ash	Nitrogen	Ether Extract	Total Extractable Polyphenols
-	-	Control	924	68.6	17.9	17.7	5.92
Alaria esculenta	Spring	AS	902	86.4	19.8	16.6	5.73
	Autumn	AA	901	72.1	18.4	17.8	6.29
Laminaria digitata	Spring	LS	910	81.9	19.4	13.8	5.64
	Autumn	LA	911	109	17.6	14.4	5.98
Pelvetiacanaliculata	Spring	PS	904	77.9	18.6	15.8	5.69
	Autumn	PA	903	85.9	17.6	15.8	6.10
Saccharinalatissima	Spring	SS	905	97.9	19.5	15.9	4.82
	Autumn	SA	905	79.5	17.8	16.9	6.58
Mastocarpusstellatus	Spring	MS	904	80.9	19.4	14.8	4.75
	Autumn	MA	905	79.7	18.1	15.5	5.83
Palmariapalmata	Spring	PPS	907	76.7	18.7	15.5	4.99
	Autumn	PPA	908	80.5	17.6	15.8	4.79
Porphyra sp.	Spring	POS	902	70.6	17.8	15.9	5.65
	Autumn	POA	901	64.1	18.1	15.8	5.96
Cladophorarupestris	Spring	CS	902	60.2	18.7	16.4	4.99
	Autumn	CA	900	70.8	17.8	16.6	5.62

As shown in Table 6, the diets including *Palmaria palmata* collected in autumn, and *Porphyra sp* and *Cladophora rupestris* collected in spring and autumn had greater ($p < 0.05$) potential gas production values (A) compared with the rest of the diets, including the control one. All the diets including seaweeds, except that with *Palmaria palmata* collected in autumn, had lower ($p < 0.05$) fractional rates of gas production and AGPR than the control.

Table 7 shows the in vitro fermentation parameters of the experimental diets. There were no differences ($p \geq 0.152$) in total VFA production, minor VFA molar proportions and NH_3-N concentrations. Compared with the control, diets including spring-harvested *Alaria esculenta, Saccharina latissima, Palmaria palmata, Laminaria digitata, Pelvetia canaliculata,* and *Mastocarpus stellatus* from both seasons had greater ($p < 0.05$) acetate proportions. All the diets except that including autumn-harvested *Alaria esculenta* had lowers ($p < 0.05$) propionate molar proportions than the control. Butyrate molar proportions were lowest for the diets with *Alaria esculenta, Laminaria digitata* and *Saccharina latissima* and greatest for the diets with *Porphyra sp.* and *Cladophora rupestris*, with the control diet having an intermediate value. Most diets including seaweeds had greater ($p < 0.05$) acetate/propionate ratios than the control diet, except those including autumn-harvested *Alaria esculenta, Laminaria digitata, Saccharina latissima* and *Palmaria palmata*. All the diets with autumn-harvested seaweeds had lower CH_4 production than the control diet.

Table 6. Parameters of gas production kinetics (A, c and AGPR) after 144 h of in vitro incubation of diets containing 50% of oat hay and 50% of concentrate either including no seaweeds (control) or different seaweed species harvested in spring and autumn in northern Norway [1].

Seaweed Species	Harvesting Season	Concentrate	A (ml)	c (h^{-1})	AGPR (ml/h)
-	-	Control	138 ^{a}	0.050 ^{b}	4.98 ^{b}
Alariaesculenta	Spring	AS	134 ^{a}	0.044 ^{a}	4.30 ^{a}
	Autumn	AA	138 ^{a}	0.043 ^{a}	4.28 ^{a}
Laminaria digitata	Spring	LS	131 ^{a}	0.040 ^{a}	3.78 ^{a}
	Autumn	LA	133 ^{a}	0.042 ^{a}	4.03 ^{a}
Pelvetiacanaliculata	Spring	PS	136 ^{a}	0.041 ^{a}	4.02 ^{a}
	Autumn	PA	129 ^{a}	0.041 ^{a}	3.82 ^{a}
Saccharinalatissima	Spring	SS	133 ^{a}	0.043 ^{a}	4.13 ^{a}
	Autumn	SA	137 ^{a}	0.043 ^{a}	4.25 ^{a}
Mastocarpusstellatus	Spring	MS	135 ^{a}	0.044 ^{a}	4.28 ^{a}
	Autumn	MA	131 ^{a}	0.043 ^{a}	4.03 ^{a}
Palmariapalmata	Spring	PPS	135 ^{a}	0.045 ^{a}	4.38 ^{a}
	Autumn	PPA	145 ^{b}	0.047 ^{ab}	4.92 ^{b}
Porphyra sp.	Spring	POS	147 ^{b}	0.041 ^{a}	4.35 ^{a}
	Autumn	POA	148 ^{b}	0.042 ^{a}	4.48 ^{a}
Cladophorarupestris	Spring	CS	146 ^{b}	0.041 ^{a}	4.35 ^{a}
	Autumn	CA	149 ^{b}	0.040 ^{a}	4.30 ^{a}
p value			<0.001	0.033	0.215
SEM			0.56	0.0014	0.098

[a-b] For each parameter, the mean values for each diet not sharing the same superscript differ ($p < 0.05$). [1] A: asymptotic gas production; c: rate of gas production; AGPR: average gas production rate. The values of *lag* were 0 for all samples. Data are expressed per 0.5 g DM fermented.

Table 7. Fermentation parameters after 24 h of in vitro incubation of diets containing 50% of oat hay and 50% of concentrate either including no seaweeds (control) or different seaweed species harvested in spring and autumn in northern Norway [1].

Seaweed Species	Harvesting Season	Concentrate	VFA (mmol/g DM)	Molar Proportions (mol/100 mol)						Acetate/Propionate (mol/mol)	NH_3-N (mg/100 mL)	CH_4 (mL/g DM)	CH_4/VFA (mL/mmol)
				Acetate	Propionate	Butyrate	Isobutyrate	Isovalerate	Valerate				
-	-	Control	5.06	62.8 ^{a}	21.2 ^{b}	11.9 ^{b}	0.98	1.20	1.79	2.96 ^{a}	10.2	64.6 ^{ab}	12.8
Alaria esculenta	Spring	AS	4.88	65.3 ^{b}	19.3 ^{a}	11.3 ^{a}	0.99	1.38	1.73	3.39 ^{b}	9.88	65.4 ^{b}	13.4
	Autumn	AA	5.02	63.8 ^{a}	22.5 ^{c}	10.4 ^{a}	0.88	1.03	1.49	2.84 ^{a}	7.09	63.2 ^{ab}	12.4
Laminaria digitata	Spring	LS	4.86	65.8 ^{b}	19.8 ^{a}	10.5 ^{a}	0.97	1.32	1.65	3.33 ^{b}	9.84	64.0 ^{ab}	13.2
	Autumn	LA	4.86	64.6 ^{b}	21.8 ^{a}	9.93 ^{a}	0.91	1.14	1.63	2.96 ^{a}	7.57	61.9 ^{a}	12.7
Pelvetia canaliculata	Spring	PS	4.80	64.8 ^{b}	19.4 ^{a}	11.9 ^{b}	0.95	1.31	1.65	3.34 ^{b}	10.1	63.6 ^{ab}	13.2
	Autumn	PA	5.06	65.0 ^{b}	19.7 ^{a}	11.0 ^{b}	1.04	1.36	1.81	3.30 ^{b}	8.10	66.6 ^{b}	13.2
Saccharina latissima	Spring	SS	4.86	65.2 ^{b}	20.3 ^{a}	10.6 ^{a}	0.98	1.31	1.65	3.22 ^{b}	9.49	63.1 ^{ab}	13.0
	Autumn	SA	5.22	63.6 ^{a}	21.8 ^{a}	10.6 ^{a}	0.99	1.30	1.68	2.91 ^{a}	7.83	65.0 ^{b}	12.4
Mastocarpus stellatus	Spring	MS	4.90	64.6 ^{b}	19.4 ^{a}	12.0 ^{b}	1.02	1.35	1.58	3.32 ^{b}	10.3	65.5 ^{b}	13.4
	Autumn	MA	4.66	64.7 ^{b}	19.5 ^{a}	11.9 ^{b}	0.99	1.32	1.64	3.31 ^{b}	9.51	61.6 ^{a}	13.2
Palmaria palmata	Spring	PPS	5.26	64.3 ^{b}	19.7 ^{a}	12.0 ^{b}	1.03	1.31	1.64	3.26 ^{b}	10.4	68.4 ^{b}	13.0
	Autumn	PPA	5.26	63.8 ^{a}	20.6 ^{a}	11.9 ^{b}	0.96	1.22	1.51	3.10 ^{ab}	8.87	67.7 ^{b}	12.9
Porphyra sp.	Spring	POS	5.16	63.6 ^{a}	19.9 ^{a}	12.6 ^{c}	0.95	1.30	1.62	3.19 ^{b}	9.34	67.2 ^{b}	13.0
	Autumn	POA	5.18	63.8 ^{a}	19.5 ^{a}	12.6 ^{c}	1.03	1.36	1.66	3.27 ^{b}	8.61	66.1 ^{b}	12.8
Cladophora rupestris	Spring	CS	5.00	63.6 ^{a}	20.0 ^{a}	12.5 ^{c}	0.95	1.26	1.67	3.19 ^{b}	8.44	65.2 ^{b}	13.0
	Autumn	CA	5.24	63.8 ^{a}	19.6 ^{a}	12.8 ^{c}	0.98	1.30	1.53	3.26 ^{b}	8.95	67.5 ^{b}	13.2
p-value			0.152	<0.001	0.008	<0.001	0.767	0.861	0.636	0.033	0.960	0.049	0.569
SEM			0.042	0.088	0.129	0.063	0.011	0.025	0.023	0.023	0.349	1.59	0.52

$^{a–c}$ For each parameter, the mean values for each diet not sharing the same superscript differ ($p < 0.05$).

4. Discussion

4.1. Chemical Composition and In Vitro Fermentation of Seaweeds

The low DM and high ash content of seaweeds are frequently reported as the main limitations to their use in ruminant diets [7,8]. Both DM and ash contents were similar to those reported for the same seaweeds and others (*Ruppia maritima*, *Ulva lactuca* and *Chaetomorpha linum*) in previous studies [7,23]. In accordance with Tayyab et al. [7], the ash content of seaweeds was greater in spring than in autumn, and the values were greater than those found in conventional feeds used in ruminant nutrition (Table 2). As previously reported [7,8,24,25]. The N content was highly variable, and it was greater in spring-harvested seaweeds than in those collected in autumn. This has been attributed to high sunlight conditions that increase the photosynthesis and nutrient assimilation and to greater N concentration in water during spring compared with autumn [24]. Both *Porphyra sp* and *Cladophora rupestris* showed an N content greater than that in the commercial concentrate used as reference in our study (Table 2; 23.0 g/kg DM), but other seaweeds had an N content similar to that in the oat hay or even lower, especially those harvested in autumn. High-protein seaweeds may be used as an alternative to conventional high-protein feeds, such as soybean meal, and recent studies [25] showed that some amino acids in *Laminaria* and *Mastocarpus* species were protected against rumen degradation, making them potential sources of by-pass protein. In agreement with previous studies [8,9,26], brown seaweeds had, in general, a greater TEP content than both red and green seaweeds, and TEP content was lower in spring-harvested seaweeds than in those collected in autumn. Brown seaweeds are rich in phlorotannins [27], which seem to be different from the tannins in terrestrial plants, but their effect on ruminants is still unknown. Polyphenols have been reported to reduce protein degradation in the rumen, but they can also reduce the fibre degradation by decreasing the attachment of microbes to feed particles [3]. The negative relationships (n = 16) observed between the TEP content and $TDMD_{144}$ (r = 0.732; p = 0.001), $TDMD_{24}$ (r = 0.503; p = 0.047), and total VFA concentrations (r = 0.478; p = 0.061) indicates a negative effect of TEP on the in vitro rumen degradation of seaweeds. However, there were no correlations between TEP content and any of the gas production parameters, which supports the idea that gas measurement should be combined with measurements of feed degradability for a better interpretation of polyphenols effects, as pointed out by Makkar [3].

The high variability observed in the potential gas production values (A) of seaweeds reflects the differences in their potential degradation in the rumen. In fact, a positive relationship between A and $TDMD_{144}$ (r = 0.510; p = 0.044; n = 16) was detected. The lowest A and $TDMD_{144}$ values were observed for *Pelvetia canaliculata*, which agrees with the low DM degradability values reported for this seaweed by Tayyab et al. [7] using the in situ technique in dairy cows and by Molina-Alcaide et al. [8] in 24-h in vitro incubations with sheep ruminal fluid. The greatest A and $TDMD_{144}$ values were observed for *Palmaria palmata* and *Saccharina latissima*, which is in agreement with the high ruminal degradability observed in previous studies for both seaweeds [7,8].

A 24-h incubation period was chosen for the in vitro incubations in our study, as this rumen retention time can be found in goats and sheep fed at moderate levels of intake [28,29]. In agreement with the results of the gas production study, *Pelvetia canaliculata* promoted the lowest total VFA production, which was only 0.41 of that observed for barley straw, and *Palmaria palmata* and *Saccharina latissima* had the greatest values, which were 1.3 and 0.75 of those observed for the concentrate, respectively. Total VFA production for *Porphyra sp.* and *Cladophora rupestris* was similar to that for barley straw, whereas the fermentation of *Alaria esculenta* and *Laminaria digitata* promoted a VFA production only slightly lower than that from fermentation of medium-quality forage such as the oat hay used in our study. These results show that seaweeds can be fermented in the rumen to a variable extent. Although the collecting season had a marked influence on the chemical composition of seaweeds, no differences between seasons were observed in total VFA production. This agrees with the lack of differences between the two harvesting seasons in the ruminal degradability of the protein of nine seaweed species observed by Gaillard et al. [25], despite the marked differences detected in protein content.

There were pronounced differences among seaweed species with regard to VFA profile. *Alaria esculenta, Laminaria digitata, Saccharina latissima* and *Palmaria palmata* harvested in autumn had high propionate proportions (≥32.7%) and their acetate/propionate ratio (1.47:1.70) was similar to that observed in ruminants fed diets based on high-cereal concentrates [30,31]. Conversely, seaweeds harvested in spring, except *Palmaria palmata*, had acetate/propionate ratios (3.13:4.73) similar or even greater than those observed for the oat hay and barley straw used as reference, and the values were similar to those reported in forage-fed ruminants [32–34]. High variations between seaweed species in the in vitro VFA profile have also been previously observed [8,13,14].

The degradation of some amino acids produces branched-chain VFA, and therefore, they can be used as an index of protein degradation [35]. *Cladophora rupestris* and *Porphyra sp.* had the greatest N content (37.1 and 55.4 g/kg DM, respectively) and also the greatest proportions of minor VFA (calculated as the sum of isobyutyrate, isovalerate and valerate; 9.55% and 7.83%), whereas *Alaria esculenta* had the lowest proportions of minor VFA (2.47%) despite having an intermediate N content (18.2 g/kg DM). As pointed out by Hume [36], the interpretation of isoacids proportions is difficult because they are captured and used by the cellulolytic bacteria and the analyzed concentrations are the balance between the N produced from degradation and the N used by the bacteria to synthesize microbial protein in the rumen. Despite this, in our study the proportions of minor VFA were positively correlated with the N content of seaweeds (r = 0.730; $p = 0.001$; n = 16).The N content was also positively correlated with NH_3-N concentrations (r = 0.952; $p < 0.001$; n = 16), which reflects the balance between the NH_3-N produced by protein degradation and that captured by ruminal microorganisms. The NH_3-N concentrations for most of the seaweeds were above the level limiting in vitro ruminal microbial growth (5 mg/100 mL) [37], but concentrations for autumn-harvested *Alaria esculenta, Laminaria digitata, Saccharina latissima and Palmaria palmata* were clearly below this level (≤ 1.58 mg/100 mL), suggesting a possible limitation of microbial growth. These seaweeds had both low N content (ranging from 6.03 g/kg DM in *Saccharina latissima* to 14.6 g/kg DM in *Palmaria palmata*) and low proportions of minor VFA (1.20 in *Alaria esculenta* to 2.72% in *Saccharina latissima*), which would indicate low protein degradation. Interestingly, these seaweed samples promoted a high-propionate fermentation pattern (≥32.7% propionate), suggesting that the low NH_3-N concentrations could also have been due to a high NH_3-N capture by ruminal microorganisms, as was reported to occur in ruminants fed diets based on high-cereal concentrates [31,38].

The production of CH_4 from seaweed fermentation was highly variable, but the positive correlation observed between CH_4 and total VFA production (r = 0.881; $p < 0.001$; n = 16) suggests that the observed differences can be partly explained by the amount of substrate fermented, as both VFA and CH_4 derive from organic matter fermentation [8]. Several studies have investigated the possible antimethanogenic effect of marine seaweeds, with controversial results. Belanche et al. [12] observed no changes in in vitro CH_4 emissions when *Laminaria digitata* or *Ascophyllum nodosum* were included in the diet at 50 g/kg DM. However, Kinley et al. [14] and Machado et al. [39] observed an antimethanogenic effect of *Asparagopsis taxiformis* included in the diet at 20 g/kg, and Machado et al. [39] observed similar effects for a freshwater/brackish alga *Oedogonium* sp. at greater doses (>500 g/kg). The CH_4/total VFA ratio in the seaweeds (Table 4) was similar or slightly lower than that of the concentrate used as reference (10.3 mL/mmol), except for *Pelvetia canaliculata* (5.05 mL/mmol), *Porphyra sp.* (11.7 mL/mmol) and *Cladophora rupestris* (12.6 mL/mmol). The greater CH_4/VFA ratio observed in *Porphyra* sp. and *Cladophora rupestris* might be related to their high N content, as it has been shown that protein fermentation also contributes to CH_4 formation [40].

4.2. Chemical Composition and In Vitro Fermentation of Experimental Diets

The level of seaweed inclusion in the concentrates was chosen from its N content and degradability with the aim that all diets had a similar N content [7]. However, a maximum of 200 g of seaweed per kg concentrate was set up following the recommendations of Rjiba-Ktita et al. [23], who observed that inclusion levels of different seaweed species greater than 200 g/kg reduced the rate and extent of degradation of the mixture. In addition, a minimum of 93 g of soyabean meal per kg concentrate was fixed to guarantee the supply of essential amino acids (mainly lysine) for the host ruminant. *Alaria esculenta, Laminaria digitata, Pelvetia canaliculata* and *Saccharina latissima* were included as energy sources and therefore, they replaced different amounts of wheat bran, corn and wheat in the concentrate. *Mastocarpus stellatus* and *Palmaria palmata* were included as sources of both energy and protein, and therefore, they replaced different amounts of wheat bran, corn, soyabean meal and sunflower meal. Finally, *Porphyra* sp. and *Cladophora rupestris* were considered as protein sources and replaced both soyabean meal and sunflower meal.

The slightly lower N content observed in the diets including autumn-seaweed compared with those including spring-seaweed is consistent with the lower N content of the autumn seaweeds, as both spring and autumn samples of each seaweed were included in the same proportion in the diet (Table 1). The inclusion of seaweeds in the diet resulted in lower *c* and AGPR values than those in the control diet, which indicates that seaweeds were slower fermented than the conventional feeds (wheat, corn, soyabean meal, sunflower meal) they replaced in the concentrate. It has to be taken into account that the differences observed among diets in fermentation parameters are not only due to the inclusion of seaweeds, but also to the different proportions of each feed included in the corresponding concentrate. The diet including *Palmaria palmata* collected during autumn was the only seaweed that had *c* and AGPR values similar to those in the control diet, which was due to their rapid fermentation rate. As indicated by the values of the potential gas production (A), the inclusion of seaweeds in the concentrates at the level used in this study did not reduce the extent of fermentation, and in some cases (autumn-harvested *Palmaria palmata* and *Porphyra* sp. and *Cladophora* collected in both spring and autumn), even confirmed it.

The lack of negative effects of the seaweeds on the in vitro degradation of the diets was confirmed by the absence of differences among diets in total VFA production. In contrast, there were some differences among diets in the VFA profile, and acetate/propionate ratio was greater than that in the control diet for all seaweeds except *Alaria esculenta, Laminaria digitata, Saccharina latissima*, and *Palmaria palmata* collected in autumn. These results are in agreement with the low acetate/propionate ratios observed in the fermentation of these seaweeds, which was similar to those observed for ruminants fed high-cereal diets. The lack of differences among diets in NH_3-N concentrations and minor VFA proportions is in accordance with the similar N contents in all the diets and also indicates similar protein degradability in all the diets.

There were some differences among diets in CH_4 production, and the diets containing *Laminaria digitata* and *Mastocarpus stellatus* collected in autumn showed the lower values. The ratio CH_4/VFA can be used as an indicator of the efficiency of ruminal fermentation, as CH_4 is an energy loss to the host animal and VFA is used as an energy source and as substrates for the synthesis of other compounds [41]. The similar values of this ratio observed for all diets ($p = 0.569$) indicate that the observed differences in CH_4 production were mostly due to the amount of substrate fermented. The positive correlation observed between CH_4 and total VFA production ($r = 0.816$; $p < 0.001$; $n = 17$) supports this hypothesis. As discussed above, differences among diets in both CH_4 and VFA production are not only due to the inclusion of seaweeds, but also to the different feed ingredients in the concentrate. These results indicate that none of the tested seaweeds had a noticeable antimethanogenic effect.

5. Conclusions

The composition of the seaweeds was variable depending on both species and the harvesting season, with seaweeds collected in autumn having less N and ash and more polyphenols than spring-harvested seaweeds. The brown seaweeds studied are sources of energy, whereas *Porphyra sp.* and *Cladophora rupestris* are good protein sources and can be used as substitutes for conventional protein feeds. Seaweeds differed in their ruminal fermentation pattern and autumn-harvested *Alaria esculenta*,

Laminaria digitata, Saccharina latissima and *Palmaria palmata* were similar to conventional high-starch feeds used in ruminant feeding. The inclusion of variable levels of seaweeds in the concentrate of a diet (up to 200 g/kg concentrate) produced only subtle effects on in vitro ruminal fermentation.

Author Contributions: E.M.-A. and V.L. obtained the funding; E.M.-A. conceived the experiments; A.d.l.M. performed the in vitro trials, analyzed the samples and did data calculations; E.M.-A. and M.D.C. did the statistical analysis and wrote the draft; M.Y.R. and M.N.-G. and M.R.W. were responsible for collection and processing of seaweeds; All authors provided advice, revised progress of the manuscript and approved the final manuscript.

Acknowledgments: Thanks to J. Fernandez for technical assistance.

References

1. Makkar, H.P.S. Review: Feed demand landscape and implications of food-not feed strategy for food security and climate change. *Animal* **2018**, *12*, 1744–1754. [CrossRef]
2. Makkar, H.P.S.; Tran, G.; Heuze, V.; Giger-Reverdin, S.; Lessire, M.; Lebas, F.; Ankers, P. Seaweeds for livestock diets: A review. *Anim. Feed Sci. Technol.* **2016**, *212*, 1–17. [CrossRef]
3. Makkar, H.P.S. Effects and fate of tannins in ruminant animals adaptation to tannins, and strategies to overcome detrimental effects of feeding tannin-rich diets. *Small Rumin. Res.* **2003**, *49*, 241–256. [CrossRef]
4. Roleda, M.Y.; Hurd, C.L. Seaweed nutrient physiology: Application of concepts to aquaculture and bioremediation. *Phycologia* **2019**, *58*, 552–562. [CrossRef]
5. Hanley, T.A.; Mckendrick, J.D. Potential nutritional limitations for black-tailed deer in a spruce hemlock forest southeastern Alaska. *J. Wildl. Manag.* **1985**, *49*, 103–114. [CrossRef]
6. Maehre, H.K.; Edvinsen, G.K.; Eilertsen, K.-E.; Elvevoll, E.O. Heat treatment increases the protein bioaccessibility in the red seaweed dulse (*Palmaria palmata*), but not in the brown seaweed winged kelp (*Alaria esculenta*). *J. Appl. Phycol.* **2016**, *28*, 581–590. [CrossRef]
7. Tayyab, U.; Novoa-Garrido, M.; Roleda, M.Y.; Lind, V.; Weisbjerg, M.R. Ruminal and intestinal protein degradability of various seaweed species measured in situ in dairy cows. *Anim. Feed Sci. Technol.* **2016**, *213*, 44–54. [CrossRef]
8. Molina Alcaide, E.; Carro, M.D.; Roleda, M.Y.; Weisbjerg, M.R.; Lind, V.; Novoa Garrido, M. In vitro ruminal fermentation and methane production of different seaweed species. *Anim. Feed Sci. Technol.* **2017**, *228*, 1–12. [CrossRef]
9. Roleda, M.Y.; Marfaing, H.; Desnica, N.; Jonsdottir, R.; Skjermo, J.; Rebours, C.; Nitschke, U. Variations in polyphenol and heavy meal contents of wild-harvested and cultivated seaweed bulk biomass: Health risk assessment and implication for food applications. *Food Control* **2019**, *95*, 121–134. [CrossRef]
10. Applegate, R.D.; Gray, P.B. Nutritional value of algae to ruminants. *Rangifer* **1995**, *15*, 15–18. [CrossRef]
11. Belanche, A.; Ramos-Morales, E.; Newbold, C.J. In vitro screening of natural feed additives from crustaceans, diatoms, algae and plant extracts to manipulate rumen fermentation. *J. Sci. Food Agric.* **2016**, *96*, 3069–3078. [CrossRef] [PubMed]
12. Belanche, A.; Jones, E.; Parveen, I.; Newbold, C.J. A metagenomics approach to evaluate the impact of dietary supplementation with ascophyllum nodosum or laminaria digitata on rumen function in rusitec fermenters. *Front. Microbiol.* **2016**, *7*, 299. [CrossRef] [PubMed]
13. Kinley, R.D.; Fredeen, A.H. In vitro evaluation of feeding North Atlantic stormtoss seaweeds on ruminal digestion. *J. Appl. Phycol.* **2015**, *27*, 282–289. [CrossRef]
14. Kinley, R.D.; De Nys, R.; Vucko, M.J.; Machado, L.; Tomkins, N.W. The red macroalgae Asparagopsis taxiformis is a potent natural antimethanogenic that reduces methane production during in vitro fermentation with rumen fluid. *Anim. Prod. Sci.* **2016**, *56*, 282–289. [CrossRef]
15. Bay-Larsen, I.; Risvoll, C.; Vestrum, I.; Bjorkhaug, H. Local protein sources in animal feed—Perceptions among arctic sheep farmers. *J. Rural Stud.* **2018**, *59*, 98–110. [CrossRef]
16. Prieto, C.; Aguilera, J.F.; Lara, L.; Fonolla, J. Protein and energy requirements for maintenance of indigenous Granadina goats. *Br. J. Nutr.* **1990**, *63*, 155–163. [CrossRef]

17. Goering, M.K.; Van Soest, P.J. Forage Fiber Analysis (Apparatus, Reagents, Procedures and Some Applications). In *Agricultural Handbook*; Agriculture Handbook No. 379; Agricultural Research Services: Washington, DC, USA, 1970.
18. Van Soest, P.J.; Win, R.; Moor, L. Estimation of the true digestibility of forages by the in vitro digestion of cell walls. In Proceedings of the 10th International Grassland Congress, Helsinki, Finland, 7–16 July 1966; pp. 438–441.
19. Association of Official Analytical Chemists (AOAC). *Official Methods of Analysis*, 18th ed.; AOAC International: Gaithersburg, MD, USA, 2005.
20. Mertens, D.R. Gravimetric determination of amylase-treated neutral detergent fiber in feeds with refluxing in beakers or crucibles: Collaborative study. *J. AOAC Int.* **2002**, *85*, 1217–1240. [CrossRef]
21. Julkunen-Tiito, R. Phenolics constituents in the leaves of northern willows: Methods for the analysis of certain phenolics. *J. Microbiol. Biotechnol. Food Sci.* **1985**, *33*, 213–217. [CrossRef]
22. Weatherburn, M.W. Phenol-hypochlorite reaction for determination of ammonia. *Anal. Chem.* **1967**, *39*, 971–974. [CrossRef]
23. Rjiba-Ktita, S.; Chermiti, A.; Bodas, R.; France, J.; Lopez, S. Aquatic plants and macroalgae as potential feed ingredients in ruminant diets. *J. Appl. Phycol.* **2017**, *29*, 449–458. [CrossRef]
24. Rodde, R.S.H.; Varum, K.M.; Larsen, B.A.; Myklestad, S.M. Seasonal and geographical variation in the chemical composition of the red alga *Palmaria palmata* (L.). Kuntze. *Bot. Mar.* **2004**, *47*, 125–133. [CrossRef]
25. Gaillard, C.; Bhatti, H.S.; Novoa-Garrido, M.; Lind, V.; Roleda, M.Y.; Weisbjerg, M.R. Amino acid profiles of nine seaweed species and their in situ degradability in dairy cows. *Anim. Feed Sci. Technol.* **2018**, *241*, 210–222. [CrossRef]
26. Mabeau, S.; Fleurence, J. Seaweed in food products: Biochemical and nutritional aspects. *Trends Food Sci. Technol.* **1993**, *4*, 103–107. [CrossRef]
27. Targett, N.M.; Arnold, T.M. Minireview—Predicting the effects of brown algal phlorotannins on marine.herbivores in tropical and temperate oceans. *J. Phycol.* **1998**, *34*, 195–205. [CrossRef]
28. Isac, M.D.; Garcia, M.A.; Aguilera, J.F.; Molina Alcaide, E. A comparative study of nutrient digestibility, kinetics of digestion and passage and rumen fermentation pattern in goats and sheep offered medium quality forages at the maintenance level of feeding. *Arch. Anim. Nutr.* **1994**, *46*, 37–50. [CrossRef] [PubMed]
29. Ranilla, M.J.; Lopez, S.; Giraldez, F.J.; Valdes, C.; Carro, M.D. Comparative digestibility and digesta flow kinetics in two breeds of sheep. *Anim. Sci.* **1998**, *66*, 389–396. [CrossRef]
30. Carro, M.D.; Ranilla, M.J.; Giraldez, F.J.; Mantecon, A.R. Effects of malate supplementation on feed intake, digestibility, microbial protein synthesis and plasma metabolites in lambs fed a high-concentrate diet. *J. Anim. Sci.* **2006**, *84*, 405–410. [CrossRef]
31. Carrasco, C.; Fuentaja, A.; Medel, P.; Carro, M.D. Effect of malate form (acid or disodium/calcium salt) supplementation on performance, ruminal parameters and blood metabolites of feedlot cattle. *Anim. Feed Sci. Technol.* **2012**, *176*, 140–149. [CrossRef]
32. Martinez, M.E.; Ranilla, M.J.; Tejido, M.L.; Ramos, S.; Carro, M.D. Comparison of Fermentation of Diets of Variable Composition in the Rumen of Sheep and Rusitec Fermenters: I. Digestibility, Fermentation Parameters and Efficiency of Microbial Protein Synthesis. *J. Dairy Sci.* **2010**, *93*, 3684–3698. [CrossRef]
33. Romero-Huelva, M.; Ramos-Morales, E.; Molina-Alcaide, E. Nutrient utilization, ruminal fermentation, microbial abundances, and milk yield and composition in dairy goats fed dietsincluding tomato and cucumber waste fruits. *J. Dairy Sci.* **2012**, *95*, 6015–6026. [CrossRef]
34. Romero-Huelva, M.; Ramirez-Fenosa, M.A.; Planelles-Gonzalez, R.; Garcia-Casado, P.; Molina-Alcaide, E. Can by-products replace conventional ingredients in concentrate of dairy goat diet? *J. Dairy Sci.* **2017**, *100*, 4500–4512. [CrossRef] [PubMed]
35. Wallace, R.J.; Cotta, M.A. Metabolism of nitrogen-containing compounds. In *The Rumen Microbial Ecosystem*; Elsevier Applied Science: London, UK, 1988; pp. 217–250.
36. Hume, I.D. Synthesis of microbial protein in the rumen. II. A response to higher volatile fatty acids. *Austr. J. Agric. Res.* **1970**, *21*, 297–304. [CrossRef]
37. Satter, L.D.; Slyter, L.L. Effect of ammonia concentration on rumen microbial protein production in vitro. *Br. J. Nutr.* **1974**, *32*, 199–208. [CrossRef] [PubMed]

38. Carrasco, C.; Carro, M.D.; Fuentaja, A.; Medel, P. Performance, carcass and ruminal fermentation characteristics of heifers fed concentrates differing in energy level and cereal type (corn vs. wheat). *Span. J. Agric. Res.* **2017**, *15*, e0606. [CrossRef]
39. Machado, L.; Magnusson, M.; Paul, N.A.; Kinley, R.; Nys, R.; Tomkins, N. Identification of bioactives from the red algae *Asparagopsis taxiformis* that promote antimethanogenic activity in vitro. *J. Appl. Phycol.* **2016**, *28*, 3117–3126. [CrossRef]
40. Haro, A.N.; Carro, M.D.; De Evan, T.; Gonzalez, J. Protecting protein against rumen degradation could contribute to reduce methane production. *J. Anim. Physiol. Anim. Nutr.* **2018**, *102*, 1482–1487. [CrossRef]
41. Vanegas, J.L.; Gonzalez, J.; Carro, M.D. Influence of protein fermentation and carbohydrate source on in vitro methane production. *J. Anim. Physiol. Anim. Nutr.* **2017**, *101*, e288–e296. [CrossRef]

14

Comparison of Rabbit, Kitten and Mammal Milk Replacer Efficiencies in Early Weaning Rabbits

Panthiphaporn Chankuang [1], Achira Linlawan [1], Kawisara Junda [1], Chittikan Kuditthalerd [1], Tuksaorn Suwanprateep [1], Attawit Kovitvadhi [2,*], Pipatpong Chundang [2], Pornchai Sanyathitiseree [3] and Chaowaphan Yinharnmingmongkol [4]

1 Faculty of Veterinary Medicine, Kasetsart University, Bangkok 10900, Thailand; panthiphaporn.c@ku.th (P.C.); achira.l@ku.th (A.L.); kawisara.ju@ku.th (K.J.); chittikan.k@ku.th (C.K.); tuksaorn.s@ku.th (T.S.)

2 Department of Physiology, Faculty of Veterinary Medicine, Kasetsart University, Bangkok 10900, Thailand; pichandang@gmail.com

3 Department of Large Animal and Wildlife Clinical Science, Faculty of Veterinary Medicine, Kasetsart University, Nakorn Pathom 73140, Thailand; fvetpos@ku.ac.th

4 Animal Space Pet Hospital, Bangkok 10170, Thailand; taaoy@hotmail.com

* Correspondence: fvetawk@ku.ac.th;

Simple Summary: A milk replacer must be given as the main diet to young rabbits that are separated from their mothers before they reach weaning age (31–35 days). This procedure, which is a rescue protocol, allows them to survive. Moreover, the early separation of young rabbits before weaning prevents negative consequences in lactating rabbits, which is beneficial to pet rabbit producers. Kitten (KMR®, Pet-Ag Inc., Hampshire, IL, USA: KMR) or mammal (Zoologic® Milk matrix 30/52, Pet-Ag Inc., Hampshire, IL, USA: MMR) milk replacers have generally been suggested for use in rabbits; however, rabbit milk has a unique composition. Therefore, a rabbit milk replacer (RMR) was formulated in this study for comparison with these commercial products. Early weaned rabbits at 18 days of age were fed daily with RMR, KMR or MMR until 36 days after birth, while a commercial pelleted diet and water were provided at an amount exceeding the normal intake. The results indicated that it is possible to use RMR as a milk replacer for rabbits without serious adverse consequences. However, the RMR group presented a lower trend in nutrient digestibility than the other groups, although there was no statistical significant difference. Therefore, prebiotics and/or probiotics should be added to RMR formulations to improve this parameter.

Abstract: Early weaned rabbits should be fed using a milk replacer in order to survive. Therefore, a rabbit milk replacer (RMR) was developed and compared with a kitten milk replacer (KMR®: KMR) and a mammal milk replacer (Zoologic® Milk matrix 30/52: MMR). Thirty-six native crossbred rabbits aged 18 days were divided into three experimental groups (six replicates/group, two rabbits/replicate), fed RMR, KMR or MMR daily until they were 36 days old and euthanized at 38 days, while a complete pelleted diet and water were provided *ad libitum*. No statistically significant differences were observed in growth performance parameters, water intake, faecal weight, nutrient digestibility, internal organ weight, caecal pH, caecal cellulose activity, number of faecal pellets and amount of crude protein intake ($p > 0.05$). Caecal amylase activity in the KMR group and caecal protease activity in the RMR group were higher than in the MMR group ($p < 0.05$). The villus height and crypt depth of the MMR group were greater than in the RMR and KMR group ($p < 0.05$). In conclusion, it is possible to feed RMR to early weaning rabbits without serious adverse effects. However, probiotics and/or prebiotics should be supplemented in milk replacers and their benefits studied.

Keywords: digestibility; enzyme activity; gut histology; milk replacer; rabbit

1. Introduction

The size of the pet market has increased sharply in recent years and was estimated to be around 131.7 billion US dollars in 2016 [1]. Moreover, the compounded annual growth rate of the global pet care market was forecast to be 4.9% between 2018 and 2025 due to changes in the new generation's lifestyle, such as living alone or child-free marriage [1]. Nevertheless, humans still need interactions with living things, which add social, medical, emotional and physical benefits to their lives [2]. Companion animals are one solution that can offer these benefits [2]. Although rabbits are not as popular as dogs and cats, they occupy third place among companion animals because they are clean, quiet, non-harmful and require little space [3].

Generally, rabbits are weaned at 31–35 days of age by rabbit producers in Thailand and other countries [4,5] and begin to consume pelleted diets around 18 days of age [6]. Milk replacer has been suggested as a means of feeding early weaned rabbits and orphaned rabbits and of solving the problem of female rabbits that do not produce milk, this being rabbits' major nutrient source for survival [7] as well as preventing gastrointestinal disease [8]. The early separation of young rabbits from their mother can prevent a negative energy balance due to lactation, which supports a higher production yield and reduces disease transmission from the mother to young rabbits, of benefit for pet rabbit producers [4,9]. In addition, the smallest rabbit breed (Netherland Dwarf) can produce around 4–6 kittens per litter; however, the milk yield of this breed is not sufficient to support their kittens, leading to a high mortality rate among young rabbits [5]. Furthermore, milk replacer can be used to rescue unweaned wild rabbits [8]. Therefore, the use of a milk replacer provides one solution to these problems.

Commercial rabbit milk replacer remains lacking or unavailable in some countries. For this reason, kitten milk replacer (KMR) has been suggested as a substitute [7]. However, although kitten milk replacer can be used in early weaning rabbits, its growth performance has been found to be inferior to rabbit milk for rabbits [8]. In addition, two milk replacer formulas using mixtures of kitten milk replacers—Fox Valley Ultraboost and/or Fox Valley 32/40 (Fox Valley Animal Nutrition, INC., Lakemoor, IL, USA)—have been used to rescue young desert cottontail (*Sylvilagus audubonii*) and eastern cottontail rabbits (*S. floridanus*), for which the mortality rate was 26–59% [8]. The high concentration of nutrients (fat, protein and energy), the near absence of lactose, the high proportion of medium-chain saturated fatty acids with bacteriostatic properties (C8:0 and C10:0) and the short milking period required constitute the unique characteristics of rabbit milk and feeding behaviour [6,10]. Therefore, milk replacer for rabbits should be formulated respecting the properties of real rabbit milk [10]. Moreover, cost-effectiveness is another problem for rabbit producers, owners and wildlife rescue center [8]. Therefore, this study aimed to compare the efficiency of a developed rabbit milk replacer with two commercial products (kitten and mammal milk replacers) based on the growth performance and health status of early weaning rabbits (18 days old).

2. Materials and Methods

2.1. Ethics Statement

This study was conducted following standard guidelines at the animal experimental unit, Faculty of Veterinary Medicine (Kasetsart University, Bangkok, Thailand) and was approved by the Institutional Animal Care and Use Committee of Kasetsart University, Bangkok, Thailand (ACKU62-VET-037).

2.2. Animals, Diets, Milk Replacer Preparation and Experimental Design

Thirty-six 18-day-old native crossbreed rabbits with initial body weights of 134 ± 6.31 g/head (mean ± standard deviation) were taken from a local rabbit farm (Saha farm, Kanchanaburi, Thailand). Rabbits were randomly separated into three experimental groups with equal numbers of each sex (six replicates per group and two rabbits per replicate) containing: (1) rabbits fed rabbit milk replacer, which was formulated in this study (RMR); (2) rabbits fed kitten milk replacer (KMR®; Pet-Ag Inc., Hampshire, IL, USA); and (3) rabbits fed mammal milk replacer (Zoologic® Milk matrix 30/52; Pet-Ag

Inc., Hampshire, IL, USA; MMR). Rabbits were placed in a stainless cage (35 cm × 35 cm × 35 cm) with controlled room temperature, light and humidity at 20 ± 2 °C, 16L:8D and 75 ± 10%, respectively. The experiment was conducted for 20 days until the rabbits reached 38 days of age. At the end of the experiments, one rabbit per replicate was euthanized by intraperitoneal injection with pentobarbital sodium at 100 mg/kg (Nembutal, Ceva corporate, France) [11] and the samples were collected for further analysis, while another rabbit of each replicate was returned to the farm and reared until it reached 60 days of age.

The formulation of the RMR, including the chemical composition of the milk replacer, complete pelleted diet and rabbit milk, is illustrated in Table 1. Every day, rabbits were fed 10 mL at 38 °C of a freshly prepared mixture of milk replacer powder and clean water using a sterile syringe at 7:00 until they reached weaning age (36 days), while clean water and complete commercial rabbit diets (Lee Feed Mill, Publ. Co., Ltd., Phetchaburi, Thailand) were provided *ad libitum* throughout the experiment. The KMR and MMR powders were diluted with warmed water at a 7:13 ratio and homogenized. The RMR consisted of two parts: a hydrogenated palm fat part and a mixed dried powder part, containing all ingredients except hydrogenated palm fat and polyoxyethylene (80) sorbitan monooleate. For RMR preparation, hydrogenated palm fat was heated in an 800 W microwave for 60 s, changing from solid to liquid form as a result. The dried mixed powder part was then mixed with warm water. Subsequently, the two parts were homogenized and polyoxyethylene (80) sorbitan monooleate was added as an emulsifier. The ratio of mixed dried powder to hydrogenated palm fat to warm water was 24.5:17.5:78. The dilution ratio was selected based on equal dry matter content between the milk replacers and the solubility of the milk mixture.

Table 1. Ingredients and chemical components of different milk replacers, rabbit milk and diet.

Items	Artificial Milk Replacers			Rabbit Diet [a]	Rabbit Milk [b]
	RMR	KMR	MMR		
Ingredients (%)					
Sodium caseinate	43.8	-	-	-	-
Skimmed milk powder	7.73	-	-	-	-
Hydrogenated palm fat	41.5	-	-	-	-
Monodicalcium phosphate	3.13	-	-	-	-
Limestone	2.30	-	-	-	-
Salt	0.94	-	-	-	-
Premix [c]	0.50	-	-	-	-
Polyoxyethylene (80) sorbitan monooleate	0.10	-	-	-	-
Chemical composition					
Dry matter (%FM)	95.8	96.2	95.2	90.7	29.8
Crude ash (%DM)	9.77	6.66	8.27	5.69	7.38
Crude protein (%DM)	41.4	48.5	30.1	16.1	41.3
Ether extract (%DM)	43.8	22.9	52.5	2.42	43.3
Crude fiber (%DM)	ND	ND	ND	25.2	ND
Nitrogen free extract (%DM) [d]	5.03	22.0	9.13	50.6	8.05
Metabolizable energy content (kcal/100g DM) [e]	535	441	584	254	540

RMR = Rabbit milk replacer which was performed in this study, KMR = Kitten milk replacer (KMR®, Pet-Ag Inc., Hampshire, IL, USA), MMR = Mammal milk replacer (Zoologic® Milk matrix 30/52, Pet-Ag Inc., Hampshire, IL, USA), FM = Fresh matter, DM = Dry matter, ND = Not detect; [a] A commercial pelleted diet for rabbits (Lee Feed Mill, Publ. Co., Ltd., Phetchaburi, Thailand); [b] Chemical composition of rabbit milk [10]; [c] Vitamin and mineral premix (Topmix-B111, Top Feed Mills Co., Ltd., Pathumthani, Thailand) were supplied per kilogram of diets at 4,800,000 IU of vitamin A; 1,200,000 IU of vitamin D3; 6000 IU of vitamin E; 600 mg of vitamin K; 600 mg of vitamin B1; 2200 mg of vitamin B2; 10,000 mg of vitamin B3; 800 mg of vitamin B6; 4 mg of vitamin B12; 48 mg of biotin; 4800 mg of Calcium pantothenate acid; 200 mg of folic acid; 24,000 mg of Zn, 16,000 mg of Fe; 32,000 mg of Mn; 32,000 mg of Cu; 200 mg of I; 40 mg of Se; 40 mg of Co; [d] Calculation [12]; [e] Calculation based on Atwater system [13].

2.3. *Perfomance, Digestibility and Faecal Evaluation*

The animals were weighed at 18, 24, 30 and 36 days of age, whereas average daily feed intake (ADFI), average daily weight gain (ADG), feed conversion ratio (FCR), water intake and weight of faeces output were evaluated at 19–24, 25–30 and 31–36 days of age. The apparent digestibility of dry matter, organic matter, ether extract and crude protein was conducted at 23–27 and 31–35 days of age and contained six replicates/groups. The procedures for feeding, faecal collection, chemical analysis and calculation were in accordance with [14]. Briefly, feed intake was measured during the period of the digestibility trial. Faeces were removed from the cage at 9:00 on the first day of the digestibility trial. Subsequently, all faeces on a net under the cage were collected at 9:00 for four days. The faeces were weighted immediately after collection, put in a sterile plastic bag and kept at −20 °C for further chemical composition analysis following the procedure of [14]. Another study was conducted, where the amount of daily faecal pellet excretion was measured by counting the dried faecal pellets between 19 and 36 days old from photos.

2.4. Internal Organs, Gut Histology and Caecal pH

The internal organ weight and the body weight of the euthanized rabbits were determined. The duodenal part of the small intestine was fixed in 10% buffered formalin for further villus morphometric evaluation. Briefly, small pieces of middle duodenum after fixation were processed, embedded in paraffin, sectioned at 7-µm thicknesses by means of a rotary microtome (Leica RM2155; Leica Instruments GmbH, Nussloch, Germany) and stained by haematoxylin and eosin method. Villi height and crypt depth were evaluated under a microscope using an image analysis programme (Image Pro Plus; Media Cybernetics, Bethesda, MD, USA). Caecal pH was measured directly using a Crison MicropH 2001 pH meter (Crison Instruments, Barcelona, Spain). The caecal content was immediately placed in sterile plastic tubes under ice for enzyme preservation and kept at −20 °C for further analysis of caecal enzyme activity.

2.5. Caecal Enzyme Activity

The crude enzyme extracted from the caecal content was extracted by homogenized caecal content with phosphate buffer solution (pH 7) at a 1:5 ratio (*w*/*v*). The homogenates were centrifuged at 18,000× *g* for 30 min at 4 °C to obtain the supernatant used to evaluate amylase, protease and cellulase activity. Amylase and cellulase activity were assayed according to [15,16] using 5% soluble starch and 1% carboxyl-methyl cellulose (CMC; medium viscosity) as the substrate, respectively. One hundred microlitres of crude enzyme extract were added to activate the digestion of the substrates. The products of the carbohydrate-digestive enzymes were stained using 1% dinitrosalicylic acid (DNS) and measured using a spectrophotometer at 540 nm against a linear range of maltose standards for amylase and glucose standards for cellulase. Protease activity was assayed according to the method described by [17] using 0.6% casein as the substrate. The product of the protein-digesting enzyme was measured spectrophotometrically at 660 nm against a linear range of tyrosine. The activity of the observed digestive enzymes was expressed as U.

2.6. Crude Protein Assessment

Each rabbit's feed intake between 19–24, 25–30 and 31–36 days old and the amount of crude protein in the milk replacer and diet were used as information to calculate the amount of crude protein intake.

2.7. Statistical Analysis

The results of this study are represented as the mean and pooled standard error of the mean. A completely randomized design was employed in this study. Therefore, one-way analysis of variance (ANOVA) was used to compare the different types of milk replacers (fixed factors) for internal organ characteristics, caecal pH, caecal digestive enzyme activities and duodenal histology, whereas the growth performances, water intake, faeces excretion, apparent digestibility, number of faecal pellets and amount of protein intake were analyzed by two-way mixed analysis of variance, with treatment groups or age serving as the between-subjects or the within-subjects factor, respectively. Duncan's multiple range test was used for post hoc analysis. Differences were considered statistically significant at $p < 0.05$. All statistical analyses in this study were performed with R-statistic software using the Rcmdr package [18].

3. Results

The effects on performance, apparent digestibility, amount of faeces excretion and crude protein intake from rabbits fed the different milk replacers are shown in Table 2. No statistically significant differences between the groups and the interactions between the studied factors (groups and age) for all parameters in Table 2 were apparent ($p > 0.05$). The age increment was correlated with increased body weight, ADFI, ADG, FCR, water intake, faeces excretion and crude protein intake ($p < 0.05$), whereas apparent digestibility did not affect dry matter, organic matter or ether extract ($p > 0.05$). However, the crude protein digestibility of rabbits at 31–35 days old was lower than rabbits at 23–27 days old ($p < 0.05$). The rabbits fed KMR and MMR displayed higher nutrient digestibility in both age ranges compared with rabbits fed RMR, but there was no statistically significant difference ($p > 0.05$). Rabbits in RMR, KMR and MMR were received the crude protein from milk daily at 2.23, 2.61 and 1.62 g dry matter/head. No deaths, morbidities or clinical signs were observed in rabbits during the experimental period. Moreover, a rabbit in each replicate was not euthanized at the end of the experiment and remained alive until it reached two months of age. In addition, there were no problems of milk perception and palatability in any group in this experiment, because the rabbits sucked milk directly from the syringe without any force feeding.

Table 2. Effect of different milk replacers on rabbit performances, apparent digestibility and crude protein intake.

Parameters	Factors							SEM	*p*-Value		
	Artificial Milk Replacers (AMR)			Age (days)					AMR	Age	AMR * Age
	RMR	KMR	MMR	0	6	12	18				
BW (g/head)	210	198	202	134 ^{a}	157 ^{b}	214 ^{c}	308 ^{d}	9.086	0.38	0.001	0.94
				19–24	25–30	31–36	-				
ADFI (g/head/day)	22.1	26.7	22.1	7.47 ^{a}	22.1 ^{b}	39.6 ^{c}	-	2.239	0.59	0.001	0.56
ADG (g/day)	9.62	9.33	8.52	4.40 ^{a}	10.8 ^{b}	14.6 ^{c}	-	0.808	0.84	0.001	0.77
FCR	2.39	2.62	2.74	1.91 ^{a}	2.21 ^{a}	2.75 ^{b}	-	0.115	0.44	0.01	0.56
Water intake (g/head/day)	41.8	30.9	34.2	10.1 ^{a}	31.9 ^{b}	68.0 ^{c}	-	3.902	0.1	0.001	0.38
Faeces excretion (g/head/day)	15	14.3	15.4	1.29 ^{a}	5.49 ^{b}	8.14 ^{c}	-	0.936	0.81	0.001	0.32
Crude protein intake (g/head/day) [1]											
Diet	3.72	3.73	3.9	1.20 ^{a}	3.56 ^{b}	6.37 ^{c}	-	0.361	0.59	0.001	0.56
Diet and milk	5.95	6.34	5.52	3.39 ^{a}	5.65 ^{b}	8.56 ^{c}	-	0.364	0.11	0.001	0.56
				23–27	31–35	-	-				
Apparent digestibility (%)											
Dry matter	59.6	63.3	63.5	60.5	63.4	-	-	1.126	0.29	0.22	0.96
Organic matter	60.7	65.1	65.2	62.8	64.2	-	-	1.123	0.18	0.51	0.85
Ether extract	68.3	73.8	70.7	71.3	70.3	-	-	2.418	0.75	0.88	0.97
Crude protein	76.5	80.8	81.1	81.5 ^{b}	77.1 ^{a}	-	-	1.007	0.07	0.03	0.34

RMR = Rabbit milk replacer, which was used in this study, KMR = Kitten milk replacer (KMR®, Pet-Ag Inc., Hampshire, IL, USA), MMR = Mammal milk replacer (Zoologic® Milk matrix 30/52, Pet-Ag Inc., Hampshire, IL, USA), SEM = pooled standard error of mean, BW = Body weight, ADFI = Average daily feed intake, ADG = Average daily weight gain, FCR = Feed conversion ratio; $^{a, b, c, d}$ The differences in superscript letter in the same row represented statistical significant differences ($p < 0.05$); [1] The rabbits in the RMR, KMR and MMR groups received crude protein from milk at 2.23, 2.61 and 1.62 g/head/day, respectively.

The consequences for internal organ weight, caecal pH, duodenal wall histology and digestive enzyme activities between the groups are compared in Table 3. The weight of each internal organ was not statistically significantly different between the groups ($p > 0.05$). Caecal pH was not influenced by the differences in milk replacers ($p > 0.05$). Caecal amylase activity in the MMR group was lower than in the KMR group ($p < 0.05$), whereas greater caecal protease activity was observed in the RMR group compared to the MMR group ($p < 0.05$). Cellulase activity was not affected by the treatments ($p > 0.05$). Respectively, the shortest villus and the shallowest crypt depth were found in the RMR and the KMR groups compared to the MMR group ($p < 0.05$).

Table 3. Effect of different milk replacers on internal organ weight, caecal pH, intestinal villi morphology and digestive enzyme activity.

Parameters	Artificial Milk Replacers (AMR)			SEM	p-Value
	RMR	KMR	MMR		
	Internal organs characteristics (g/live body weight)				
Liver	4.46	4.00	3.81	0.162	0.25
Spleen	0.14	0.10	0.10	0.015	0.51
Kidney	1.24	1.18	1.12	0.046	0.61
Thoracic organs [1]	1.07	1.09	1.15	0.063	0.38
Pancreas	0.06	0.07	0.05	0.007	0.99
Full stomach	7.40	7.44	7.22	0.555	0.79
Stomach wall	2.09	2.01	1.99	0.061	0.65
Intestinal organs [2]	7.38	6.87	7.37	0.242	0.54
Full caecum	16.3	16.1	16.2	0.422	0.97
Caecal wall	2.75	2.49	2.34	0.134	0.48
Caecal pH	6.50	6.30	6.38	0.081	0.63
	Caecal digestive enzyme activities (U)				
Amylase	12.2 ^{ab}	18.5 ^{b}	10.4 ^{a}	1.41	0.04
Protease ($\times 10^{-1}$)	7.26 ^{b}	6.84 ^{ab}	5.79 ^{a}	0.253	0.04
Cellulase	3.71	3.50	3.60	0.057	0.31
	Duodenal villi morphology (μm)				
Villus height	320 ^{a}	361 ^{ab}	380 ^{b}	8.50	0.04
Villus crypt	66.2 ^{ab}	62.6 ^{a}	68.4 ^{b}	0.737	0.007

RMR = Rabbit milk replacer, which was used in this study, KMR = Kitten milk replacer (KMR®, Pet-Ag Inc., Hampshire, IL, USA), MMR = Mammal milk replacer (Zoologic® Milk matrix 30/52, Pet-Ag Inc., Hampshire, IL, USA), SEM = pooled standard error of mean, BW = Body weight, FI = Feed intake, ADG = Average daily weight gain, FCR = Feed conversion ratio; $^{a, b}$ The differences in superscript letter in the same row represented statistical significant differences ($p < 0.05$); [1] Thoracic organs includes lungs and heart; [2] Intestinal organs includes stomach, small intestine, large intestine, caecum and rectum with content.

The average number of faeces pellets from two rabbits of each replicate in the experiments is illustrated in Figure 1 and Table A1. The graph shows a steady increase in the number of faecal pellets with increasing age ($p < 0.001$); however, a sharp drop occurred in all study groups at 35 days of age, followed by another increase. The largest number of faecal pellets existed in the MMR group compared to the RMR group ($p < 0.05$; Appendix A Table A1), with the KMR group between them ($p > 0.05$). A significant interaction between the fixed factors (age and treatment group) was identified ($p < 0.05$). Generally, the same increasing trend was observed in all groups, except for the sharp rise in the number of faecal pellets in the MMR, RMR and KMR groups at 22–23, 24–26 and 34–36 days of age, respectively.

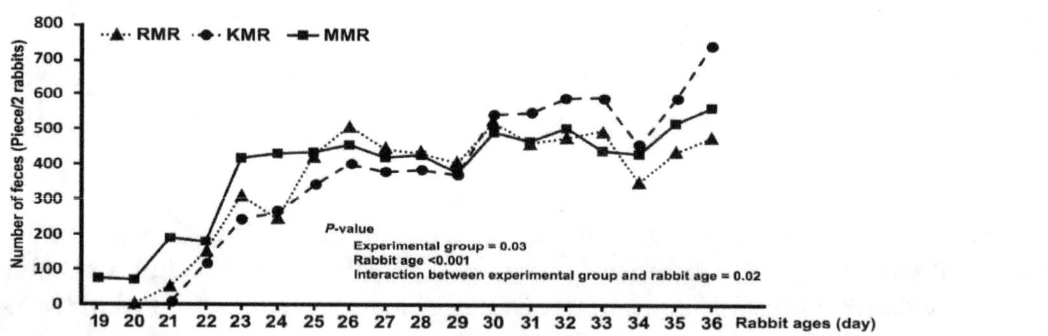

Figure 1. Effect of different milk replacers on number of faecal pellets (RMR, rabbit milk replacer in this study; KMR, kitten milk replacer, KMR®, Pet-Ag Inc., Hampshire, IL, USA; and MMR, mammal milk replacer, Zoologic® Milk matrix 30/52, Pet-Ag Inc., Hampshire, IL, USA).

4. Discussion

The RMR was formulated according to the profile of rabbit milk; therefore, its chemical composition was the most similar to the composition of rabbit milk [10]. KMR contained a higher proportion of crude protein and was lower in fat and energy than RMR and rabbit milk. On the other hand, MMR was lower in crude protein and higher in fat and energy than RMR and rabbit milk. A high density of nutrients and energy was the unique characteristic of rabbit milk, which contained respectively around four and three times higher proportions of protein and lipids than cow's milk [10]. A short milking time is a common nursing behaviour, explaining the high density of rabbit milk [6]. The amount of milk replacer fed to young rabbits was calculated on the basis of stomach capacity. Milk replacer in powder form was used for all formulations in this study because a highly concentrated milk mixture can be formulated from dried powder but not in liquid form [8,9]. Rabbit milk protein comprises around 70% and 30% casein and whey protein, respectively [10]. Therefore, casein served as a major protein ingredient in the milk replacer formulation for RMR and the two other commercial milk replacers, whereas dried skimmed milk powder represented another protein source, which was used in a lower proportion than casein. Respectively, either whey or milk protein concentrate was supplemented in KMR and MMR, whereas in RMR they were not. A low level of lactose is present in rabbit milk; therefore, cow's or goat's milk is limited as a milk replacer formulation [10]. Moreover, lactase activity in rabbits decreases with age and does not respond to the lactose concentration in the diet. Therefore, excessive lactose intake can lead to a digestive disorder [6]. Differences in the type of raw protein source and quantity influence the diversity of the amino acid profile. Although the amino acid profile was not evaluated in this study, the intake of amino acids would have been sufficient for rabbits because casein, considered an ideal protein, was used as the main ingredient and the percentage of crude protein in all milk replacers was higher than nutrient requirements [4,6].

Fat in rabbit milk represents the major energy sources for rabbits [10], whereas excessive starch intake promotes digestive problems and increases the mortality rate of young rabbits [4]. The highest nitrogen-free extract was present in the KMR formulation; however, no adverse effects were observed in this group. Medium-chain fatty acids, mainly caprylic (C8:0) and capric acid (C10:0), were the major components of fatty acids in rabbit milk, comprising around 50% of total fatty acids [10]. Vegetable oil or hydrogenated palm oil served as the only lipid sources in milk replacers, comprising a high proportion of polyunsaturated fatty acids; however, they still contained these medium-chain fatty acids. Another function of caprylic and capric acid was their antibacterial properties, which maintained microbial community development and prevented pathogen invasion. Supplementation with these medium-chain fatty acids should confer health benefits on rabbits. Nevertheless, no health problem occurred in this study, although these medium-chain fatty acids were not supplied in the formulations. The essential fatty acids were supplied in sufficient amounts to fulfil the nutritional requirements of all the study groups, as there was a very high amount of fat in all the formulations [6]. However, an antioxidant such as tocopherol should be supplemented to prevent lipid oxidation. The minerals and vitamins in RMR were higher than the minimum requirements in rabbits [6]. In the early period of the experiment, the rabbits received nutrients and energy from the milk replacers in high proportions compared to the pelleted diet, which represented only around 1/3 of the total intake in the first period. Subsequently, a higher intake of the pelleted diet became the main nutrient and energy source. Growth performances did not differ between treatment groups, and no morbidity or mortality of rabbits occurred during or after the experimental period. Moreover, the rabbits' final body weights were similar to those of their counterparts reared with their mother at the same age, as reported in the study of [5]. These findings demonstrate the potential of using milk replacers in early weaned rabbits. However, the low energy and nutrient requirements of the rabbits in this study were characteristics of a native breed with a slow growth rate compared to hybrid or commercial meat rabbits [5]. Therefore, milk replacers can certainly be used in local breed or pet rabbits, possibly offering benefits to veterinarian and pet rabbit producers. Rabbits raised in intensive rearing systems and/or with high growth rates may be studied further as to which milk replacer can support their growth performances and the maintenance of a healthy condition. Another study [9] reported lower growth performances in commercial meat rabbits fed KMR compared with rabbits fed rabbit milk.

A previous report has shown that the gut microbial community of rabbits plays an important role relating to the efficiency of nutrient fermentation, productive performance and heath condition [19]. A sterile gut is observed in rabbits immediately after birth. Subsequently, microbes slowly and continuously colonize the rabbit gut and the amount of antimicrobial substances in rabbit milk and nutrients have major effects on the development of gut functions and the microbial community [10,19]. A large variation in the bacterial community was observed between young rabbits at an early age. However, the composition of the bacterial community between rabbits became more similar with increasing age, especially with pelleted diets [13,19]. To our knowledge, the caecum is a major organ of bacterial fermentation in rabbits, providing around 50–60% of daily energy requirements. The caecal environment greatly affects the microbial community in rabbits, especially in terms of pH, being influenced in a major way by fermentable products called volatile fatty acids. A low nutrient fermentation efficiency can result in a higher value of caecal pH, increasing the risk of a digestive disorder [6]. The caecal pH of the RMR group seemed to be higher than that of the other groups. However, the caecal pH of the RMR group was lower than 6.73, insufficient to promote a gut health problem [6]. Moreover, there was no statistically significant difference in caecal pH, growth performance, digestibility, morbidity and mortality during the experiments, supporting the possibility of using RMR in rabbits.

Caecal enzyme activity was another indicator representing microbial activity. The amount of caecal amylase and protease in the RMR and KMR groups was higher than in the MMR group. Fermentation products from probiotic bacteria and prebiotics play an important role in supporting the development of a normal flora in rabbits [20]. Supplementation with fermentation products from several normal flora bacteria (*Lactobacillus* sp., *Enterococcus faecium*, *Bifidobacterium bifidum*, *Pediococcus adicilactici*) and prebiotics (fructooligosaccharide) in the KMR formulation contributed to higher enzyme activity, whereas RMR and MMR did not contain these supplements. Such findings are in accordance with another research study [8]. A higher survival rate was observed in the milk replacer with probiotics and prebiotics (35.3%) compared to without these additions (21.3%) in infant cottontail rabbits in which the chemical composition of these milk replacers were similar [8]. Thus, probiotics and/or prebiotics may have been the key factor determining this result. The unsuitable chemical components of KMR with respect to rabbit milk and the slow growth rate of bacteria in the first period could have been the cause of the lower feed intake in the first period in the KMR group. Subsequently, feed intake in the KMR group was higher than in the others at the end of the experiment, as a consequence of the full development of microbes. The closeness of the chemical component of RMR to that of rabbit milk may have promoted the appropriate substrate for early microbial colonization, leading to higher caecal enzyme activity. None of prebiotic supplement and difference in chemical composition in MMR could be the cause of the lower enzyme activity in the caecum compared to other groups. Therefore, supplementation with probiotics and/or substrates for a normal flora in feed formulation with the appropriate chemical components (i.e., in line with rabbit milk) may represent the best procedure to achieve good development of the microbial community in early weaned rabbits. However, a microbial community analysis should be performed in a future study to confirm this hypothesis.

No statistically significant difference in growth performance was observed in this study. However, a trend existed that can be explained in detail. The lowest FCR was observed in the RMR group as a consequence of the lower ADFI and the higher ADG. This may have been due to the appropriate crude protein and fat proportion in the RMR, which promoted lower adaptation following the milk replacer in the first period. However, the highest ADFI was observed in the KMR group as the lowest energy density because the feed intake was stimulated by a physiological function until the animal had obtained its daily energy requirement [6]. Therefore, a lower pelleted diet intake could be observed in the group already receiving a high-density milk replacer in the RMR and MMR groups. The digestibility of dry matter and nutrients in the RMR group tended to be lower than in the other groups. The supplementation of prebiotics and substrates for microflora in KMR can support the microbial community and facilitate increased digestibility. In addition, MMR contained the highest energy density, which bacteria can use as energy sources to produce butyrate. Intestinal epithelial cells can utilize such short-chain fatty acids as an energy source, promoting proliferation, differentiation and gut immunity [21]. Thereby, the greatest villus height and villus crypt depth were seen in the MMR group, enabling better digestion and absorption and providing higher nutrient digestibility RMR as a result [22]. Furthermore, the amount and characteristics of hard faeces can be used to indicate digestive health [7]. The total weight of faecal excretion did not differ between groups, but the largest number of faecal pellets was found in MMR, followed by KMR and RMR. Moreover, no soft faeces were found under the rabbit cage, indicating that crude protein intake did not exceed their requirement [19].

Lower growth performance and survival rates were reported in the group fed with KMR compared with the group fed with rabbit milk from lactating does [8]. Unfortunately, this study did not compare the milk replacers with rabbit milk. Early separation at 14 days of age may have been the cause of the negative consequences found in the study of [8], whereas separation at 18 days of age did not result in any serious adverse outcomes here. In addition, a high mortality rate was observed in the study of [8] because the rabbits were too young, were injured and were experiencing high levels of stress due to being wild rabbits.

5. Conclusions

Differences in efficiency between the RMR developed in this study and commercial milk replacers (KMR and MMR) used in 18-day-old rabbits were revealed in the current study. Based on the results, it was possible to use RMR as a milk replacer for 18-day-old rabbits and to wean at 36 days of age, this not providing any adverse consequences for final body weight, ADG, FCR, ADFI, nutrient digestibility, internal organ characteristics, caecal pH, amount of faeces excretion and crude protein intake. Lower nutrient digestibility was observed in the RMR group without statistically significant differences. Therefore, probiotics and/or prebiotics can be supplemented in formulations to promote a suitable microbial community and to provide benefits in terms of growth performance and nutrient digestibility.

Author Contributions: Conceptualization, A.K.; methodology, A.K.; investigation, P.C. (Panthiphaporn Chankuang), A.L., K.J., C.K., T.S., P.C. (Pipatpong Chundang) and A.K.; resources, P.S. and C.Y.; data curation, A.K.; writing—original draft preparation, A.K.; writing—review and editing, P.C. (Panthiphaporn Chankuang), A.L., K.J., C.K., T.S., P.C. (Pipatpong Chundang), A.K., P.S. and C.Y.; project administration, A.K.; funding acquisition, A.K. All authors have read and agreed to the published version of the manuscript.

Acknowledgments: The authors would like to thank Faculty of Veterinary Medicine, Veterinary Teaching Hospital Kamphaengsaen campus (Kasetsart University, Nakhon Pathom, Thailand) and Mr.Saha Bairak (Saha farm, Kanchanaburi, Thailand) for providing the rabbit milk replacer and rabbits, respectively. Moreover, we would like to acknowledge Thanaporn Sriprathardtrakul, Jiraphat Wangka and Sirapoom Narktap for the support in rabbit rearing and feeding.

Appendix A

Table A1. Effect of different milk replacers on number of faecal pellets.

Artificial Milk Replacers	Age (days)	Number of Feces
RMR [A]		
	19	0 [a]
	20	0 [a]
	21	6 [ab]
	22	117 [b]
	23	244 [cd]
	24	263 [c]

Table A1. *Cont.*

Artificial Milk Replacers	Age (days)	Number of Feces
	25	338 [def]
	26	401 [efgh]
	27	379 [defg]
	28	385 [defg]
	29	371 [cde]
	30	543 [hi]
	31	548 [fgh]
	32	590 [hi]
	33	592 [ghi]
	34	454 [def]
	35	587 [ghi]
	36	742 [i]
KMR [AB]		
	19	75.6
	20	70.8
	21	191
	22	180
	23	418
	24	432
	25	435
	26	456
	27	421
	28	428
	29	378
	30	493
	31	467
	32	505
	33	441
	34	431
	35	519
	36	563
MMR [B]		
	19	0
	20	3.33
	21	52.3
	22	151
	23	311
	24	245
	25	426
	26	509
	27	444
	28	432
	29	406
	30	520
	31	463
	32	476
	33	497
	34	350
	35	436
	36	477
SEM		12.02
p-Value		
	Groups	0.03
	Age	0.001
	Groups * Age	0.02

RMR = Rabbit milk replacer in this study, KMR = Kitten milk replacer (KMR®; Pet-Ag Inc., Hampshire, IL, USA), MMR = Mammal milk replacer (Zoologic® Milk matrix 30/52; Pet-Ag Inc., Hampshire, IL, USA; SEM = Pooled standard error of mean; [a–i, A, B] The differences in superscript capital or lower case letter in the same column represent statistical significant differences of groups or age, respectively ($p < 0.05$).

References

1. Grand View Research. Pet Care Market Size, Share & Trends Analysis Report by Pet Type (Dog, Cat, Fish, Bird), by Products, Competitive Landscape, and Segment Forecasts, 2018–2025. Available online: https://www.grandviewresearch.com/industry-analysis/pet-care-market (accessed on 20 April 2020).
2. Morley, C.; Fook, J. The importance of pet loss and some implications for services. *Mortality* **2005**, *10*, 127–143. [CrossRef]
3. Rooney, N.J.; Blackwell, E.J.; Mullan, S.M.; Saunders, R.; Baker, P.E.; Hill, J.M.; Sealey, C.E.; Turner, M.J.; DE Held, S. The current state of welfare, housing and husbandry of the English pet rabbit population. *BMC Res. Notes* **2014**, *7*, 942–954. [CrossRef] [PubMed]
4. Gutiérrez, I.; Espinosa, A.; Gracía, J.; Carabano, R.; De Blas, J.C. Effect of levels of starch, fiber, and lactose on digestion and growth performance of early-weaned rabbits. *J. Anim. Sci.* **2002**, *80*, 1029–1037. [CrossRef] [PubMed]
5. Kovitvadhi, A.; Chandang, P.; Tirawattanawanich, C.; Kasemsuwan, S.; Rukkwamsuk, T. The study on cluster and obstacles of meat rabbit production in Thailand. *J. Mahanakorn. Vet. Med.* **2019**, *14*, 115–127.
6. De Blas, C.; Wiseman, J. *Nutrition of the Rabbit*, 2nd ed.; CAB International: Oxfordshire, UK, 2010; pp. 1–19, 180–185, 229.
7. Varga, M. *Textbook of Rabbit Medicine*, 2nd ed.; Butterworth Heinemann: London, UK, 2014; pp. 84–87.
8. Paul, G.; Friend, D.G. Comparison of outcomes using two milk replacer formulars based on commercially available products in two species of infant cottontail rabbits. *J. Wildl. Rehabil.* **2017**, *37*, 13–19.
9. Ferguson, F.A.; Lukefahr, S.D.; McNitt, J.L. A technical note on artificial milk feeding of rabbit kits weaned at 14 days. *World Rabbit Sci.* **1997**, *5*, 65–70. [CrossRef]
10. Maertens, L.; Lebas, F.; Szendrö, Z.S. Rabbit milk: A review of quantity, quality and non-dietary affecting factors. *World Rabbit Sci.* **2006**, *14*, 205–230. [CrossRef]
11. AVMA (American Veterinary Medical Association). *AVMA Guidelines for the Euthanasia of Animals: 2020 Edition*; American Veterinary Medical Association: Schaumburg, IL, USA, 2020; pp. 63–64.
12. AOAC (Association of Official Analytical Chemists). *Official Method of Analysis*, 18th ed.; Association of Official Analytical Chemists: Rockville, MD, USA, 2006.
13. AFFCO (Association of American Feed Control Officials). *2020 Official Publication*; Association of American Feed Control Officials Inc.: Champaign, IL, USA, 2020.
14. Kovitvadhi, A.; Gasco, L.; Ferrocino, I.; Rotolo, L.; Dabbou, S.; Malfatto, V.; Gai, F.; Peiretti, P.G.; Falzone, M.; Vignolini, C.; et al. Effect of purple loosestrife (*Lythrum salicaria*) diet supplementation in rabbit nutrition on performance, digestibility, health and meat quality. *Animal* **2016**, *10*, 10–18. [CrossRef]
15. Areekijseree, M.; Engkagul, A.; Kovitvadhi, U.; Thongpan, A.; Mingmuang, M.; Pakkong, P.; Rungruangsak-Torrissen, K. Temperature and pH characteristics of amylase and proteinase of adult fresh water pearl mussel, *Hyriopsis* (*Hyriopsis*) *bialatus*, Simpson (1900). *Aquaculture* **2004**, *234*, 575–578. [CrossRef]
16. Vatanparast, M.; Hosseininaveh, V.; Ghadamyari, M.; Minoo-Sajjadian, S. Plant cell wall degrading enzymes, pectinase and cellulase, in the digestive system of the red palm weevil, *Rhynchophorus ferrugineus* (Coleoptera: Curculionidae). *Plant Prot. Sci.* **2014**, *50*, 190–198. [CrossRef]
17. Folin, O.; Ciocalteu, V. On tyrosine and tryptophan determinations in proteins. *J. Biol. Chem.* **1927**, *73*, 627–650.
18. R: A Language and Environment for Statistical Computing. Available online: https://www.R-project.org (accessed on 19 March 2020).
19. Combes, S.; Fortun-Lamothe, L.; Cauquil, L.; Gidenne, T. Engineering the rabbit digestive ecosystem to improve digestive health and efficacy. *Animal* **2013**, *7*, 1429–1439. [CrossRef] [PubMed]
20. Falcão-e-Cunha, L.; Castro-Solla, L.; Maertens, L.; Marounek, M.; Pinheiro, V.; Freire, J.; Mourão, J.L. Alternatives to antibiotic growth promoters in rabbit feeding: A review. *World Rabbit Sci.* **2007**, *15*, 127–140. [CrossRef]
21. Kinoshita, M.; Suzuki, Y.; Saito, Y. Butyrate reduces colonic paracellular permeability by enhancing PPARγ activation. *Biochem. Biophys. Res. Commun.* **2002**, *293*, 827–831. [CrossRef]
22. Makovicky, P.; Tumova, E.; Volek, Z.; Makovicky, P.; Vodicka, P. Histological aspects of the small intestine under variable feed restriction: The effects of short and intense restriction on a growing rabbit model. *Exp. Med.* **2014**, *8*, 1623–1627. [CrossRef]

15

In Vitro Techniques using the Daisy ^{II} Incubator for the Assessment of Digestibility

Sonia Tassone [1,*], Riccardo Fortina [1] and Pier Giorgio Peiretti [2]

[1] Department of Agriculture, Forestry, and Food Sciences, University of Turin, 10095 Grugliasco, Italy; riccardo.fortina@unito.it

[2] Institute of Sciences of Food Production, National Research Council, 10095 Grugliasco, Italy; piergiorgio.peiretti@ispa.cnr.it

* Correspondence: sonia.tassone@unito.it;

Simple Summary: The Ankom Daisy^{II} incubator (AD^{II}; Ankom Technology Corporation Fairport, NY, USA) has gained acceptance as an alternative to traditional in vitro procedures. It reduces the labour requirement and increases the number of determinations that can be completed by a single operator. The apparatus allows for the simultaneous incubation of several feedstuffs in sealed polyester bags in the same incubation vessel, which is rotated continuously at 39.5 °C. With this method, the material that disappears from the bag during incubation is considered digestible. The method, which was first developed to predict the digestibility of feedstuffs for ruminants, has been modified and adapted to improve its accuracy and prediction capacity. Modifications used by various researchers include the use of different inocula, buffer solutions, and sample weights. Recently, attempts have been made to adapt the method to determine nutrient digestibility of feedstuff in non-ruminant animals, including pets.

Abstract: This review summarises the use of the Ankom Daisy^{II} incubator (AD^{II}; Ankom Technology Corporation Fairport, NY, USA), as presented in studies on digestibility, and its extension to other species apart from ruminants, from its introduction until today. This technique has been modified and adapted to allow for different types of investigations to be conducted. Researchers have studied and tested different procedures, and the main sources of variation have been found to be: the inoculum source, sample size, sample preparation, and bag type. In vitro digestibility methods, applied to the AD^{II} incubator, have been reviewed, the precision and accuracy of the method using the AD^{II} incubator have been dealt with, and comparisons with other methods have been made. Moreover, some hypotheses on the possible evolutions of this technology in non-ruminants, including pets, have been described. To date, there are no standardised protocols for the collection, storage, and transportation of rumen fluid or faeces. There is also still a need to standardise the procedures for washing the bags after digestion. Moreover, some performance metrics of the instrument (such as the reliability of the rotation mechanism of the jars) still require improvement.

Keywords: in vitro digestibility; inoculum; rumen fluid; faeces; enzyme; Ankom Daisy^{II} incubator

---

1. Introduction

The in vitro digestion method was first developed as an alternative to the costly, labour-intensive, time consuming, and ethically difficult in vivo method to predict nutrient digestibility in ruminants.

The first method, described by Tilley and Terry [1] as a two-stage rumen fluid–pepsin technique (TT), provided satisfactory estimates of in vivo apparent digestibility [2], although some authors found that the TT was just accurate for fresh grasses and not for silages or straw [3–5]. Van Soest et al. [6] (VS) and Goering and Van Soest [7] (GVS) modified the TT by replacing the acid–pepsin step with a neutral detergent digestion step; this version of the method is faster and more accurate than the original TT, and it is able to estimate the in vitro true digestibility of feedstuffs on the basis of the undigested cell-wall constituents.

In an attempt to overcome problems related to the variability of the rumen fluid [8], Czerkawski and Breckenridge [9] developed a continuous-culture system using an apparatus described by Gray et al. [10] and by Aafjes and Nijhof [11] as a starting point: the "RUmen SImulation TEChnique" (Rusitec), which is still successfully used to generate inocula for in vitro studies [12–14].

Other in vitro methods have been developed to estimate the digestibility of feedstuff. Menke and Steingass [15] proposed to measure the gas produced during fermentation and feed composition data to estimate the energy content of feeds. Theodorou et al. [16], considering previous studies [17,18], developed an in vitro method to measure the accumulation of head-space gas; this method was then revised by other authors, who used computerised pressure sensors to monitor the gaseous products of the microbial metabolism and found a clear linear relationship between the disappearance of neutral detergent fibre (NDF) and the production of gas [19,20].

The need for a piece of apparatus that would be capable of automating traditional in vitro digestibility analysis and resolving some analytical errors such as those pertaining to sample handling and manual filtration steps led to the development of the Ankom Daisy^{II} incubator (AD^{II}; Ankom Technology Corporation Fairport, NY, USA).

This review summarises the use of the AD^{II} incubator—from its introduction until today—in digestibility studies on ruminants, compares and correlates it with other digestibility procedures, and discusses the sources of variability of the results and the extension of this technology to other non-ruminant species. Finally, some hypotheses on the future evolution and development of this technology and on the standardisation of the procedure are presented

2. The Ankom Daisy^{II} Incubator

The AD^{II} incubator started out as a project for a Canadian customer and was introduced to the public in 1994 as a wooden and somewhat fragile cabinet [21]. In 1997, a new model was made with a more resistant metal cabinet, exactly as in the currently marketed form (Figure 1). AD^{II} is essentially based on the in vivo simulation of digestion. With this device, it is possible to simultaneously analyse up to 92 samples in a thermostatically controlled chamber that contains four rotating digestion jars. The temperature inside the chamber is maintained at 39 ± 0.5 °C by a heat controller; a timer allows each incubation period to be set. Samples are weighed in F57 filter bags (25 μm pore size) (Ankom Technology Corporation Fairport, NY, USA) and put into the jars (up to 23/jar) together with the inoculum (rumen fluid, faeces, or enzymes) and a buffer solution. Each of the four glass jars, placed on the rotation racks inside the incubator, contains a perforated agitator baffle that divides the internal volume into two parts and allows for the free movement of the digestion medium. The bags are weighed before and after a specific period of incubation, and the material that has disappeared is considered digestible dry matter. The AD^{II} incubator offers advantages in time, efficiency and labour requirements over conventional methods, such as the Tilley and Terry method and the Van Soest method. Because of its design, the AD^{II} design is capable of testing a large number of samples [22–24]. It has been identified as an easy, inexpensive, and efficient instrument for the prediction of the digestibility of several feedstuffs and diets [8,22]. However, compared to other techniques (such as the batch culture technique, the use of the Ankom Gas Production System or the Rumen Simulation Technique), the AD^{II} incubator has been demonstrated to give higher values at different incubation times [25].

One application of the AD^{II} incubator is the estimation of neutral detergent fibre digestibility (NDFD) at single time points (such as 30 or 48 h) [26].

Attempts to address the variability of results have involved the assessment of the vessel type and the sealing, venting, and gassing procedures [27]; the comparisons of different types of fibre-bag and the use of sodium sulphite for long incubation periods [28]; the development of specific in vitro methods to determine indigested NDF and to estimate the individual pool sizes and rates of digestion for application for diet formulation purposes [29]; the evaluation of the storage times and temperatures of rumen fluid before its transfer to the incubation flask [30]; the effects of the priming techniques of rumen fluid [31,32]; comparisons with in situ and various in vitro methods [33,34]; and the quantification of two pools of digestible NDF (fast and slowly digested) with a minimal number of fermentation time points [35].

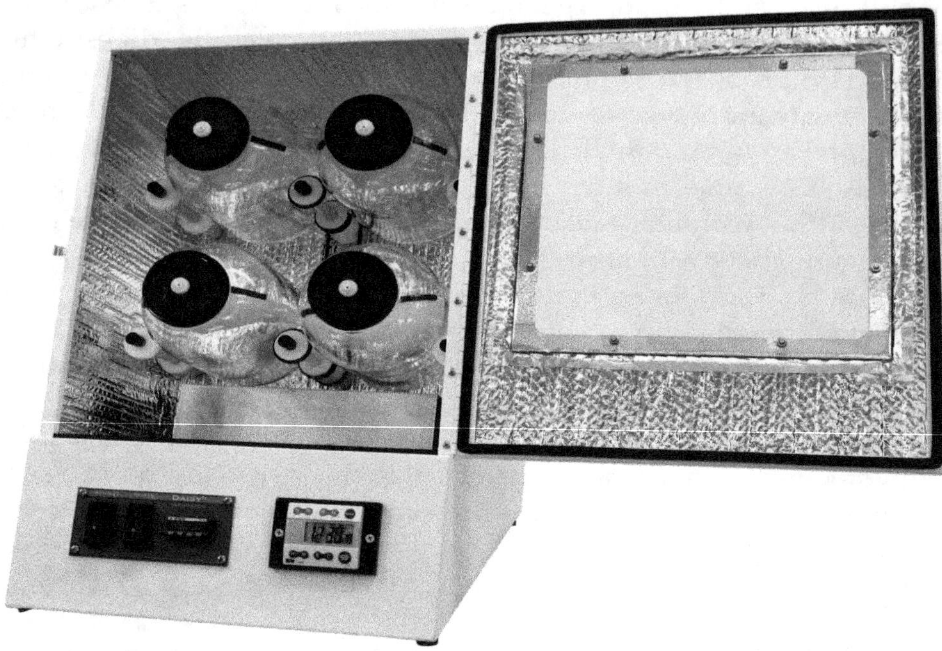

Figure 1. The Daisy^{II} incubator (Ankom Technology Corporation Fairport, New York, NY, USA).

Recently, the AD^{II} incubator has been used for the in vitro long-term ruminal digestion (240 h) of undigested NDF (uNDF) [28]. To estimate the kinetics of NDF degradation, longer time intervals are essential, especially when using complex models. Complex models may require inputs of fast, slow, and indigestible NDF pools [35,36], which can be determined with ease when using the AD^{II} incubator.

A list of practical recommendations on the use of AD^{II} incubator and a list of the main problems concerning the use of the instrument that require further study are reported in Table S1.

3. Inoculum Applied to the Use of the Ankom Daisy^{II} Incubator

Inoculum is very important for in vitro fermentation studies, but it also represents the greatest source of uncontrolled variation in fermentation systems. The inoculum has to create a similar environment to that of the digestive tract [37], but its digestive capacity may be influenced by the animal species, breed, and individual and within animal variations from time to time [38]. The characteristics and quality of the inoculum is not a specific problem of the AD^{II} incubator. As there is a lack of specific information on the AD^{II} incubator and some authors have studied inoculum for in vitro analysis, we reported their experience with other digestibility systems, because this information may also be useful for Ankom Daisy^{II}.

3.1. Rumen Fluid

As for other systems, the most frequently used inoculum source in the AD^{II} incubator is rumen fluid (RF). The necessity of fistulated and cannulated animals to provide this inoculum raises a number of practical problems, e.g., the need for surgical facilities, constant care to avoid infections, and the costs associated with the long-term maintenance of these animals. Moreover, the use of cannulated animals for this purpose has been criticised on ethical grounds.

Different solutions, in which there is no need to use cannulated animals, have been studied to resolve cost issues and ethical concerns about the well-being of animals. RF can be obtained via the oesophagus, thereby avoiding the need for cannulation, but such samples are often contaminated with saliva, and their collection causes considerable stress to the host animal. Moreover, as a result of the placement of the sampling device, the samples may not be representative of the entire rumen contents [37]. A very different approach with more details on this matter can be found in a paper by Ramos-Morales et al. [39]. These authors assessed in vivo trials conducted with ruminally cannulated sheep and goats to validate the use of stomach probing as an alternative to rumen cannulation in small ruminants with the aim of detecting any differences in ruminal fermentation and in the microbial community between species, diets, and sampling times.

A more ethically acceptable approach that reduces stress and alleviates the suffering of animals by avoiding an invasive procedure is the collection of RF at slaughtering [40]. Alba et al. [41] verified, through the use of an AD^{II} incubator, that the rumen inoculum obtained from slaughtered cattle can be used to replace the use of cannulated animals and that this approach is a viable alternative to digestibility analysis.

This method is accepted by the Rumen Microbial Genomics Network [42] for microbiota studies and has been mentioned as an alternative to sampling via cannula [43].

A supplemental video of the sampling procedures of RF at slaughtering is available online [44]. These procedures involve the collection of the rumen content into plastic bags a few minutes after slaughtering; the rumen content is squeezed, and the RF is filtered and collected into pre-heated plastic bottles. The presence of oxygen is avoided by squeezing the bottles while closing them; the rumen fluid is transported to the lab (max. 1.5 h time) at a temperature of 39–42 °C.

The effects of the source of inoculum with various combinations of donor cow diets generally vary to a great extent [45]. The results of a trial conducted by Holden et al. [22] showed that the source of inoculum affected in vitro dry matter digestibility (DMD). A grass hay donor cow diet resulted in lower digestibility values than a corn silage-based, total mixed ration donor cow diet for alfalfa hay, grass hay, steam flaked corn, and dry ground corn. No influence of the donor diet was found for mixed haylage, corn silage, grain mixture, or high moisture shelled corn. King and Plaizier [46] found that the source of inoculum (steers or cows) did not affect apparent or true DMD to any great extent. They also found that forage digestibility was similar when using the RF from sheep and from cattle [23]. Ammar et al. [47], using an AD^{II} device, found that the RF of sheep and goats was similar under the conditions of the experiment when all the donor animals were fed the same diet and were maintained under the same conditions.

Robinson et al. [30] examined the influence of storage time and temperature on the ability of rumen microorganism to degrade NDF. They reported that within-day delays of up to 6.5 h between the time of collection of rumen inoculum and the time of the initiation of the in vitro incubation had no impact on the measured 48 h digestion of NDF if the RF was maintained at 39 °C under anaerobic conditions during the delay. Similarly, the RF of sheep, preserved for up to 6 h in crushed ice, had no effect on any fermentation parameters [48]. Another possible RF storage system for in vitro incubation is short-term refrigeration [41]. Chaudhry and Mohamed [49] tested thawed RF from frozen rumen contents (stored at −20 °C for 4 w) against fresh RF from the same slaughtered cattle. Though the thawed RF had a lower degradation than the fresh one, it could be used to predict in vitro digestibility, as the values were closely correlated ($R^2 = 0.95$). However, it was still necessary to test its suitability for routine use. Hervas et al. [48] instead found a reduction in fermentative activity as a result of freezing (24 h). Spanghero et al. [14] recently compared inoculum collected at slaughtering with RF samples obtained from a continuous fermenter that were fresh, refrigerated at 4 °C, chilled at −80 °C, and freeze-dried. They evaluated the fermentability by measuring the NDF, crude protein degradability, and gas production. They confirmed that short-term refrigeration is a valuable technique to manage RF, whereas methods based on low temperatures significantly reduce the *Fibrobacter succinogenes*, which are very important for fibre degradation. Denek et al. [50] studied the preservation of microorganisms with a cryoprotectant under different deep-frozen conditions. They showed that RF treated with 5% dimethyl sulphoxide and frozen in liquid nitrogen gave similar results to fresh RF, but they also showed that the incubation time needed to be increased to 72 h to measure the digestibility of roughages. Belanche et al. [51] assessed the relevance of different factors (the diet of the donor animal, the fermentation substrate, microbial fraction, and the inoculum preservation method) to maximize the rumen inoculum activity, and they found that the highest microbial numbers and in vitro fermentation rates were recorded for fresh RF sampled after 3 h from donor animals fed a high concentrate diet.

As far as the microbial population that develops in an AD^{II} incubator is concerned, Soto et al. [52] showed the variations such a population underwent during the incubation process, and they compared the results with those of a Wheaton bottle and a single-flow continuous-culture fermenters using the same goat RF. In an AD^{II} incubator, they monitored the different microbial groups (bacteria, archaea, fungi, and protozoa) for 48 h by means of real time-PCR and terminal-restriction fragment length polymorphism. They observed a general decrease in the microbial population and important changes in microbiota profile, as the methanogens population increased. A similar trend was observed for the Wheaton bottle at 72 h, but there was also a growth of fibrolytic bacteria. However, the continuous-culture fermenters kept the rumen microbiota similar to that sampled from the rumen.

Spanghero et al. [14] found that the fermentation liquid from rumen continuous-fermenters can be used to generate inoculum for in vitro purposes.

Problems can arise for microorganisms, regarding the preparation of inoculum [37], connected with feed particles, the use of multiple layers of cheese cloth, and/or the use of some physical methods (e.g., the Stomacher method or the maceration of the rumen content in a food processor), which may destroy cell integrity.

3.2. *Faecal Inocula*

Fresh faeces (FF) have been used as an alternative source of ruminal inoculum in many experiments [41]. All these studies have demonstrated that bovine faeces may be used as microbial inocula for in vitro digestion and gas production, but this use has some limitations, such as a lower enzymatic activity than RF [53–55]. According to Akhter et al. [56], cattle faeces could also be used as an alternative to sheep RF.

Tufarelli et al. [57] tested faecal samples of yaks (*Bos grunniens*) as an alternative microbial inoculum source and compared them with RF, which was used as a control. They found that a faecal extract could be utilised instead of RF to estimate in vitro digestibility and that an AD^{II} incubator, with faecal liquor, is able to simply assess the adaptation capability of ruminant species to a pasture. These results were confirmed using camel faeces as a source of inoculum for AD^{II} [58].

Bovine FF may be used to replace bovine RF for incubation times no lower than 48 h [59]. Chiaravalli et al. [60] utilised an AD^{II} incubator to estimate the undigestible NDF of seven substrates using three different inocula (one rumen and two faeces) and considering two incubation times (240 and 360 h). The undigestible NDF results showed that faecal inoculum could be used to replace RF for long incubation times and that faeces can be used as an inoculum for end-point measurements.

The diet of an animal can change its microbial population. Guzmán and Sager [61] compared the microbial inoculum collected from a rumen-fistulated Aberdeen Angus steer fed with alfalfa hay and then with low quality digit hay (*Digitaria eriantha*), as well as the faeces collected from the same animal to evaluate the substrate, inoculum, and digestibility interaction. Using both inoculum sources, the true DMD was found to be affected by the diet of the donor animal, and the RF values ranked higher in the runs. Moreover, Kim et al. [62] suggested considering the diet, because it has an important effect on faecal microbiota, in particular when a forage-based diet is compared with a concentrate.

Faeces have also been extensively used as inoculum for in vitro incubation trials on monogastrics. Lowman et al. [63] were the first to demonstrate that equine faeces can be used as a source of microbial inoculum and that the faecal microflora of equines can remain viable for several hours after excretion. Other authors have confirmed these results. Earing et al. [64] demonstrated that the in vitro methodologies developed for the AD^{II} incubator could produce accurate estimates of in vivo equine apparent DMD and NDFD when equine faeces were used as the inoculum source. They evaluated three incubation periods in their study: 30, 48, and 72 h. Though the 30 and 48 h in vitro estimates were consistently less accurate than the in vivo estimates, they ranked diets in the same order as the in vivo method, and the 72-h period provided the most similar digestibility estimates to the in vivo data. Tassone et al. [65] evaluated the use of the AD^{II} incubator for the apparent and true DMD and NDFD measurements of feedstuffs considering four incubation times (30, 48, 60, and 72 h) using donkey faeces as a source of microbial inoculum. All the digestibility parameters increased significantly after 30–72 h of incubation, with average coefficients of variation for repeatability and reproducibility of 3.4% and 7.3% for apparent DMD; 1.7% and 4.3% for true DMD; and 6.6% and 14.6% for NDFD, respectively.

Table 1 summarises the references pertaining to rumen fluid and fresh faeces inocula applied to the AD^{II} incubator.

Table 1. Rumen fluid (RF) and fresh faeces (FF) inocula applied to the Ankom $Daisy^{II}$ incubator (AD^{II}).

Inoculum	Species	Sample Type	Notes	Ref.
RF	Dairy cattle	10 feeds	Variability of the dry matter digestibility for different donor cow diets as sources of inoculum	[22]
RF	Dairy cattle	By-products	AD^{II} vs. gas production, RF from slaughtered or cannulated cows	[41]
RF	Steers and Dairy cattle	Grains, total mixed ration, silages	Effect of the RF on the apparent and true dry matter digestibility DMD	[46]
RF	Sheep and Goats	Leaves, flowers and fruits of 5 browse plant species	Comparison of the true DMD and gas production kinetics with RF from animals fed the same diet	[47]
FF	Yaks	Forage produced at high altitude	Faeces vs. RF for a comparative digestibility trial	[57]
FF	Sheep and Camels	Fodder species from an arid environment	RF from sheep, and faeces from camels: comparative digestibility trial	[58]
FF-RF	Cattle	Feeds with different neutral detergent fibre (NDF) contents	NDF digestibility and undigested NDF measured with RF and 2 FF from cows fed different diets	[60]
FF-RF	Steers	35-day regrowth alfalfa hay	Comparative evaluation of the true dry matter digestibility; steers fed alfalfa or digit grass	[61]
FF	Horses	4 dietary treatments (hays or hay + oat)	Comparative evaluation of in vivo vs. in vitro DM and NDF digestibility	[64]
FF	Donkeys	7 common feeds for donkeys	Evaluation of the apparent and true DMD and neutral detergent fibre digestibility (NDFD) at 4 incubation times (30, 48, 60, and 72 h)	[65]

3.3. *Enzymatic Inoculum*

Enzymatic methodologies, in which microbial inoculum is eliminated, were developed to avoid problems associated with variations in rumen fluid over time [37]. This approach can be recommended because it offers an improved standardisation of the methodology, a reduction in the variations that may be attributed to the inoculum source and preparation, and a reduced dependence on surgically modified animals as rumen fluid donors [66]. However, the attempt to use enzymes instead of rumen fluid or other inocula have resulted in problems of variability in their preparation [67], and very little work has been done to optimise enzyme activities or incubation conditions. Though there are no available studies on ruminant digestibility in which enzymes were used in an AD^{II} incubator, many authors have already used enzymes in digestibility studies on pigs [68], rabbits [69,70], and dogs [71].

4. Sample Size, Sample Weight and Bag Type

The sample bags in the AD^{II} incubator constantly rotate in jars (0.95 rpm), and the internal septum leads to the complete immersion of the bags at every spin of the jar; in this way, gases do not accumulate inside the bag, and samples are prevented from floating freely in the flask. The continuous shaking of samples produced significantly higher digestibility results than when shaking occurred only twice daily [72]. As reported by Alende et al. [25], the use of filter bags may be advantageous, because filtration and recovery have been mentioned as sources of variability of the digestibility coefficients. Additionally, jars positioned horizontally render a higher digestibility than vertically placed ones. Holden et al. [22] found no significant differences when grains and forages were incubated in the same digestion vessel.

The first and most extensively used AD^{II} incubator bag is the F57 bag. The F57 bag is made up of an extruded polyethylene fibre with a three-dimensional filtration matrix that facilitates the maximum flow of a solution, thereby obtaining the best substrate interaction and minimum particle loss. The F57 filter bag has an approximately 25 μm pore size, is 50 mm long and 50 mm wide at the open top, and tapers to a bottom width of 30 mm. Sample processing, particularly concerning the grind size, interacts with the pore size of the bag and affects the extent of feed disappearance [73]. The ratio of the sample size to the bag surface area, suggested by Vanzant et al. [74] to increase the accuracy of degradability predictions relative to in vivo ruminal disappearance, is 10 mg/cm^2.

In previous studies, sample sizes of both 0.25 g [28,30] and 0.5 g [33,34] were used in conjunction with Ankom procedures [75]. Coblentz and Akins [76] compared the NDF digestibility values of triticale forages determined with the AD^{II} device, and they considered two sample sizes (0.25 and 0.50 g) and incubation periods of 12, 24, 30, 48, 144, and 240 h. The results were compared with those obtained from a commercial laboratory that used a traditional methodology. With the 0.25 g sample size, the linear equations between the Ankom and the traditional methods did not show differences both 30 and 48 h. There was less agreement, particularly for the 30 h incubation, when a sample of 0.50 g sample was used. The NDF digestibility values were generally greater for the 0.25 g sample size when using the Ankom methods, especially for incubation times of 24, 30, and 48 h.

Cattani et al. [77] evaluated what sample size (0.25 or 0.50 g/bag) allowed for a better correlation to be achieved between the NDFD and true DMD values obtained with the AD^{II} and a conventional batch culture technique. The regressions between the mean values, provided for the various feeds by the two methods, for the NDF and true DMD, had R^2 values of 0.75 and of 0.92,and an RSD (relative standard deviation) of 10.9% and of 4.8%, respectively, for the 0.50 g/bag size. The corresponding regressions for NDFD and true DMD showed R^2 values of 0.94 and of 0.98 and an RSD of 3.0% and of 1.3%, respectively, for the 0.25 g/bag size. This screening analysis therefore indicated that the reduction of the sample size from 0.50 to 0.25 g of feed sample/bag (corresponding to 12 and 6 mg/cm^2 of bag surface), when using an AD^{II} device, allowed for more closely correlated and less variable estimates of NDFD and true DMD to be obtained than those provided by the batch culture technique.

A recent work that evaluated the rate kinetics of triticale forages considered 0.3 g samples sealed within fibre bags as a procedural compromise between the 0.25 g sample size recommended for short incubation times and the necessity of ensuring that an adequate amount of residue remained after a long digestion time (144 and 240 h) [78].

The critics of the Ankom bag method have indicated the potential loss of small indigestible particles through its pores and that any method should decrease the loss of small particles without restricting access to the protozoa and bacterial populations. Ankom recommends F58 for crude fibre, neutral, and acid detergent fibre analyses. A pore sizes of <10 μm can restrict the number of protozoa and bacteria that enter digestion bags, so a smaller bag pore size than that of F58 is not advisable. Wilman and Adesogan [23] verified that soluble matter from samples high in soluble substances is able to escape from F57, thereby influencing the microbial population and increasing cell wall degradation in any samples low in soluble substances that are in the same jar. Valentine et al. [28] compared Ankom F57 bags (25 μm) with F58 bags (8–10 μm pore size) to measure undigested NDF after 240 h of incubation and found that both had significant effects on lowering undegraded NDFom values. In conventional procedures, smaller pore size filters generally tend to have greater average undegraded NDFom values than methods with larger pore size filters. They expected a similar finding, because potentially undigested NDF may be retained by finer filters, whereas potentially indigestible and digestible NDF may inadvertently escape from a coarser filter. They found when using the same technique for in vitro analysis, that Ankom F57 and F58 gave similar digestion rate results.

Adesogan [79] tested alternative bags to Ankom F57. He determined the in vitro apparent dry matter digestibility of the feed samples in an AD^{II} incubator using Ankom F57 bags and dacron bags with pore sizes of 30 and 50 μm, with or without a 5 g glass ball placed in the bags to ensure submersion in the media. He obtained different digestibility estimates when the alternative bags were used instead of the F57 bags, but the Ankom bags gave a more precise prediction of conventionally measured digestibility estimates than the alternative bags. Using Ankom bags ensures more standardised and repeatable results. The characteristics of alternative bags should be disclosed whenever they are used, instead of F57 bags, to estimate digestibility. Anassori et al. [80] also used dacron bags (pore size of 50 μm) in an AD^{II} to measure the organic matter digestibility (OMD) of forage-based sheep diets supplemented with raw garlic, garlic oil, and monensin. They compared AD^{II} with the TT and gas production. The values obtained with the AD^{II} method were always higher than those obtained with the TT and (for diets containing garlic oil) with in vitro gas production methods. According to the authors, in the AD^{II} procedure, a proportion of non-digestible fine particles may have been removed during incubation, boiling, and rinsing, thus reducing the weight of the residue and increasing the estimate of digestibility compared to that obtained with other methods.

Table 2 summarises the references pertaining to the sample size and bag type applied to the AD^{II} incubator.

Table 2. Sample size and bag type applied to the Ankom $Daisy^{II}$ incubator (AD^{II}).

Sample Size (g)	Bag Type	Sample Type	Notes	Ref.
0.25	F57	Forages and plant parts	Particle breakdown: 0.5, 1.0, and 1.5 mm	[23]
0.25	F57 and F58	Temperate and tropical grasses and legumes	uNDF after 240; effect of Na_2SO_3	[28]
0.25	Polyethylene polyester polymer bags	Low- and high-quality forages and grains	Different time delays and storage time between the collection of RF and the analysis	[30]
0.25	F57 and dacron bags (pore size: 0.30 and 0.50 μm)	Dried samples + 5 g glass balls	DMD of feed samples with alternatives to F57 and weighted to ensure submersion in the media	[79]
0.25	Dacron bags (pore size: 0.50 μm)	5 feeds + garlic or garlic oil vs. Monensin	Sheep RF, the effect of inclusion on organic matter digestibility (OMD)	[80]
0.30	F57	Triticale	Short and long (240 h) incubation times	[78]
0.50	5 × 3 cm pore size 0.45 μm	Pastures, forages and by-products	Comparison of in situ DM and NDF degradation kinetics	[33]
0.50	F0285 (pore size 0.25 μm)	Corn silage	Comparison of in vitro and in situ estimates of indigestible NDF at 2 fermentation end points (120 and 288 h)	[34]
0.25 0.50	F57	Triticale	Comparison of NDFD with 2 sample sizes	[76]
0.25 0.50	F57	7 feeds	Correlation with a conventional batch culture	[77]

5. Buffer Solutions and in Vitro Digestibility Methods Applied to the Ankom Daisy^{II} Incubator

Many methods and buffer solutions that are used to study in vitro digestibility, first for ruminants and then for monogastrics, have also been applied to the AD^{II} incubator.

A buffer solution (either phosphate, carbonate, or both) is used during incubation to control the pH and to supply nutrients for the inoculum microorganisms. Without a buffer, the short chain of fatty acids would lower the pH [81]. As authors have reported, only phosphate buffers do not require preparation under CO_2. The references of the different buffer solutions used for in vitro digestibility analysis are briefly reported in Table 3. However, a comparison of buffer solutions is still lacking. In 2000, Figueiredo et al. [72] compared buffers that had been described by Marten and Barnes [82] with those that had been described by Minson and McLeod [83], and the authors verified that the solutions could replace each other.

Table 3. Different buffer solutions used in in vitro digestibility trials with the Ankom Daisy^{II} incubator (AD^{II}) in different animal species.

Buffer Solution References	AD^{II} References	Animal Species
[82]	[24]	Ruminants
	[61]	Ruminants
	[64]	Horses
	[65]	Donkeys
[83]	[72]	Ruminants
[84]	[85]	Ruminants
[86]	[25]	Ruminants

6. Precision and Accuracy of the Method Using the Ankom Daisy^{II} Incubator

The utilisation and the diffusion of AD^{II} to study in vitro digestibility is a result of the reliability and accuracy of the method.

Damiran et al. [87] found a coefficient of variation (CV) of 4.7% for DMD measured with AD^{II} and a CV of 12.2% for NDFD. A CV of <1% was observed between sample replicates in other laboratories for the in vitro true digestibility values, but this coefficient normally ranged between 1–3% [21]. However, it is a little higher for NDFD analysis and typically ranges from 2.0–4.5%, depending on the type. Corn silage samples are always a little more variable. If any sample has a CV of over 5%, it should be re-analysed. Figueiredo et al. [72] verified a good reproducibility when measuring digestibility with AD^{II}. They reported a low coefficient of variation (CV = 2.65%) between jars and within jars, with values of 3.92, 2.13., 6.12, and 1.94 for jar numbers of 1, 2, 3, and 4, respectively. Tagliapietra et al. [88], in situ and in vitro, studied the rumen fluid of 11 feeds collected by means of oro-ruminal suction from intact donor cows. The reproducibility coefficient of the DMD for AD^{II} was 96.0%. The DMD values were underestimated when filter bags were considered, compared to in situ-nylon bags and in vitro conventional bottles. Nevertheless, it was possible to overcome the lower repeatability provided by the filter bags by increasing the number of replicates: three filter bags led to approximately the same standard error as the mean of 2.5 nylon bags and the mean of 2 conventional bottle measurements. The results showed a direct proportionality between the DMD values obtained in situ and in vitro with different techniques (in situ nylon vs. in vitro conventional bottles and in situ synthetic filter bags vs. AD^{II}).

Spanghero et al. [89] studied the NDF degradability of 18 hays considering different incubation times (2, 4, 8, 16, 24, 48, and 72 h) and found that the variability (CV) of the AD^{II} incubator (including jar repeatability) was 2.8%—that is, a similar value to that generally found for some chemical analyses of feedstuffs [90] and one that is lower than that obtained for in situ measurements (including low repeatability, CV: 3.7%).

Spanghero et al. [91] also evaluated the precision of the AD^{II} device in measuring the in vitro NDF degradability of 162 hay samples from permanent Austrian grasslands. The obtained results showed a within forage standard error of 2.8%. This limited repeatability of the measurement was attributed to various sources of variability (bag porosity, dimensions, amount of substrate, etc.), but not to the different jar positions in the fermenter, because the average values obtained after five incubations for the different jars were not statistically different.

Spanghero et al. [92] also investigated the precision and accuracy of the AD^{II} incubator for NDFD analysis and the accuracy and reproducibility of the associated calculated net energy of lactation. Five laboratories analysed 10 fibrous feed samples each; the fermentation times in the AD^{II} incubator were 30 and 48 h. The precision was measured as the standard deviation (SD) of the reproducibility (SR) and repeatability (Sr) of the between and within laboratory variability. Extending the fermentation time from 30 to 48 h increased the NDFD values (from 42% to 54%) and improved the NDFD precision, in terms of both Sr (12% and 7% for 30 and 48 h, respectively) and SR (17% and 10% for 30 and 48 h, respectively). The 48-h period of incubation improved the accuracy and reproducibility of the calculated net energy of lactation.

The accuracy and precision of NDFD, determined after short or long-time intervals, has recently been of considerable research and industry interest, as the relative consistency of the results.

Cişmileanu and Toma [93] studied the repeatability, reproducibility, and accuracy of an AD^{II} incubator using a new version of the TT. The stages of the method were similar to those of the traditional version: one stage with buffered rumen liquid and one stage with pepsin–HCl. An alfalfa hay sample was tested to establish the OMD by means of the in vivo method, and it was then considered as an internal control feed with a known digestibility. The authors observed that the coefficient of variability was 1.11% for repeatability and 1.85% for reproducibility. The accuracy was the same as that obtained with the conventional method.

Moreover, even if the AD^{II} incubator is fully functional, sometimes the jars do not rotate correctly and suffer from slowdowns, stops, and starts [94]. Some structural adjustments are therefore necessary to better exploit the potential of the AD^{II} incubator and to implement its diffusion and use.

Table 4 summarises the references pertaining to the precision and accuracy of the method using the Ankom Daisy^{II} incubator.

Table 4. Precision and accuracy of the method using the Ankom Daisy^{II} incubator (AD^{II}).

Parameters	Notes	Ref.
DMD and NDFD by means of the two-stage rumen fluid–pepsin technique (TT), the AD^{II} incubator and in situ; 0.25 and 0.50 sample size; 1 and 2 mm grinding size	The digestibility values estimated means of the by AD^{II} incubator and in situ techniques were correlated (R^2 = 0.58–0.88) with values estimated by means of conventional in vitro and in vivo techniques. In most cases, the AD^{II} incubator and in situ techniques overestimated DMD and NDFD	[87]
In vitro DMD vs. Minson and McLeod technique [83]	Good reproducibility between and within the jars in the AD^{II} incubator	[72]
In situ (2 different filter bags) and TDMD (traditional bottles or the AD^{II} incubator)	The AD^{II} incubator underestimated the TDMD values but there was direct proportionality between the in situ and in vitro DMD values	[88]
NDFD of 18 hays	The variability was similar to that of some chemical analysis and lower than the in situ measurements	[89]
NDFD of 162 hays	Similar average values	[91]
NDFD and the associated calculated net energy lactation (NEl) of 10 fibrous feeds; 5 laboratories	Improved NDFD precision and improved accuracy and reproducibility of the calculated NEl for an extended fermentation time (48 h)	[92]
Validation of a modified TT by achieved by testing the repeatability and reproducibility of the new TT as well as the correlation with a previous version of the method	Good repeatability and reproducibility achieved when using the new version of the TT with the AD^{II} incubator; the same accuracy was achieved as that of the conventional method	[93]

7. Comparison with other Methods

Many methods are available to measure in vitro digestibility, but only a few articles have compared the results obtained using an AD^{II} incubator with the results of other procedures [25].

The first results on digestibility in ruminants obtained using an AD^{II} incubator were presented by Komarek et al. [95] in 1994 at the National Conference on forage quality in Lincoln (USA) [96]. The following year, Ayangbile et al. [97] showed that there were no differences between DMD data obtained from an AD^{II} incubator and data obtained by means of the conventional Tilley and Terry methods [1,7]. Traxler et al. [98] determined the true DMD on four forages for different incubation times (48, 72, and 144 h), and even though the conventional Van Soest method [6] was found to be more efficient, the results basically confirmed the conclusions of Ayangbile et al. [97].

Cohen et al. [99] incubated corn silage samples in tubes according to the GVS method [7] and in an AD^{II} incubator at different times using unwashed F57 bags or F57 bags washed in acetone before being filled. The NDFD measured with the AD^{II} incubator was lower than that in the tubes, probably because of the retention of gas and acid end products within the bags, and the values of the washed filter bags were similar to those obtained by shaking the tubes. Traxler [100] instead noted very few differences between the AD^{II} incubator and the GVS method [7].

Over time, other studies have confirmed that the AD^{II} incubator can be used to predict the DMD digestibility of forages, grains, and mixed rations for ruminants [7,22–24,26,73,87,101].

Ammar et al. [102] compared the TT and VS methods [6] using an AD^{II} incubator for leguminous shrub species. The medium was prepared according to the VS method. After incubation in a buffered rumen fluid, samples were either subjected to a 48 h pepsin–HCl digestion (TT) or gently rinsed and extracted with a neutral detergent solution at 100 °C, as described in the VS method. The apparent digestibility was generally lower than the true digestibility, and the differences were always significant, particularly in leaves.

The same author [103] used the VS method applied to the Ankom technique [104] to obtain the in vitro digestibility of the stems and leaves of grasses and legumes taken from the first and subsequent cuts of a permanent meadow. In this experiment, rumen fluid was withdrawn from adult sheep.

Gargallo et al. [85] verified the use of an AD^{II} incubator to determine the intestinal digestion of crude protein using Calsamiglia and Stern's three-step procedure (TSP) [84]. Four tests were conducted to study the effect of the type of pepsin, the type of bags, the amount of sample, and the number of bags per jar on the estimated intestinal digestion using the AD^{II} incubator and the TSP techniques on soybean meal samples, heated at different temperatures, and with 12 protein supplements. The results showed that the intestinal digestion of soybean meal and the 12 protein supplements from the TSP and the AD^{II} incubator (with R510) were closely correlated. The amount of sample per bag and the number of bags per jar did not affect the estimates, and up to 30 bags (Ankom R510) with 5 g of sample could be used in each jar of an AD^{II} incubator to estimate the intestinal digestion of the proteins in ruminants.

In 2017, Cişmileanu and Toma [93] successfully validated a new version of the TT applied to AD^{II}, in which the stages of the traditional procedure were maintained. Two stages, the first one with buffered rumen liquid and the second with the pepsin–HCl solution, were considered.

Holden et al. [22] compared a modification of the TT and the AD^{II} incubator techniques to determine DMD, considering sources of inoculum from two different donor cow diets, as well as all the forage and total mixed rations. Their results showed that the AD^{II} incubator did not affect the digestibility values of the forages or grains to any great extent, as well as that the source of inoculum could affect DMD.

Wilman and Adesogan [23] compared the TT and an the AD^{II} incubator to estimate apparent and true DMD, apparent and true OMD, and NDFD. The analysed forage samples comprised 72 combinations of two forage species (*Lolium multiflorum* and *Medicago sativa*), three plant parts, three degrees of particle breakdown, two field replicates with rumen fluid from sheep, and two field replicates with rumen fluid from cattle. It was found that the sieve size used when milling did not influence the true OMD. However, small differences were observed between the two forage species: the standard errors and coefficients of variation were higher for the AD^{II} incubator (mean: 4.0%) than for the TT (mean: 2.7%). When they used the TT, they found it was possible to more precisely predict the true digestibility than the apparent digestibility from the AD^{II} incubator results; the difference between apparent and true digestibility, when estimated using the AD^{II} incubator, appeared unrealistically low. The estimated digestibility was similar when rumen fluid from sheep and from cattle was used. In conclusion, the TT gives more precise results than the AD^{II} incubator, albeit at the cost of requiring more labour. Mabjeesh et al. [73] performed the same comparison (AD^{II} vs. TT) on 17 concentrates and protein supplements, and they obtained a satisfactory relationship (R^2 = 0.81), even though the AD^{II} incubator gave higher values for some energy concentrate and protein supplements.

Ricci et al. [105] compared the precision and accuracy of in vitro ruminal DM degradability using the TT, an AD^{II} incubator, and the gas-production technique to estimate the in vivo DM digestibility of tall wheatgrass, hay, and haylage. The goodness-of-fit of all the techniques with the in vivo DM digestibility and the relationships between them were evaluated by means of a simple linear regression analysis. The Pearson correlation coefficient (ρ) was used to evaluate the strength of the association between the observed and in vitro estimated data. The concordance correlation coefficient (ρc) was used as a single indicator to integrate both precision and accuracy (Cb). This indicator (scaled between 0 and 1) is a reproducibility index that evaluates the agreement between two sets of data by measuring the shift in location from the concordance line (the 45° line through the origin) in the observed versus predicted plot. Cb is a bias correction factor that indicates how far the best fit line deviates from the concordance line. Linear relationships were observed between the in vivo and the TT, AD^{II}, and gas production values. The TT had the highest correlation (0.98), and this was followed by the gas-production technique (0.97) and then by AD^{II} (0.96). However, the TT exhibited the lowest accuracy (ρc = 0.341), and AD^{II} exhibited the highest (ρc = 0.850). The regression analysis showed an overestimation of the in vivo dry matter digestibility above 48.8% for AD^{II} and an underestimation below this value. AD^{II} is faster and more accurate than the other techniques, and it therefore appears to be the most suitable for in vitro digestion trials. Figueiredo et al. [72] compared the AD^{II} technique with Minson and McLeod's technique [83], (they modified the TT in 1972) and found higher values when they used the AD^{II} procedure.

Some authors have conducted comparison between an AD^{II} incubator and in situ system. Robinson et al. [30] reported higher NDFD values at 48 h with an AD^{II} incubator. Spanghero et al. [92] showed that the results of an AD^{II} incubator were closely correlated with the results of an in situ method (R^2 = 0.98). Spanghero et al. [89] compared the NDF degradability of 18 hays, measured by means of an in situ method (nylon bag technique) and the AD^{II} incubator. The incubation times were 2, 4, 8, 16, 24, 48, and 72 h. The NDFD values obtained in situ and in vitro with the AD^{II} incubator after 48 h of incubation were closely correlated (R^2 = 0.94). In another study [91], they verified that the NDF degradability of 162 hay samples measured in an AD^{II} incubator was 25–30% higher than the effective in situ values. The regression analysis between the in vitro and in situ NDFD values showed a medium degree of correlation and a low level of accuracy.

Tagliapietra et al. [88] compared four in situ methods with nylon bags and filter bags, as well as in vitro with conventional individual bottles or AD^{II}, to measure the DMD of 11 feeds. The reproducibility coefficients of the dry matter digestibility were 97.9%, 95.1%, 98.8%, and 96.0% for the in situ-nylon, filter bags, conventional bottles, and AD^{II}, respectively. The in situ and in vitro filter bags underestimated the dry matter digestibility values compared to the in situ-nylon bags and conventional bottles. They concluded that in vitro estimates of dry matter digestibility at 48 h with AD^{II}, using rumen fluid collected from intact cows, can produce similar values to those obtained in situ. The filter bags underestimated the dry matter digestibility values compared to the in situ-nylon bags and conventional bottles. However, it was possible to overcome the lower repeatability provided by the filter bags by increasing the number of replicates: three filter bags gave approximately the same standard error as the mean of 2.5 nylon bags and the mean of two CB measurements. The results showed a direct proportionality between the dry matter digestibility values obtained in situ and in vitro with different techniques (in situ-nylon vs. conventional bottles and in situ-filter vs. AD^{II}).

Alende et al. [25] compared three different DMD methods (AD^{II} incubator, batch culture, and Ankom gas production) considering four incubation times (12, 24, 36, and 48 h); the results obtained at 24 h were compared with those obtained from dual-flow, continuous-culture fermenters. The results showed that different methods yield different DMD values. When the incubation time was longer than 12 h, the predicted DMD from the AD^{II} incubator was greater than when the gas production and the batch culture methods were used. The apparent DM digestibility, estimated using the continuous culture fermenter, was similar to that obtained from the batch culture and gas production, but it was lower than that of the AD^{II} incubator. Damiran et al. [87] concluded that the AD^{II} technique is able to accurately predict in vivo and the in situ DMD. Table 5 summarises the references pertaining to comparisons with other methods.

Table 5. Comparison of the Ankom Daisy^{II} incubator (AD^{II}) with other digestibility methods.

Methods	Results (Referred to AD^{II} Technique)	Ref.
TT	True DMD: no differences	[97]
VS	True DMD considering 3 incubation times: the AD^{II} technique was less efficient but there were no significant differences	[98]
GVS	NDFD at different times: AD^{II} always lower than GVS; better results with F57 washed in acetone	[99]
GVS	NDFD: very few differences	[100]
TT, VS	Apparent and true DMD: significant differences	[102,103]
TSP	Intestinal digestibility of crude protein (R510 filter bags, up to 5 g sample): results closely results	[85]
TT	Validation of a modified TT with the AD^{II} technique	[93]
TT	Similar digestion values; the source of inoculum may affect DMD	[22]
TT	Apparent and true DMD, apparent and true OMD, NDFD. The TT gives more precise results but requires more labour	[23]
TT	Good agreement, but the AD^{II} technique gave higher values for some feeds	[73]
TT, gas production, in vivo	The results of 3 in vitro techniques (AD^{II}, TT and gas production) were highly correlated with in vivo; AD^{II} technique is faster and more accurate	[105]
Minson and McLeod [83]	Higher digestibility values were obtained with AD^{II}	[72]
In situ	NDFD was closely correlated	[92]
In situ	NDFD was 25–30% higher than in situ; a medium degree of correlation and low accuracy were achieved	[91]
In situ	Incubation at different times. The digestible NDF values were closely correlated at 48 h incubation, but the AD^{II} values of the NDFD were higher than the in situ values.	[89]
Different in situ and in vitro techniques	Lower reproducibility coefficients for AD^{II} than the other techniques; direct proportionality was observed between the in situ and in vitro DMD for different techniques	[88]
Batch culture, gas production	The AD^{II} dry matter digestibility values were higher than the gas production and batch culture values for longer incubation times than 12 h	[25]
In vivo, in situ, TT	The AD^{II} technique accurately predicted the in vivo DMD but overestimated in situ DMD; AD^{II} less accurately correlated with the TT	[87]

TT = Tilley and Terry; VS = Van Soest; GVS = Goering Van Soest, TSP = three-step procedure.

8. Use of Daisy^{II} Incubator for Non-Ruminants

8.1. Horses

The in vivo standard and the inert marker methods are optimal for the determination and assessment of the digestibility of horse feeds, but they are time consuming. The use of in vitro fermentation procedures, such as enzyme-based essays, for the prediction of pre-caecal starch digestibility [106], and the gas production technique, developed for ruminants [15] using either caecal fluid [107] or faeces as inocula [108] to study diet digestion and fermentative end products has become increasingly more popular in equine nutrition. Abdouli and Attia [109] developed a simple in vitro method that is suitable for both concentrates and forages and that combines both the pre-caecal and hind gut digestion processes. These authors focused on the duration needed to establish feed pre-digestion by pepsin–amylase and its subsequent effect on gas production and organic matter digestibility using horse faeces as a source of microbial inoculum, and they compared the results with those from low-to-high-starch and protein feeds. They concluded that this procedure should be extended and validated with a large array of feeds with known digestibility values, because the enzymatic pre-digestion treatment effects varied between samples (non-pre-digested hay, barley grain, and soybean meal). Equine faeces is a suitable source of microbial inoculum for in vitro gas production studies, and the evaluated in vitro batch culture technique showed a considerable potential for the routine prediction of the nutritive value of a wide range of equine feedstuffs [79].

Lattimer et al. [110,111] studied the effects of *Saccharomyces cerevisiae* on the in vitro fermentation of a high concentrate or high-fibre diet for horses using equine faeces as an in vitro inoculum source in an AD^{II} incubator. These authors demonstrated that the use of 0.25-g samples may yield more accurate and less varied estimates of DM digestibility. Furthermore, the DM digestibility values for the in vivo and in vitro were similar, and they concluded that the AD^{II} incubator could be used to predict the DM digestibility of diets. Earing et al. [64], evaluating the in vitro digestion of four different diets using the AD^{II} incubator, recently confirmed that equine faeces are a suitable source of microbial inoculum for in vitro digestibility studies on horses. They found comparable DM digestibility for diets consisting of timothy hay, timothy hay with oats, and alfalfa hay with oats between in vitro and in vivo methods, while different digestibility values were observed between the two methods for an alfalfa hay diet. These authors stated that further research is needed, using a wider range of forages and methods, to determine whether in vitro and in vivo digestibility methods produce similar results for horses and to establish in vitro digestibility as a viable technique for estimating digestibility in horses.

Blažková et al. [112] compared the in vivo DM digestibility of corn silage for horses with that obtained using equine faeces in an AD^{II} incubator. These authors concluded that DM digestibility is only comparable with data on ruminants, and they showed that horses have a lower DM digestibility of corn silage than ruminants. Moreover, they demonstrated that equine faeces are a suitable source of microbial inoculum for in vitro digestibility.

8.2. Donkeys

Despite the increasing interest in donkeys, studies on this species are very limited. Tassone et al. [65] demonstrated that donkey digestibility can be predicted, with a high repeatability and reproducibility, using an AD^{II} incubator, a closed-system fermentation apparatus, and donkey faeces as a source of microbial inoculum. Moreover, these authors observed that the digestibility of different feeds for donkeys needs different incubation times.

8.3. Camelids

In vitro TTs that use camel rumen liquor as an inoculum require fistulated animals to provide this inoculum [113,114]. Rumen fluid can also be obtained, for the same purpose, from slaughtered dromedaries. Lifa et al. [115] therefore investigated the suitability of this rumen fluid with the aim of evaluating the in vitro degradation characteristics of highly fermentable industrial by-products (citrus, tomato, and apple), fibrous forages, and their mixtures. They concluded that rumen fluid extracted from slaughtered dromedaries is a valuable tool for determining the in vitro degradation of camel feeds. None of these experiments on camelids were conducted using an AD^{II} incubator.

The successful use of a liquid suspension of camel faeces, as an alternative inoculum for an in vitro AD^{II} incubator, yielded valid in vitro estimates of the DM, NDF, and ADF (acid detergent fibre) digestibility of forages and grains and could make it unnecessary to resort to fistulated animals (particularly in tropical countries) to obtain inoculum; this could solve some practical problems, such as the constant care needed to avoid infections and the costs associated with the long-term maintenance of donor animals, as well as ethical considerations and the necessity of surgical facilities [58].

Laudadio et al. [58] evaluated the in vitro digestibility of the fodder species browsed by camels in pastures in an arid region of Southern Tunisia using an AD^{II} incubator. They used different sources of faecal liquor, collected from camels, healthy mature sheep, and goats, as alternative microbial inoculum sources to test the nutrient digestibility of these forages, as well as rumen liquor, collected from sheep, as a control for the in vitro AD^{II} incubator. These authors stated that the similarity of the different repetitions for all the fodders in the estimation of nutrient digestibility in the AD^{II} incubator reflects its accuracy, making it comparable with traditional methods in regard to digestibility. They concluded that the AD^{II} incubator is appropriate for the determination of the in vitro digestibility of nutrients when using camel faecal liquor, which could be used instead of rumen fluid to estimate the in vitro digestibility of forages.

8.4. Rabbits

An AD^{II} incubator was also used in rabbit studies to determine the in vitro insoluble fibre [116] and in vitro digestibility of rabbit feedstuffs [69,70,117–120]. Abad et al. [69] adapted the in vitro digestion procedure proposed by Carabaño et al. [121] and compared the quantifications of soluble fibre in rabbit feedstuffs using different chemical and in vitro approaches. The method was modified using Ankom filter bags, which were placed in an AD^{II} incubator jar rather than in crucibles (reference method) to facilitate sample filtering. No difference was observed when crucibles and Ankom bags were used (both in single or collective digestion) for two-step pepsin/pancreatin in vitro DM digestibility, corrected for ash and protein. The correlations obtained for in vitro DM digestibility were higher (0.99) than those reported by Vogel et al. [24], who studied the in vitro DM digestibility of forages for ruminants (0.92). The latter authors reported higher in vitro digestibility when using Ankom bags than when using crucibles (0.602 vs. 0.563, respectively), whereas Abad et al. [69] found much less of a difference.

Ferreira et al. [70], in order to evaluate the potential use of dried or autoclaved sugarcane bagasse and enriched or non-enriched with vinasse in the diets of growing rabbits and to determine their in vitro dry matter digestibility, modified the last step of the Abad et al. method [69] using a caecal contents diluted at a ratio of 1:1 (w/v) with a buffered mineral solution [122] as inoculum. Ferreira et al. used the same method to determine the in vitro dry matter digestibility of rabbit diets supplemented with macaúba seed cake meal [117] or with tropical ingredients, co-products, and by-products [118].

The Ramos et al. method [123], which is based on that of Boisen et al. [124], in which Ankom bags are used, and which, in turn, was modified by Abad et al. [69], was used to determine the in vitro dry matter digestibility of rabbit diets supplemented with co-products derived from olive cake [119] or with citrus co-products [120].

8.5. Guinea Pigs

López et al. [125] used an AD^{II} incubator to compare two types of "in vitro" digestibility assays, using commercial enzymes and guinea pig caecal liquor with the in vivo assay to identify the assay that resembled the in vivo response the most, and they found that the optimal in vitro method to use for comparisons with the in vivo test is the caecal liquor technique because it presents a smaller difference in results.

8.6. Pigs

Several in vitro feed digestibility estimation methods have been developed and can be divided into three groups, that is single-, two-, or three-step models that simulate gastric digestion, gastric/small intestinal digestion, and gastric/small intestinal/large intestinal digestion, respectively [126]. The Boisen and Fernandez [127] in vitro gastric-ileal digestion procedure was been adapted for use in an AD^{II} incubator and it allows for the simultaneous incubation of different pig feedstuffs in sealed polyester bags (5 × 10 cm bags; R510, Ankom Technology, Macedon, NY) in the same incubation vessel [68].

Fushai [128] determined, with an AD^{II} incubator, the in vitro digestibility of growing pig diets supplemented with exogenous enzymes. Each feed was digested in pepsin, followed by pancreatin, with the recovery of the fibrous residues. The pepsin–pancreatin fibre extracts were digested, by means of Viscozyme and Roxazyme, in a third step to complete the simulated pig gastro-intestinal digestion process.

Torres-Pitarch et al. [129] determined the in vitro ileal digestibility of pig diets by means of a two-step in vitro incubation procedure, adapted from that of Akinsola [68] using an AD^{II} incubator at 39 °C with samples incubated inside Ankom F57 bags. The first step, which simulated the digestion in the stomach, was that of enzymatic hydrolysis with a pepsin solution at pH 2.0 and 39 °C for 5 h, and the second step involved hydrolysis with a multi-enzyme pancreatin at pH 6.8 and 39 °C for 17 h.

Pahm [130] compared the use of an AD^{II} incubator with three Huang et al. [131] in vitro procedures using cellulase in the third step to that of Boisen and Fernandez [127] using Viscozyme or faecal inoculum in the third step. When using the AD^{II} incubator, these authors concluded that, of the three evaluated in vitro procedures, that of increasing the incubation length of the Boisen and Fernandez [127] using Viscozyme in the third step was the one that improved the sensitivity of the assay the most, and it provided a better R^2 between the dry matter digestibility and apparent total tract digestibility of the gross energy, and between the dry matter digestibility and digestible energy, than the procedures that used cellulase or faecal inoculum.

Youssef and Kamphues [132] analysed a commercial swine diet, with lignocellulose A and B, by means an AD^{II} incubator, to determine its in vitro dry matter digestibility, using the fresh faeces of pigs as the inoculum source. The fermentation rates of the tested ingredients were evaluated using the caecum contents of swine as inoculum precursors, and these were then compared with that obtained with faeces inocula. The in vitro results were confirmed in vivo by testing the digestibility rate of the most digestible product of the lignocellulose ingredients. These authors found that the use of faeces/excreta liquor provided a valid estimate of the fermentation or digestibility of feeds, and they concluded that this procedure could be an effective way of approximating the digestibility of pig diets.

8.7. Dogs

Candellone et al. [71] recently performed in vitro analyses of dog pet food using the methods proposed by Hervera et al. [133] and Biagi et al. [134] utilizing Ankom bags and an AD^{II} incubator. They concluded that the two in vitro methods slightly overestimated the digestibility coefficients of the considered dog diets, when compared with the in vivo digestibility values. The in vitro method proposed by Hervera et al. [133] and utilized in this study yielded values closer to the in vivo results, in line with Hervera et al. [135], who showed a higher accuracy approach of in vivo crude protein apparent digestibility (R^2 = 0.81) and in vivo digestible energy (R^2 = 0.94), respectively.

9. Conclusions

This review summarised the use of the AD^{II} incubator in studies on digestibility in ruminants, as well as its extension to non-ruminants. From its introduction until today, the AD^{II} incubator has proved to be able to allow for the analysis of multiple feedstuffs, to improve the precision and reproducibility of an assay, and to reduce the time and costs of analysis. DMD values from AD^{II} and in situ techniques may be higher than those obtained in vivo [104], but both systems allow for the true digestibility of feedstuffs to be estimated, while the in vivo values only refer to the apparent digestibility.

Even though the use of the AD^{II} incubator is by now standardised, there is still a need for further research, as reported in Table S1, to summarise some practical recommendations concerning the correct use of the AD^{II} incubator. To date, there are no standardised protocols for the collection, storage, and transportation of the rumen fluid or faeces. There is also a need to standardise the procedures for washing the bags after digestion. A major problem is the type of inoculum, which is the main source of variability of the system. Some performance metrics of the instrument (such as the reliability of the rotation mechanism of the jars) also require improvement.

The authors verified the need for caution when comparing data obtained from different methods, because they can yield different results [25]. Table S2 reports the variability of the AD^{II} instrument for 48 h of incubation, as well as the coefficient of variability (CV, %) within and between laboratory, runs, jars, and samples. Table S3 shows the correlation between AD^{II} and in vivo, in situ, and Tilley and Terry digestibility, as well as the respective linear equations.

The authors also verified that there is a lack of a standard terminology in studies and, as such, propose the use of the acronyms reported in Table S4 to make the language homogeneous.

Some potential developments and evolutions in the use of the AD^{II} incubator were also described. Created and developed for digestibility studies on ruminants, before being extended to monogastric and other non-ruminant species, this technology, in the future, could in fact be used for human digestibility studies or to obtain more detailed knowledge on the nutraceutical function of some feeds.

Supplementary Materials:
Table S1: Practical recommendations on the use of the Ankom $Daisy^{II}$ incubator (AD^{II}), Table S2: Variability (CV, %) of the Ankom $Daisy^{II}$ incubator (AD^{II}) after 48 h of incubation, Table S3: Linear equation between the Ankom $Daisy^{II}$ incubator (AD^{II}) at 48 h and other digestibility systems, Table S4: Acronyms for digestibility trials.

Author Contributions: Conceptualization, P.G.P. and S.T.; writing—original draft preparation, P.G.P., R.F., and S.T.; writing—review and editing, P.G.P., R.F., and S.T. All authors have read and agreed to the published version of the manuscript.

Acknowledgments: The authors would like to thank M. Jones for the linguistic revision of the manuscript.

References

1. Tilley, J.M.A.; Terry, R.A. A two-stage technique for the in vitro digestion of forage crops. *Grass Forage Sci.* **1963**, *18*, 104–111. [CrossRef]
2. Van Soest, P.J. *Nutritional Ecology of the Ruminant*, 2nd ed.; Cornell University Press: Ithaca, NY, USA, 1994.
3. Klopfenstein, T.J.; Krause, V.E.; Jones, M.J.; Woods, W. Chemical treatment of low quality forages. *J. Anim. Sci.* **1972**, *35*, 418–422. [CrossRef]
4. Givens, D.I.; Cottyn, B.G.; Dewey, P.J.S.; Steg, A. A comparison of the neutral detergent-cellulase method with other laboratory methods for predicting the digestibility in vivo of maize silages from three European countries. *Anim. Feed Sci. Technol.* **1995**, *54*, 55–64. [CrossRef]
5. Adesogan, A.T.; Givens, D.I.; Owen, E. Prediction of the in vivo digestibility of whole crop wheat from in vitro digestibility, chemical composition, in situ rumen degradability, in vitro gas production and near infrared reflectance spectroscopy. *Anim. Feed Sci. Technol.* **1998**, *74*, 259–272. [CrossRef]
6. Van Soest, P.J.; Wine, R.H.; Moore, L.A. Estimation of the true digestibility of forages by the in vitro digestion of cell walls. In Proceedings of the 10th International Grassland Congress, Finnish Grassland Association, Helsinki, Finland, 7–8 July 1966; pp. 438–441.
7. Goering, M.K.; Van Soest, P.J. *Forage Fiber Analysis (Apparatus, Reagents, Procedures and some Applications)*; Agricultural Handbook No. 379; USDA: Washington, DC, USA, 1970.

8. Adesogan, A.T. What are feeds worth? A critical evaluation of selected nutritive value methods. In Proceedings of the 13th Annual Florida Ruminant Nutrition Symposium, Gainesville, FL, USA, 11–12 January 2002; pp. 33–47.
9. Czerkawski, J.W.; Breckenridge, G. Design and development of a long-term rumen simulation technique (Rusitec). *Br. J. Nutr.* **1977**, *38*, 371–384. [CrossRef]
10. Gray, F.V.; Weller, A.F.; Pilgrim, A.F.; Jones, G.E. A stringent test for the artificial rumen. *Aust. J. Agric. Res.* **1962**, *13*, 343–349. [CrossRef]
11. Aafjes, J.H.; Nijhof, J.K. A simple artificial rumen giving good production of volatile fatty acids. *Br. Vet. J.* **1967**, *123*, 436–446. [CrossRef]
12. Carro, M.D.; Ranilla, M.J.; Martin-García, A.I.; Molina-Alcaide, E. Comparison of microbial fermentation of high- and low-forage diets in Rusitec, single-flow continuous-culture fermenters and sheep rumen. *Animal* **2009**, *3*, 527–534. [CrossRef] [PubMed]
13. Martínez, M.E.; Ranilla, M.J.; Tejido, M.L.; Saro, C.; Carro, M.D. Comparison of fermentation of diets of variable composition and microbial populations in the rumen of sheep and Rusitec fermenters. II. Protozoa population and diversity of bacterial communities. *J. Dairy Sci.* **2010**, *93*, 3699–3712. [CrossRef]
14. Spanghero, M.; Chiaravalli, M.; Colombini, S.; Fabro, C.; Froldi, F.; Mason, F.; Moschini, M.; Sarnataro, C.; Schiavon, S.; Tagliapietra, F. Rumen inoculum collected from cows at slaughter or from a continuous fermenter and preserved in warm, refrigerated, chilled or freeze-dried environments for in vitro tests. *Animals* **2019**, *9*, 815. [CrossRef]
15. Menke, K.H.; Steingass, H. Estimation of the energetic feed value obtained from chemical analysis and in vitro gas production using rumen fluid. *Anim. Res. Dev.* **1988**, *28*, 7–55.
16. Theodorou, M.K.; Williams, B.A.; Dhanoa, M.S.; McAllan, A.B. A new laboratory procedure for estimating kinetic parameters associated with the digestibility of forages. In Proceedings of the International Symposium on Forage Cell Wall Structure and Digestibility, USD-ARS, Madison, WI, USA, 7–10 October 1991.
17. Hungate, R.E. *The Rumen and Its Microbes*; Academic Press: New York, NY, USA, 1966.
18. Menke, K.H.; Raab, A.; Salewski, H.; Steingass, D.; Fritz, D.; Scneider, W. The estimation of digestibility and metabolizable energy content of ruminant feedstuffs from the gas production when they are incubated with rumen liquor in vitro. *J. Agric. Sci.* **1979**, *193*, 217–225. [CrossRef]
19. Pell, A.N.; Schofield, P. Computerized monitoring of gas production to measure forage digestion in vitro. *J. Dairy Sci.* **1993**, *76*, 1063–1073. [CrossRef]
20. Schofield, P.; Pell, A.N. Measurement and kinetic analysis of the neutral detergent-soluble carbohydrate fraction of legumes and grasses. *J. Anim. Sci.* **1995**, *73*, 3455–3463. [CrossRef]
21. Layton, B.; Ankom Technology Corporation Fairport, NY, USA. Personal communication, 2019.
22. Holden, L.A. Comparison of methods of in vitro dry matter digestibility for ten feeds. *J. Dairy Sci.* **1999**, *82*, 1791–1794. [CrossRef]
23. Wilman, D.; Adesogan, A. A comparison of filter bag methods with conventional tube methods of determining the in vitro digestibility of forages. *Anim. Feed Sci. Technol.* **2000**, *84*, 33–47. [CrossRef]
24. Vogel, K.P.; Pedersen, J.F.; Masterson, S.D.; Toy, J.J. Evaluation of a filter bag system for NDF, ADF, and IVDMD forage analysis. *Crop. Sci.* **1999**, *39*, 276–279. [CrossRef]
25. Alende, M.; Lascano, G.J.; Jenkins, T.C.; Koch Pas, L.E.; Andrae, J.G. Comparison of 4 methods for determining in vitro ruminal digestibility of annual ryegrass. *Prof. Anim. Sci.* **2018**, *34*, 306–309. [CrossRef]
26. Coblentz, W.K.; Akins, M.S.; Ogden, R.K.; Bauman, L.M.; Stammer, A.J. Effects of sample size on neutral detergent fiber digestibility of triticale forages using the Ankom Daisy^{II} Incubator system. *J. Dairy Sci.* **2019**, *102*, 6987–6999. [CrossRef]
27. Hall, M.B.; Mertens, D.R. In vitro fermentation vessel type and method alter fiber digestibility estimates. *J. Dairy Sci.* **2008**, *91*, 301–307. [CrossRef]
28. Valentine, M.E.; Karayilandli, E.; Cherney, J.H.; Cherney, D.J. Comparison of in vitro long digestion methods and digestion rates for diverse forages. *Crop. Sci.* **2019**, *59*, 422–435. [CrossRef]
29. Raffrenato, E.; Ross, D.A.; Van Amburgh, M.E. Development of an in vitro method to determine rumen undigested aNDFom for use in feed evaluation. *J. Dairy Sci.* **2018**, *101*, 9888–9900. [CrossRef]
30. Robinson, P.H.; Campbell Matthews, M.; Fadel, J.G. Influence of storage time and temperature on in vitro digestion of neutral detergent fibre at 48 h, and comparison to 48 h in sacco neutral detergent fibre digestion. *Anim. Feed Sci. Technol.* **1999**, *80*, 257–266. [CrossRef]

31. Goeser, J.P.; Combs, D.K. An alternative method to assess 24-h ruminal in vitro neutral detergent fiber digestibility. *J. Dairy Sci.* **2009**, *92*, 3833–3841. [CrossRef]
32. Goeser, J.P.; Hoffman, P.C.; Combs, D.K. Modification of a rumen fluid priming technique for measuring in vitro neutral detergent fiber digestibility. *J. Dairy Sci.* **2009**, *92*, 3842–3848. [CrossRef]
33. Trujillo, A.I.; Marichal, M.D.J.; Carriquiry, M. Comparison of dry matter and neutral detergent fibre degradation of fibrous feedstuffs as determined with in situ and in vitro gravimetric procedures. *Anim. Feed Sci. Technol.* **2010**, *161*, 49–57. [CrossRef]
34. Bender, R.W.; Cook, D.E.; Combs, D.K. Comparison of in situ versus in vitro methods of fiber digestion at 120 and 288 h to quantify the indigestible neutral detergent fiber fraction of corn silage samples. *J. Dairy Sci.* **2016**, *99*, 5394–5400. [CrossRef]
35. Raffrenato, E.; Nicholson, C.F.; Van Amburgh, M.E. Development of a mathematical model to predict pool sizes and rates of digestion of 2 pools of digestible neutral detergent fiber and an undigested neutral detergent fiber fraction within various forages. *J. Dairy Sci.* **2019**, *102*, 351–364. [CrossRef]
36. Mertens, D.R. Using uNDF to predict dairy cow performance and design rations. In Proceedings of the Four-State Dairy Nutrition and Management Conference, Dubuque, IA, USA, 12–13 June 2016; pp. 12–19.
37. Mould, F.L.; Kliem, K.E.; Morgan, R.; Mauricio, R.M. In vitro microbial inoculum: A review of its function and properties. *Anim. Feed Sci. Technol.* **2005**, *123–124*, 31–50. [CrossRef]
38. López, S. In vitro and in situ techniques for estimating digestibility. In *Quantitative Aspects of Ruminant Digestion and Metabolism*, 2nd ed.; CAB International: Wallingford, UK, 2005; pp. 87–121.
39. Ramos-Morales, E.; Arco-Pérez, A.; Martín-García, A.I.; Yáñez-Ruiz, D.R.; Frutos, P.; Hervás, G. Use of stomach tubing as an alternative to rumen cannulation to study ruminal fermentation and microbiota in sheep and goats. *Anim. Feed Sci. Technol.* **2014**, *198*, 57–66. [CrossRef]
40. Beyihayo, G.A.; Omaria, R.; Namazzi, C.; Atuhaire, A. Comparison of in vitro digestibility using slaughtered and fistulated cattle as sources of inoculum. *Uganda J. Agric. Sci.* **2015**, *16*, 93–98. [CrossRef]
41. Alba, H.D.R.; Oliveira, R.L.; de Carvalho, S.T.; Ítavo, L.C.V.; Ribeiro, O.L.; do Nascimento Júnior, N.G.; Freitas, M.D.; Bezerra, L.R. Can ruminal inoculum from slaughtered cattle replace inoculum from cannulated cattle for feed evaluation research? *Semin. Ciênc. Agrár.* **2018**, *39*, 2133–2144. [CrossRef]
42. RMG Network. A Report in Support of the Rumen Microbial Genomics (RMG) Network Describing Standard Guidelines and Protocols for Data Acquisition, Analysis and Storage. Available online: http://www.rmgnetwork:user/file/37/pdf (accessed on 25 February 2020).
43. Yáñez Ruiz, D.R.; Bannink, A.; Dijkstra, J.; Kebreab, E.; Morgavi, D.P.; O'Kiely, P.; Reynolds, C.K.; Schwarm, A.; Shingfield, K.J.; Yu, Z.; et al. Design, implementation and interpretation of in vitro batch culture experiments to assess enteric methane mitigation in ruminants—A review. *Anim. Feed Sci. Technol.* **2016**, *216*, 1–18. [CrossRef]
44. Bioscreen Technologies Research Laboratories—Artificial Rumen by-Pass. Available online: https://www.youtube.com/watch?v=AAT63sytI0w (accessed on 25 February 2020).
45. Marinucci, M.T.; Dehority, B.A.; Loerch, S.C. In vitro and in vivo studies of factors affecting digestion of feeds in synthetic fiber bags. *J. Anim. Sci.* **1992**, *70*, 296–307. [CrossRef] [PubMed]
46. King, J.; Plaizier, J.C. Effects of source of rumen fluid on in vitro dry matter digestibility of feeds determined using the DAISY^{II} Incubator. *Can. J. Anim. Sci.* **2006**, *86*, 439–441. [CrossRef]
47. Ammar, H.; Lopez, S.; Andres, S.; Ranilla, M.J.; Boda, R.; Gonzalez, J.S. In vitro digestibility and fermentation kinetics of some browse plants using sheep or goat ruminal fluid as the source of inoculum. *Anim. Feed Sci. Technol.* **2008**, *147*, 90–104. [CrossRef]
48. Hervas, G.; Frutos, P.; Giraldez, F.J.; Mora, M.J.; Fernandez, B.; Mantecon, A.R. Effect of preservation on fermentative activity of rumen fluid inoculum for in vitro gas production techniques. *Anim. Feed Sci. Technol.* **2005**, *123–124*, 107–118. [CrossRef]
49. Chaudhry, A.S.; Mohamed, R.A.I. Fresh of frozen rumen contents from slaughtered cattle to estimate in vitro degradation of two contrasting feeds. *Czech. J. Anim. Sci.* **2012**, *6*, 265–273. [CrossRef]
50. Denek, N.; Can, A.; Avci, M. Frozen rumen fluid as microbial inoculum in the two-stage in vitro digestibility assay of ruminant feeds. *S. Afr. J. Anim. Sci.* **2010**, *40*, 251–256. [CrossRef]
51. Belanche, A.; Palma-Hidalgo, J.M.; Nejjam, I.; Serrano, R.; Jiménez, E.; Martín-García, I.; Yáñez-Ruiz, D.R. In vitro assessment of the factors that determine the activity of the rumen microbiota for further applications as inoculum. *J. Sci. Food Agric.* **2019**, *99*, 163–172. [CrossRef]

52. Soto, E.C.; Molina-Alcaide, E.; Khelil, H.; Yáñez-Ruiz, D.R. Ruminal microbiota developing in different in vitro simulation systems inoculated with goats rumen liquor. *Anim. Feed Sci. Technol.* **2013**, *185*, 9–18. [CrossRef]
53. Mauricio, R.M.; Owen, E.; Mould, F.L.; Givens, I.; Theodorou, M.K.; France, J.; Davi, D.R.; Dhanoa, M.S. Comparison of bovine rumen liquor and bovine faeces as inoculum for an in vitro gas production technique for evaluating forages. *Anim. Feed Sci. Technol.* **2001**, *89*, 33–48. [CrossRef]
54. Hughes, M.M.; Mlambo, V.; Lallo, C.H.O.; Jennings, P.G.A. Potency of microbial inocula from bovine faeces and rumen fluid for in vitro digestion of different tropical forage substrates. *Grass Forage Sci.* **2012**, *67*, 263–273. [CrossRef]
55. Ramin, M.; Lerose, D.; Tagliapietra, F.; Huhtanen, P. Comparison of rumen fluid inoculum vs. faecal inoculum on predicted methane production using a fully automated in vitro *gas* production system. *Livest. Sci.* **2015**, *181*, 65–71. [CrossRef]
56. Akhter, S.; Owen, E.; Theodorou, M.K.; Butler, E.A.; Minson, D.J. Bovine faeces as a source of micro-organisms for the in vitro digestibility assay of forages. *Grass Forage Sci.* **1999**, *54*, 219–226. [CrossRef]
57. Tufarelli, V.; Cazzato, E.; Ficco, A.; Laudadio, V. Assessing nutritional value and in vitro digestibility of Mediterranean pasture species using yak (*Bos grunniens*) faeces as alternative microbial inoculum in a Daisy^{II} incubator. *J. Food Agric. Environ.* **2010**, *8*, 477–481. [CrossRef]
58. Laudadio, V.; Lacalandra, G.M.; Monaco, D.; Khorchani, T.; Hammadi, M.; Tufarelli, V. Faecal liquor as alternative microbial inoculum source for in vitro (Daisy^{II}) technique to estimate the digestibility of feeds for camels. *J. Camelid Sci.* **2009**, *2*, 1–7.
59. Cone, J.W.; Van Gelder, A.H.; Bachmann, H. Influence of inoculum source on gas production profiles. *Anim. Feed Sci. Technol.* **2002**, *99*, 221–231. [CrossRef]
60. Chiaravalli, M.; Rapetti, L.; Rota Graziosi, A.; Galassi, G.; Crovetto, G.M.; Colombini, S. Comparison of faecal versus rumen inocula for the estimation of NDF digestibility. *Animals* **2019**, *9*, 928. [CrossRef]
61. Guzmán, M.L.; Sager, R.L. Ruminant Fecal Inolucum for In Vitro Feed Digestibility Analysis. 2016. Available online: https://unsl.academia.edu/RicardoSager (accessed on 25 February 2020).
62. Kim, M.; Kim, J.; Kuehn, L.A.; Bono, J.L.; Berry, E.D.; Kalchayanand, N.; Freetly, H.C.; Benson, A.K.; Wells, J.E. Investigation of bacterial diversity in the feces of cattle fed different diets. *J. Anim. Sci.* **2014**, *92*, 683–694. [CrossRef]
63. Lowman, R.S.; Theodorou, M.K.; Hyslop, J.J.; Dhanoa, M.S.; Cuddeford, D. Evaluation of an in vitro batch culture technique for estimating the in vivo digestibility and digestible energy content of equine feeds using equine faeces as the source of microbial inoculum. *Anim. Feed Sci. Technol.* **1999**, *80*, 11–27. [CrossRef]
64. Earing, J.E.; Cassill, B.D.; Hayes, S.H.; Vanzant, E.S.; Lawrence, L.M. Comparison of in vitro digestibility estimates using the Daisy^{II} incubator with in vivo digestibility estimates in horses. *J. Anim. Sci.* **2010**, *88*, 3954–3963. [CrossRef]
65. Tassone, S.; Renna, M.; Barbera, S.; Valle, E.; Fortina, R. In vitro digestibility measurement of feedstuffs in donkeys using the Daisy^{II} incubator. *J. Equine Vet. Sci.* **2019**, *75*, 122–126. [CrossRef]
66. Varadyova, Z.; Baran, M.; Zelenak, I. Comparison of two in vitro fermentation gas production methods using both rumen fluid and faecal inoculum from sheep. *Anim. Feed Sci. Technol.* **2005**, *123–124*, 81–94. [CrossRef]
67. De Boever, J.L.; Cottyn, B.G.; Andries, J.I.; Buysse, F.X.; Vanacker, J.M. The use of a cellulase technique to predict digestibility, metabolizable and net energy of forages. *Anim. Feed Sci. Technol.* **1988**, *19*, 247–260. [CrossRef]
68. Akinsola, M.P. Development of an In Vitro Technique to Determine Digestibility of High Fiber Pig Feed. Ph.D. Thesis, Tshwane University of Technology, Pretoria, South Africa, 2013.
69. Abad, R.; Ibáñez, M.A.; Carabaño, R.; García, J. Quantification of soluble fibre in feedstuffs for rabbits and evaluation of the interference between the determinations of soluble fibre and intestinal mucin. *Anim. Feed Sci. Technol.* **2013**, *182*, 61–70. [CrossRef]
70. Ferreira, F.N.A.; Ferreira, W.M.; Silva Neta, C.S.; Inácio, D.F.S.; Mota, K.C.N.; Costa Júnior, M.B.; Rocha, L.F.; Lara, L.B.; Fontes, D.O. Effect of dietary inclusion of dried or autoclaved sugarcane bagasse and vinasse on live performance and in vitro evaluations on growing rabbits. *Anim. Feed Sci. Technol.* **2017**, *230*, 87–95. [CrossRef]

71. Candellone, A.; Prola, L.; Peiretti, P.G.; Tassone, S.; Longato, E.; Pattono, D.; Russo, N.; Meineri, G. In vivo and in vitro digestibility, palatability and nutritive quality of extruded dog food based on mechanically separated chicken meat or meat by-product. *Ital. J. Anim. Sci.* **2019**, *18* (Suppl. 1), 109.
72. Figueiredo, M.; Mbhele, A.; Zondi, J. An evaluation of the Daisy II-220 technique for determining in vitro digestibility of animal feeds in comparison with the Minson & McLeod technique. *S. Afr. J. Anim. Sci.* **2000**, *30*, 45–46.
73. Mabjeesh, S.J.; Cohen, M.; Arieli, A. In vitro methods for measuring the dry matter digestibility of ruminant feedstuffs: Comparison of methods and inoculum source. *J. Dairy Sci.* **2000**, *83*, 2289–2294. [CrossRef]
74. Vanzant, E.S.; Cochran, R.C.; Titgemeyer, E.C. Standardization of in situ techniques for ruminant feedstuff evaluation. *J. Anim. Sci.* **1998**, *76*, 2717–2729. [CrossRef]
75. In Vitro True Digestibility Using the DaisyII Incubator. Available online: https://www.ankom.com/sites/default/files/document-files/Method_3_Invitro_D200_D200I.pdf (accessed on 25 February 2020).
76. Coblentz, W.; Akins, M. Comparisons of fiber digestibility for triticale forages at two different sample sizes using the Ankom Daisy Incubator II System. In Proceedings of the ADSA Annual Meeting, Cincinnati, OH, USA, 23–26 June 2019. Abstract T69.
77. Cattani, M.; Tagliapietra, F.; Bailoni, L.; Schiavon, S. In vitro rumen feed degradability assessed with Daisy^{II} and batch culture: Effect of sample size. *Ital. J. Anim. Sci.* **2009**, *8*, 169–171. [CrossRef]
78. Coblentz, W.K.; Akins, M.S.; Kalscheur, K.F.; Brink, G.E.; Cavadini, J.S. Effects of growth stage and growing degree day accumulations on triticale forages: 1) Dry matter yield, nutritive value, and in vitro dry matter disappearance. *J. Dairy Sci.* **2018**, *101*, 8965–8985. [CrossRef]
79. Adesogan, A.T. Effect of bag type on the apparent digestibility of feeds in Ankom Daisy^{II} incubators. *Anim. Feed Sci. Technol.* **2005**, *119*, 333–344. [CrossRef]
80. Anassori, E.; Dalir-Naghadeh, B.; Pirmohammadi, R.; Taghizadeh, A.; Asri-Rezaei, S.; Farahmand-Azar, S.; Besharati, M.; Tahmoozi, M. In vitro assessment of the digestibility of forage based sheep diet, supplemented with raw garlic, garlic oil and monensin. *Vet. Res. Forum* **2012**, *3*, 5–11.
81. Coles, L.T.; Moughan, P.J.; Darragh, A.J. In vitro digestion and fermentation methods incuding gas production techniques, as applied to nutritive evaluation of foods in the hindgut of humans and other simple-stomached animals. *Anim. Feed Sci. Technol.* **2005**, *123–124*, 421–444. [CrossRef]
82. Marten, G.C.; Barnes, R.F. Prediction of energy digestibility of forages in vitro rumen fermentation and fungal enzyme systems. In Proceedings of the International Workshop on Standardization of Analytical Methodology for Feeds, Ottawa, ON, Canada, 12–14 March 1979; Pigden, W.J., Balch, C.C., Graham, M., Eds.; Unipub: New York, NY, USA, 1979; pp. 61–71.
83. Minson, D.J.; Mcleod, M.N. *Division of Tropical Pastures Technical*; Paper No. 8; Commonwealth Scientific and Industrial Research Organization: Melbourne, Australia, 1972.
84. Calsamiglia, S.; Stern, M.D. A three-step in vitro procedure for estimating intestinal digestion of protein in ruminants. *J. Anim. Sci.* **1995**, *73*, 1459–1465. [CrossRef]
85. Gargallo, S.; Calsamiglia, S.; Ferret, A. A modified three-step in vitro procedure to determine intestinal digestion of proteins. *J. Anim. Sci.* **2006**, *84*, 2163–2167. [CrossRef]
86. Slyter, L.L.; Bryant, M.P.; Wolin, M.J. Effect of pH on population and fermentation in a continuously cultured rumen ecosystem. *Appl. Microbiol.* **1966**, *14*, 573–578. [CrossRef]
87. Damiran, D.; DelCurto, T.; Bohnert, D.W.; Findholt, S.L. Comparison of techniques and grinding size to estimate digestibility of forage based ruminant diets. *Anim. Feed Sci. Technol.* **2008**, *141*, 15–35. [CrossRef]
88. Tagliapietra, F.; Cattani, M.; Hindrichsen, I.K.; Hansen, H.H.; Colombini, S.; Bailoni, L.; Schiavon, S. True dry matter digestibility of feeds evaluated in situ with different bags and in vitro using rumen fluid collected from intact donor cows. *Anim. Prod. Sci.* **2012**, *52*, 338–346. [CrossRef]
89. Spanghero, M.; Boccalon, S.; Gracco, L.; Gruber, L. NDF degradability of hays measured in situ and in vitro. *Anim. Feed Sci. Technol.* **2003**, *104*, 201–208. [CrossRef]
90. Bovera, F.; Spanghero, M.; Galassi, G.; Masoero, F.; Buccioni, A. Repeatability and reproducibility of the Cornell Net Carbohydrate and Protein System analytical determinations. *Ital. J. Anim. Sci.* **2003**, *2*, 41–50. [CrossRef]
91. Spanghero, M.; Gruber, L.; Zanfi, C. Precision and accuracy of the NDF rumen degradability of hays measured by the Daisy fermenter. *Ital. J. Anim. Sci.* **2007**, *6*, 363–365. [CrossRef]

92. Spanghero, M.; Berzaghi, P.; Fortina, R.; Masoero, F.; Rapetti, L.; Zanfi, C.; Tassone, S.; Gallo, A.; Colombini, S.; Ferlito, J.C. Precision and accuracy of in vitro digestion of neutral detergent fiber and predicted net energy of lactation content of fibrous feeds. *J. Dairy Sci.* **2010**, *93*, 4855–4859. [CrossRef]
93. Cişmileanu, A.E.; Toma, S. Validation of the in vitro ruminant digestibility method applied on Daisy incubator. *Anim. Sci. Biotech.* **2017**, *50*, 8–10.
94. Ankom. Technical FAQs. Available online: https://www.ankom.com/technical-support/daisy-incubator (accessed on 15 February 2020).
95. Komarek, A.R.; Robertson, J.B.; Van Soest, P.J. Comparison of filter bag technique to conventional filtration in the Van Soest analysis of 21 feeds. In Proceedings of the National Conference on Forage Quality, Evaluation and Utilization, Lincoln, NE, USA, 13–15 April 1994.
96. Cherney, D.J.R.; Traxler, M.J.; Robertson, J.B. Use of Ankom fiber determination systems to determine digestibility. In Proceedings of the NIRS Forage and feed Testing Consortium Annual Conference, Madison, WI, USA, 19–20 February 1997.
97. Ayangbile, O.A.; Meier, J.C.; Vogel, M.K.; Robertson, J.; McElroy, A.R.; Komarek, A.R. Cryogenically protected and fresh rumen inoculum for digestibility study. In Proceedings of the 23rd Biennial Conference on Rumen Function, Chicago, IL, USA, 14–16 November 1995.
98. Traxler, M.J.; Robertson, J.B.; Van Soest, P.J.; Fox, D.G.; Pell, A.N. A comparison of methods for determining IVDMD at three time periods using the filter bag technique versus conventional methods. *J. Dairy Sci.* **1995**, *78*, 274.
99. Cohen, M.A.; Maslanka, H.E.; Kung, L. An evaluation of automated and manual in vitro methods for estimation of NDF digestion. In Proceedings of the Conference on Rumen Physiology, Chicago, IL, USA, 11–13 November 1997.
100. Traxler, M.J. Predicting the Effect of Lignin on the Extent of Digestion and the Evaluation of Alternative Intake Models for Lactating Dairy Cows Consuming High NDF Forages. Ph.D. Thesis, Cornell University, Ithaca, NY, USA, 1997.
101. Brons, E.; Plaizier, J.C. Comparisons of methods for in vitro dry matter digestibility of ruminant feeds. *Can. J. Anim. Sci.* **2005**, *85*, 243–245. [CrossRef]
102. Ammar, H.; López, S.; Gonzales, J.S.; Ranilla, M.J. Seasonal variations in the chemical composition and in vitro digestibility of some Spanish leguminous shrub species. *Anim. Feed Sci. Technol.* **2004**, *115*, 327–340. [CrossRef]
103. Ammar, H.; López, S.; Bochi-Brum, O.; Garcia, R.; Ranilla, M.J. Composition and in vitro digestibility of leaves and stems of grasses and legumes harvested from permanent mountain meadows at different stages of maturity. *J. Anim. Feed Sci.* **1999**, *8*, 599–610. [CrossRef]
104. Ankom Procedures for Fiber and In Vitro Analysis. Available online: http://www.ankom.com (accessed on 25 February 2020).
105. Ricci, P.; Romera, A.J.; Burges, J.C.; Fernández, H.H.; Cangiano, C.A. Case study: Precision and accuracy of methodologies for estimating in vitro digestibility of *Thinopyrum ponticum* (Tall Wheatgrass) hay and haylage fed to beef cattle. *Prof. Anim. Sci.* **2009**, *25*, 625–632. [CrossRef]
106. Rowe, J.; Bird, S.; Brown, W. *Safe and Effective Grain Feeding for Horses: A Report for the Rural Industries Research and Development Corporation*; RIRDC Project No. UNE-62A; RIRDC: Barton, Australia, 2001.
107. Macheboeuf, D.; Jestin, M.; Martin-Rosset, W. Utilization of the gas test method and horse faeces as a source of inoculum. *BSAP Occas. Publ.* **1998**, *22*, 187–189. [CrossRef]
108. Macheboeuf, D.; Jestin, M. Utilization of the gas test method using horse faeces as a source of inoculums. In Proceedings of the Symposium on In Vitro Techniques for Measuring Nutrient Supply to Ruminants, Penicuik, UK, 8–10 July 1997; p. 36.
109. Abdouli, H.; Attia, S.B. Evaluation of a two-stage in vitro technique for estimating digestibility of equine feeds using horse faeces as the source of microbial inoculum. *Anim. Feed Sci. Technol.* **2007**, *132*, 155–162. [CrossRef]
110. Lattimer, J.M.; Cooper, S.R.; Freeman, D.W.; Lalman, D.A. Effects of *Saccharomyces cerevisiae* on in vitro fermentation of a high concentrate or high fiber diet in horses. In Proceedings of the 19th Symposium of the Equine Science Society, Tucson, AZ, USA, 31 May–3 June 2005; pp. 168–173.

111. Lattimer, J.M.; Cooper, S.R.; Freeman, D.W.; Lalman, D.L. Effect of yeast culture on in vitro fermentation of a high-concentrate or high-fiber diet using equine fecal inoculum in a Daisy^{II} incubator. *J. Anim. Sci.* **2007**, *85*, 2484–2491. [CrossRef]
112. Blažková, K.; Homolka, P.; Maršálek, M. The corn silage digestibility by horses. *MendelNet* **2009**, *9*, 171–176.
113. Towhidi, A.; Zhandi, M. Chemical composition, in vitro digestibility and palatability of nine plant species for dromedary camels in the province of Semnan, Iran. *Egypt. J. Biol.* **2007**, *9*, 47–52.
114. Mohammadabadi, T.; Kakar, A.R. Comparison of in vitro digestibility of diets containing Subabul plant as fodder in dromedary camel and cow. *Explor. Anim. Med. Res.* **2019**, *9*, 61–66.
115. Lifa, M.; Haddi, M.L.; Tagliapietra, F.; Cattani, M.; Guadagnin, M.; Sulas, L.; Muresu, R.; Schiavon, S.; Bailoni, L.; Squartini, A. Chemical and fermentative characteristics of agricultural byproducts and their mixtures with roughages incubated with rumen fluid from slaughtered dromedaries. *Turk. J. Vet. Anim. Sci.* **2018**, *42*, 590–599. [CrossRef]
116. Abad-Guamán, R.; Carabaño, R.; Gómez-Conde, M.S.; García, J. Effect of type of fiber, site of fermentation, and method of analysis on digestibility of soluble and insoluble fiber in rabbits. *J. Anim. Sci.* **2015**, *93*, 2860–2871. [CrossRef]
117. Ferreira, W.M.; Ferreira, F.N.A.; Inácio, D.F.D.S.; Mota, K.C.D.N.; Costa Júnior, M.B.D.; Silva Neta, C.S.; Rocha, L.F.D.; Miranda, E.R.D. Effects of dietary inclusion of macaúba seed cake meal on performance, caecotrophy traits and in vitro evaluations for growing rabbits. *Arch. Anim. Nutr.* **2018**, *72*, 138–152. [CrossRef]
118. Ferreira, F.N.A.; Ferreira, W.M.; Inácio, D.F.D.S.; Neta, C.S.S.; Mota, K.C.D.N.; Costa Júnior, M.B.D.; Rocha, L.F.D.; Caicedo, W.O. In vitro digestion and fermentation characteristics of tropical ingredients, co-products and by-products with potential use in diets for rabbits. *Anim. Feed Sci. Technol.* **2019**, *252*, 1–10. [CrossRef]
119. De Blas, J.C.; Rodriguez, C.A.; Bacha, F.; Fernandez, R.; Abad-Guamán, R. Nutritive value of co-products derived from olive cake in rabbit feeding. *World Rabbit Sci.* **2015**, *23*, 255–262. [CrossRef]
120. De Blas, J.C.; Ferrer, P.; Rodríguez, C.A.; Cerisuelo, A.; García-Rebollar, P.; Calvet, S.; Farias, C. Nutritive value of citrus co-products in rabbit feeding. *World Rabbit Sci.* **2018**, *26*, 7–14. [CrossRef]
121. Carabaño, R.; Nicodemus, N.; García, J.; Xiccato, G.; Trocino, A.; Amorós, P.; Maertens, L. In vitro analysis, an accurate tool to estimate dry matter digestibility in rabbits. Intra- and inter-laboratory variability. *World Rabbit Sci.* **2008**, *16*, 195–203. [CrossRef]
122. Theodorou, M.K.; Williams, B.A.; Dhanoa, M.S.; McAllan, A.B.; France, J. A simple gas production method using a pressure transducer to determine the fermentation kinetics of ruminant feeds. *Anim. Feed Sci. Technol.* **1994**, *48*, 185–197. [CrossRef]
123. Ramos, M.A.; Carabaño, R.; Boisen, S. An in vitro method for estimating digestibility in rabbits. *J. Appl. Rabbit Res.* **1992**, *15*, 938–946.
124. Boisen, S. A model for feed evaluation based on in vitro digestibility dry matter and protein. In *In Vitro Digestion for Pigs and Poultry*; Fuller, M.F., Ed.; CAB International: Wallingford, UK, 1991; pp. 135–145.
125. López, S.; Guevara, H.; Duchi, N.; Moreno, G. Evaluation of two "in vitro" digestibility tests with the "in vivo" test of alfalfa (*Medicago sativa*) in guinea pig (*Cavia porcellus*) feeding. *Eur. Sci. J.* **2018**, *14*, 399–404. [CrossRef]
126. Chiang, C.-C.; Croom, J.; Chuang, S.-T.; Chiou, P.W.S.; Yu, B. Development of a dynamic system simulating pig gastric digestion. *Asian Aust. J. Anim. Sci.* **2008**, *21*, 1522–1528. [CrossRef]
127. Boisen, S.; Fernández, J.A. Prediction of the total tract digestibility of energy in feedstuffs and pig diets by in vitro analyses. *Anim. Feed Sci. Technol.* **1997**, *68*, 277–286. [CrossRef]
128. Fushai, F. Fermentability of Dietary Fibre and Metabolic Impacts of Including High Levels of fibrous Feed Ingredients in Maize-Soybean Growing Pig Diets Supplemented with Exogenous Enzymes. Ph.D. Thesis, Tshwane University of Technology, Pretoria, South Africa, 2014.
129. Torres-Pitarch, A.; McCormack, U.M.; Beattie, V.E.; Magowan, E.; Gardiner, G.E.; Pérez-Vendrell, A.M.; Torrallardonaf, D.; O'Dohertye, J.V.; Lawlor, P.G. Effect of phytase, carbohydrase, and protease addition to a wheat distillers dried grains with solubles and rapeseed based diet on in vitro ileal digestibility, growth, and bone mineral density of grower-finisher pigs. *Livest. Sci.* **2018**, *216*, 94–99. [CrossRef]
130. Pahm, S.F.C. In Vivo and In Vitro Disappearance of Energy and Nutrients in Novel Carbohydrates and Cereal Grains by Pigs. Ph.D. Thesis, University of Illinois, Urbana-Champaign, IL, USA, 2011.
131. Huang, G.; Sauer, W.C.; He, J.; Hwangbo, J.; Wang, X. The nutritive value of hulled and hulless barley for growing pigs. 1. Determination of energy and protein digestibility with in vivo and in vitro methods. *J. Anim. Feed Sci.* **2003**, *12*, 759–769. [CrossRef]

132. Youssef, I.M.; Kamphues, J. Fermentation of lignocellulose ingredients in vivo and in vitro via using fecal and caecal inoculums of monogastric animals (swine/turkeys). *Beni Suef Univ. J. Basic Appl. Sci.* **2018**, *7*, 407–413. [CrossRef]
133. Hervera, M.; Baucells, M.D.; Blanch, F.; Castrillo, C. Prediction of digestible energy content of extruded dog food by in vitro analyses. *J. Anim. Physiol. Anim. Nutr.* **2007**, *91*, 205–209. [CrossRef]
134. Biagi, G.; Cipollini, I.; Grandi, M.; Pinna, C.; Vecchiato, C.G.; Zaghini, G. A new in vitro method to evaluate digestibility of commercial diets for dogs. *Ital. J. Anim. Sci.* **2016**, *15*, 617–625. [CrossRef]
135. Hervera, M.; Baucells, M.D.; González, G.; Pérez, E.; Castrillo, C. Prediction of digestible protein content of dry extruded dog foods: Comparison of methods. *J. Anim. Physiol. Anim. Nutr.* **2009**, *93*, 366–372. [CrossRef] [PubMed]

16

Exploring Additive, Synergistic or Antagonistic Effects of Natural Plant Extracts on In Vitro Beef Feedlot-Type Rumen Microbial Fermentation Conditions

Ignacio Fandiño [1,†], Gonzalo Fernandez-Turren [2], Alfred Ferret [1], Diego Moya [1], Lorena Castillejos [1] and Sergio Calsamiglia [1,*]

1 Departamento de Ciència Animal i dels Aliments, Universitat Autònoma de Barcelona, Servicio de Nutrición y Bienestar Animal (SNiBA), 08193 Bellaterra, Spain; alfred.ferret@uab.cat (A.F.); diego.moya@usask.ca (D.M.); Lorena.castillejos@uab.cat (L.C.)

2 Departamento de Producción Animal, Instituto de Producción Animal, Facultad de Veterinaria, Universidad de la República, Ruta 1 km 42, CP 80100 San José, Uruguay; gonzalofernandezt@gmail.com

* Correspondence: sergio.calsamiglia@uab.cat;

† Deceased.

Simple Summary: Essential oils (EO) can be used as natural alternatives to in-feed antibiotics. Most EO products in the market are based on a combination of EO or their active molecules but prove of additivity or synergy is lacking. The effect of six EO (tea tree oil—TeTr, oregano oil—Ore, clove bud oil—Clo, thyme oil—Thy, rosemary oil—Ros and sage oil—Sag) and different mixes on in vitro microbial fermentation profile of a feedlot beef cattle type fermentation were evaluated for their additive, synergistic or antagonistic effects. Mixing TeTr with Thy, Ore or Thy + Ore modified rumen microbial fermentation profile, but the size of the effect was similar to that obtained with TeTr alone, suggesting that the effects were not additive. When Thy, Ore or Thy + Ore were mixed with Clo, most effects on rumen fermentation profile disappeared even when TeTr was part of the mix, suggesting an antagonistic interaction of Clo with Thy and Ore. Results do not support the hypothesis of additivity among the EO tested, and antagonistic effects may occur among some of them, at least in a low pH, beef-type fermentation conditions.

Abstract: Six Essential oils (EO) (tea tree oil—TeTr, oregano oil—Ore, clove bud oil—Clo, thyme oil—Thy, rosemary oil—Ros, and sage oil—Sag) in Experiment 1; and different combinations of selected oils in Experiment 2, were evaluate at four doses in an in vitro microbial fermentation system using ruminal fluid from beef cattle fed a 10:90 straw: Concentrate diet. In Experiment 1, TeTr, Ore, Clo and Thy improved rumen fermentation profile in a direction consistent with better feed utilization. In Experiment 2, TeTr mixed with Thy, Ore, Thy + Ore or Clo at 200 and 400 mg/L increased the molar proportion of propionate and decreased that of acetate, and the acetate to propionate ratio. However, the size of the effect was similar to that obtained with TeTr alone, suggesting that effects were not additive. When Thy, Ore or Thy + Ore where mixed with Clo, most effects on rumen fermentation profile disappeared, suggesting an antagonistic interaction of Clo with Thy and Ore. Results do not support the hypothesis of additivity among the EO tested, and antagonistic effects of Clo mixed with Thy or Ore were demonstrated at least in a low pH, beef-type fermentation conditions.

Keywords: essential oils; rumen microbial fermentation; synergies

1. Introduction

Essential oils (EO) are aromatic oily liquids obtained from plants that can be used as natural alternatives to in-feed antibiotics [1,2]. Many different plant EO and their active components have been tested for their effects on ruminal microbial fermentation profile [3–5]. Because of the diverse mechanisms of action and effects, Calsamiglia et al. [1] suggested that the wise selection and combination of different EO may enhance ruminal fermentation. When combinations of EO are used, the effect may be additive, synergistic or antagonistic. Most EO products for ruminants in the market are based on combination of different EO or their active molecules (i.e., CRINA-Ruminants®, DSM, Switzerland; AGOLIN®, Agolin Sa, Switzerland, XTRACT®, Pancosma, Switzerland), but proof of additivity or synergy is lacking. Some studies attributed additive and synergistic effects to phenolic and alcohol compounds. In general, compounds with similar structures exhibit additive rather than synergistic effects. Antagonistic effects have been attributed to the interaction between non-oxygenated and oxygenated monoterpene hydrocarbons [6]. All these interactions have been demonstrated in cosmetic and food industry [6–8]. However, it is surprising that no studies have been specifically designed to prove additivity or synergies of different EO on rumen microbial fermentation. The mechanism of action of most EO is mediated through their interaction with the cell membrane, and this interaction is dependent on the fermentation conditions like the source of rumen fluid, the substrate of fermentation and pH [1,5]. The additive, synergistic or antagonistic effects in EO mixes may also be dependent on these fermentation conditions. While most research in vitro has been conducted using rumen fluid from dairy cattle, forage as a main fermentation substrate and pH above 6.5, research in high concentrate feedlot-type beef fermentation conditions is more limited.

We hypothesize that the combination of different EO, in particular simple phenolic-type molecules of the family of monoterpenoids and phenylpropanes, will result in additive or synergistic effects. The objective of the present study was to evaluate the effects of different doses of six EO and the additive, synergistic or antagonistic effects of their combinations under feedlot beet-type ruminal microbial fermentation conditions in vitro.

2. Materials and Methods

2.1. Experimental Protocol and Treatments

The research protocol was approved by Campus Laboratory Animal Care Committee of the Universitat Autónoma of Barcelona (Spain) (reference CEAAH 00604). The effects of six EO and their combinations were evaluated in an in vitro batch fermentation system. Rumen fluid was obtained from four beef heifers fed a 10:90 barley straw:concentrate diet (161 g/kgcrude protein, 321 g/kg neutral detergent fiber, 168 g/kg acid detergent fiber on a dry matter basis) designed to meet or exceed nutrient recommendations [9] of growing beef cattle. The concentrate consisted of (DM basis) ground barley grain (444 g/kg), corn gluten feed (155 g/kg), ground corn grain (133 g/kg), soybean meal (72 g/kg), sunflower meal (72 g/kg), ground wheat grain (69 g/kg), palm kernel (52 g/kg), vitamin A (4600 IU/kg), vitamin D (450 IU/kg), vitamin E (7.5 mg/kg), zinc (10.5 mg/kg), iron (6 mg/kg), manganese (1.2 mg/kg), copper (0.75 mg/kg), cobalt (0.15 mg/kg), iodine (0.11 mg/kg) and selenium (0.08 mg/kg). Rumen fluid was strained through two layers of cheesecloth and mixed in a 1:1 proportion with a phosphate-bicarbonate buffer [10]. The pH of the buffered fluid was adjusted to 6.20 with hydrochloric acid. Incubations were conducted in a 75 ml tubes containing 50 ml of the diluted rumen fluid and 0.5 g of the same diet fed to the donor animals. The diet was ground through a 2 mm screen, and each tube was gassed with CO_2 before sealing with rubber corks with a gas release valve. Incubations were conducted in a water bath at 39 °C for 24 h in two consecutive periods with triplicates within period.

In Experiment 1, four doses of different EO (10, 50, 200 and 400 mg/L of the culture fluid) were evaluated. Most EO have shown to be effective at doses around 200–400 mg/L [1,3,5]. Doses were selected based on this evidence and at lower doses to show potential additivity and synergies of effects. Treatments were a negative control without EO (CTR), a positive control with Monensin at

10 mg/L (M5273 Sigma-Aldrich Chemical, St Louis, MO, USA) and six different phenolic-derived monoterpene or phenylpropane EO, of which two were oxygenated and four non-oxygenated: 1) Tea tree oil (*Melaleuca alternifolia*; TeTr; containing 42% terpinen-4-ol as active component), Oregano oil (*Origanum vulgare*; Ore; containing 53% carvacrol as active components), Clove bud oil (*Syzygium aromaticum*; Clo; containing 82% eugenol as main active component), Thyme oil (*Thymus vulgaris*; Thy; containing 47% thymol as main active component), Rosemary oil (*Rosmarinus officinalis*; Ros; containing 61% 1,8-cineole as main active component) and Sage oil (*Salvia officinalis*; Sag; containing 65% thujone as main active component). In Experiment 2, four EO selected from Experiment 1 (TeTr, Thy, Ore and Clo) were used in 13 different combinations at four concentrations (10, 50, 200 and 400 mg/L of the culture fluid; Table 1) to test additive, synergistic or antagonist effects.

Table 1. Treatments and combination of essential oils evaluated in *in vitro* fermentation.

Treatment Number	Combination of Essential Oils (%)			
	Tea-tree	Clove Bud	Thyme	Oregano
1	75	25	-	-
2	75	-	25	-
3	75	-	-	25
4	50	50	-	-
5	50	-	25	25
6	50	25	25	-
7	50	25	-	25
8	25	50	25	-
9	25	50	-	25
10	25	75	-	-
11	-	75	25	-
12	-	75	-	25
13	-	50	25	25

2.2. Sample Collection and Chemical Analyses

After 24 h, the fermented fluid pH was measured immediately (Model 507; Crison Instruments, Alella, Barcelona, Spain) and liquid samples were withdrawn from each tube to analyze ammonia-N and volatile fatty acid (VFA) concentrations. Samples for ammonia-N analysis (5 mL of fermentation fluid preserved with 1 mL of a 50 g/L orthophosphoric acid solution) were centrifuged at 15,000× *g* for 15 min at 4 °C, and the supernatant was analyzed for ammonia-N by colorimetry [11]. Samples for VFA analysis were prepared as described by Jouany [12]. Samples of fermented fluid (5 mL) were preserved with 1 mL of a solution made of 2 g/L of mercuric chloride, 35 g/L orthophosphoric acid and 2 g/L of 4-methylvaleric acid as an internal standard, and frozen. After thawing, samples were centrifuged at 3000× *g* for 30 min, and the supernatant fraction was analyzed by gas chromatography (Model 6890; Hewlett Packard, Palo Alto, CA, USA) using a polyethylene glycol nitroterephthalic acid-treated capillary column (BP21; SGE, Europe Ltd., Buckinghamshire, UK) at 275 °C in the injector and a gas flow rate of 29.9 mL/min.

2.3. Statistical Analyses

Results of the batch fermentation experiment were analyzed as a randomized complete block design using the PROC MIXED procedure (SAS, version 9.4, SAS Institute Inc., Cary, NC, USA), where the essential oils or mixes were the fixed effect and the period was considered a random effect. Because the objective was to identify the lowest dose of EO at which effects were observed, the Dunnett test was used to identify significant differences ($p < 0.05$). When required, the synergistic, additive or antagonistic effects of mixes were evaluated comparing the relative response of the mixes vs. the corresponding effects of the main essential oil using a paired t-test at a significant level of $p < 0.05$.

3. Results

Because of the large number of treatments tested, two runs in Experiment 1, and six runs in Experiment 2 were conducted. To account for possible run-to-run variation, a negative (no additive) control was used, and results are reported as percent change compared with this negative control (Tables 2 and 3).

3.1. Experiment 1 (Individual Essential Oils)

Results from Experiment 1 are shown in Table 2. Monensin was used as a positive control and always increased ($p < 0.05$) the proportion of propionate and decreased ($p < 0.05$) the molar proportion of acetate and the acetate to propionate (A:P) ratio. In most cases, monensin also reduced the molar proportion of butyrate and brach-chain VFA (BCVFA), and the concentration of ammonia-N when compared with control.

None of the EO treatments affected rumen fermentation profile at 10 mg/L compared with control. Tea tree oil at 50 and 200 mg/L decreased ($p < 0.05$) the molar proportion of acetate (−9 to −15%) and the A:P ratio (−22 to −36%), and increased ($p < 0.05$) the molar proportion of propionate (+18 to +34%) and butyrate (+14 to +21%). At 400 mg/L TeTr only increased ($p < 0.05$) the molar proportion of butyrate (+14%) compared with control. Oregano oil at 50 and 200 mg/L decreased ($p < 0.05$) the molar proportion of acetate (−9 to −11%) and the A:P ratio (−26 to −9%) and increased ($p < 0.05$) the molar proportion of propionate (+24 to +25%) and the molar proportion of butyrate (+12 to +16%), but at 400 mg/L increased ($p < 0.05$) the molar proportion of butyrate (+17%) and the pH (+2%) compared with control. Clove bud oil at 50 mg/L had no effects on rumen microbial fermentation, but at 200 and 400 mg/L increased ($p < 0.05$) the molar proportion of butyrate (+10 to +18%), and at 400 mg/L, decreased ($p < 0.05$) the molar proportion of propionate (−13%), and increased ($p < 0.05$) the A:P ratio (+15%) and the pH (+2%) compared with control. Thyme oil at 50 and 200 decreased ($p < 0.05$) the molar proportion of acetate (−6%) and the A:P ratio (−25 to −28%), and increased ($p < 0.05$) the molar proportion of propionate (+24 to +30%). At 200 and 400 mg/L Thy decreased ($p < 0.05$) the molar proportion of butyrate (−15 to −21%), but only at 400 mg/L decreased ($p < 0.05$) the total VFA concentration (−4%), the BCVFA concentration (−16%) and the concentration of ammonia-N (−17%), and increased ($p < 0.05$) the molar proportion of acetate (+5%). Rosemary oil at 50 mg/L had no effect on rumen microbial fermentation, but at 200 and 400 mg/L decreased ($p < 0.05$) the molar proportion of butyrate (−14 to −19%) and at 400 mg/L increased ($p < 0.05$) the molar proportion of acetate (−4%) and decreased the total VFA concentration (−5%) compared with control. Sage oil at 50 mg/L had no effect on rumen microbial fermentation, but at 200 mg/L decreased ($p < 0.05$) the molar proportion of butyrate (−17%) and at 400 mg/L, increased ($p < 0.05$) the molar proportion of propionate (+12%), reduced ($p < 0.05$) the A:P ratio (−13%) and the concentration of ammonia-N (−16%), and tended to reduce ($p < 0.10$) the total VFA concentration (−4%) compared with control.

Table 2. Effect of Tea Tree (TeTr), Oregano (Ore), Clove bud (Clo), Thyme (Thy), Rosmarinus officinalis (Ros), Sage oil (Sag) and Monensin (MON) on pH, ammonia-N and total and individual VFA concentrations in vitro fermentation of 10:90 forage to concentrate diet ($n = 6$).

Treatment	pH	NH_3, mg/dL	Total VFA, mM	Acetate, %	Propionate, %	Butyrate, %	Valerate, %	BCVFA [2], mM	A:P [3]
CTR	5.33	13.2	133.2	61.8	18.8	13.4	3.46	3.47	3.28
TeTr10 [a]	5.32	13.2	132.7	61.3	18.1	14.4	3.53	3.41	3.38
TeTr50 [b]	5.33	13.2	129.0	56.5 *	22.2 *	15.2 *	3.56	3.26	2.55 *
TeTr200 [c]	5.33	13.6	127.0	52.4 *	25.2 *	16.2 *	3.53	3.54	2.11 *
TeTr400 [d]	5.38	13.2	127.8	60.2	18.3	15.2 *	3.81	3.15	3.28
Ore10 [a]	5.33	13.7	135.9	59.9	19.9	14.4	3.36	3.32	3.07
Ore50 [b]	5.33	13.5	132.6	56.0 *	23.3 *	15.0 *	3.25	3.29	2.43 *
Ore200 [c]	5.33	13.6	128.7	54.8 *	23.5 *	15.5 *	3.53	3.32	2.33 *
Ore400 [d]	5.44 *	13.6	131.1	60.8	17.6	15.6 *	3.49	3.19	3.44
Clo10 [a]	5.31	13.3	135.9	60.6	19.0	14.6	3.23	3.33	3.19
Clo50 [b]	5.33	13.5	133.0	60.9	18.8	14.6	3.12	3.40	3.24
Clo200 [c]	5.33	13.5	133.1	60.3	19.2	14.7 *	3.22	3.41	3.14
Clo400 [d]	5.44 *	13.9	127.2	61.8	16.4 *	15.8 *	3.46	3.22	3.77 *
MON	5.44 *	11.1 *	130.0	53.9 *	28.2 *	12.5	3.39	2.64 *	1.99 *
SEM [1]	0.02	0.80	2.63	0.95	0.83	0.28	0.14	0.20	0.10
CTR	5.26	17.4	131.5	61.8	20.0	14.3	1.60	2.90	3.14
Thy10 [a]	5.13	17.3	130.1	63.1	20.4	13.0	1.46	2.48	3.14
Thy50 [b]	5.18	17.9	127.0	58.4 *	24.8 *	13.3	1.49	2.59	2.37 *
Thy200 [c]	5.25	16.8	127.0	58.3 *	26.0 *	12.2 *	1.37	2.59	2.26 *
Thy400 [d]	5.37	14.4 *	125.7 *	64.9 *	20.0	11.2 *	1.59	2.44 *	3.30
Ros10 [a]	5.16	16.9	128.2	63.1	20.6	12.9	1.44	2.47	3.11
Ros50 [b]	5.15	17.1	128.0	63.1	20.4	12.9	1.66	2.59	3.14
Ros200 [c]	5.31	17.2	127.1	63.7	20.6	12.3 *	1.39	2.56	3.17
Ros400 [d]	5.31	17.4	125.3 *	64.3 *	20.2	11.6 *	1.49	2.85	3.30
Sag10 [a]	5.22	17.4	128.2	63.7	20.0	12.4	1.37	3.28	3.11
Sag50 [b]	5.19	17.3	129.3	63.7	20.4	12.3	1.37	3.31	3.14
Sag200 [c]	5.13	17.4	127.5	63.7	20.8	11.8 *	1.33	3.34	3.08
Sag400 [d]	5.16	14.5 *	125.9 *	60.7	22.4 *	12.9	1.43	3.16	2.72 *
MON	5.24	11.3 *	134.1	55.9 *	25.8 *	14.6	1.73	2.84	2.20 *
SEM [1]	0.05	0.71	2.00	0.71	0.44	0.70	0.14	0.16	0.08

[1] SEM—standard error of the mean. [2] BCVFA—Branch-chained VFA. [3] A:P—Acetate to propionate ratio. * Means within a column differ from control ($p < 0.05$). [a] 10 mg/L. [b] 50 mg/L. [c] 200 mg/L. [d] 400 mg/L.

Table 3. Effect of different doses of Treatment 1 (T1: 75% Tea tree, 25% Thyme), Treatment 2 (T2: 75% Tea tree, 25% Clove bud), Tretament 3 (T3: 75% Tea tree, 25% Oregano), Treatment 4 (T4; 50% Tea tree, 50% Clove bud), Treatment 5 (T5: 50% Tea tree, 25% Thyme, 25% Oregano), Treatment 6 (T6: 50% Tea tree, 25% Clove bud, 25% Thyme) and Monensin (MON) on pH, ammonia-N and total and individual VFA concentrations in vitro fermentation of 10:90 forage to concentrate diet ($n = 6$).

Treatment	pH	NH_3, mg/dL	VFA, mM	Acetate, %	Propionate, %	Butyrate, %	Valerate, %	BCVFA [2], mM	A:P [3]
CTR	5.33	27.5	130.8	62.0	21.0	14.0	1.08	2.47	2.94
T1 (10) [a]	5.32	25.2	129.1	61.5	21.3	14.3	1.05	2.54	2.91
T1 (50) [b]	5.30	25.2	128.2	61.2	21.3	14.7	1.10	2.41	2.87
T1 (200) [c]	5.30	26.5	128.6	58.6 *	23.6 *	14.3	1.20	2.36	2.48 *
T1 (400) [d]	5.28+	25.2	126.5	58.3 *	24.4 *	14.4	1.14	2.40	2.39 *
MON	5.29+	25.2 *	123.5 +	58.9 *	25.9 *	12.2 +	1.29	2.10 *	2.28 *
SEM [1]	0.01	1.34	2.70	0.68	0.53	0.51	0.09	0.05	0.09
CTR	5.38	27.0	116.9	60.1	20.8	15.4	1.24	2.64	2.89
T2 (10) [a]	5.39	24.9	116.7	60.1	21.5	14.6	1.30	2.69	2.82
T2 (50) [b]	5.36	26.3	116.3	59.6	21.5	15.6	1.19	2.58	2.80
T2 (200) [c]	5.38	26.1	114.6	59.6	21.7	15.0	1.38	2.60	2.75 *
T2 (400) [d]	5.37	27.0	115.0	56.9 *	24.2 *	15.4	1.36	2.59	2.36 *
T3 (10) [a]	5.36	27.3	115.7	60.0	20.6	15.6	1.20	2.64	2.88
T3 (50) [b]	5.38	26.6	116.2	59.7	21.5	15.3	1.33	2.62	2.78
T3 (200) [c]	5.36	26.0	114.9	59.3	21.5	14.8	1.16	2.59	2.81
T3 (400) [d]	5.37	26.7	116.5	57.3 *	24.0 *	15.1	1.48	2.58	2.40 *
MON	5.36	22.7	114.2	57.5 *	26.5 *	2.03 *	0.98	2.29 *	2.17 *
SEM[1]	0.01	1.34	2.70	0.68	0.53	0.51	0.09	0.05	0.09
CTR	5.40	24.9	126.0	63.8	18.1	14.6	1.09	2.95	3.53
T4 (10) [a]	5.39	25.6	127.2	62.7	19.0	14.7	1.10	3.01	3.30
T4 (50) [b]	5.34	25.9	127.2	62.3	19.6	14.5	1.26	3.04	3.19
T4 (200) [c]	5.38	24.9	127.2	63.2	18.7	14.6	1.18	2.90	3.41
T4 (400) [d]	5.31	23.0	125.3	59.8 *	22.5 *	14.1	1.14	3.10	2.66 *
T5 (10) [a]	5.37	24.7	124.9	62.5	19.6	14.6	1.11	2.74	3.20
T5 (50) [b]	5.33	25.1	127.2	63.0	18.7	14.9	1.14	2.84	3.37
T5 (200) [c]	5.38	24.2	125.6	60.2 *	21.9 *	14.1	1.28	2.98	2.75 *
T5 (400) [d]	5.35	21.4	127.2	58.8 *	22.7 *	15.2	1.20	2.93	2.60 *
T6 (10) [a]	5.41	27.6	129.2	61.6	20.6	14.9	1.22	2.28	3.02
T6 (50) [b]	5.37	27.7	125.9	61.1	20.8	14.1	1.20	2.37	3.05
T6 (200) [c]	5.39	28.1	128.6	62.1	21.2	14.7	1.21	2.34	2.92
T6 (400) [d]	5.38	27.1	127.2	59.3 *	21.0	14.6	1.28	2.39	2.58
MON	5.37	24.1 *	123.5 +	59.6 *	25.1 *	12.3 +	1.31	2.03 *	2.38 *
SEM [1]	0.01	1.12	2.70	0.90	0.69	0.48	0.10	0.08	0.13

[1] SEM—standard error of the mean. [2] BCVFA—Branch-chained VFA. [3] A:P—Acetate to propionate ratio. * Means within a column differ from control ($p < 0.05$). [a] 10 mg/L. [b] 50 mg/L. [c] 200 mg/L. [d] 400 mg/L.

3.2. Experiment 2 (Combination of Essential Oils)

Treatments 7 to 13 had no effect on rumen fermentation profile and are not reported. The effects of treatments 1 to 6 are shown in Table 3. Monensin reduced ($p < 0.05$) the molar proportion of acetate, the molar proportion of butyrate, the BCVFA concentrations, the A:P ratio and the concentration of ammonia-N, and increased ($p < 0.05$) the molar proportion of propionate compared with control. At 10 and 50 mg/L, none of the treatments had effects on rumen fermentation profile compared with control.

Treatment 1 at 200 and 400 mg/L decreased ($p < 0.05$) the molar proportion of acetate (−5 to −6%) and the A:P ratio (−16 to −19%), and increased ($p < 0.05$) the molar proportion of propionate (+12 to +16%). Treatment 1 at 400 mg/L tended to reduce ($p < 0.10$) the pH compared with control Treatment 2 only at 400 mg/L decreased ($p < 0.05$) the molar proportion of acetate (−5%) and the A:P ratio (−18%), and increased ($p < 0.05$) the propionate proportion (+16%) when compared with control. Treatment 3 only at 400 mg/L decreased ($p < 0.05$) the molar proportion of acetate (−4%) and the A:P ratio (−17%), and increased ($p < 0.05$) the molar proportion of propionate (+15%) compared with control. Treatment 4 had no effect at 200 mg/L but at 400 mg/L decreased ($p < 0.05$) the molar proportion of acetate (-6%) and the A:P ratio (−25%), and increased ($p < 0.05$) the molar proportion of propionate (+24%) compared with control. Treatment 5 at 200 and 400 mg/L decreased ($p < 0.05$) the molar proportion of acetate (−6 to −8%) and the A:P ratio (−22 to +26%), and increased ($p < 0.05$) the molar proportion of propionate (+21 to +25 %) when compared with control. Treatment 6 had no effect at 200 mg/L, and only the molar proportion of acetate decreased (−5%; $p < 0.05$) at 400 mg/L compared with the control. When mixes modified rumen microbial fermentation (T1 to T6), the relative effects of the mixes of EO were compared with the closest doses of TeTr oil in Experiment 1, and the size of the effect increased, suggesting no additive or synergistic effects.

4. Discussion

In the last two decades, there has been a wealth of data produced by research on the effect of EO on rumen microbial fermentation, most of it in in vitro conditions [1,2]. However, most of it has focused on dairy cattle, where fermentation was conducted in high forage diets at pH 6.5–7.0. There are few reports on the effects of EO on ruminal microbial fermentation in high concentrate diets fed to feedlot beef cattle [5,13,14]. The antimicrobial activity of EO is affected by pH, and ruminal pH in cattle fed high-concentrate diets is different from that of dairy cattle diets and often below 6.0 [5]. These effects may also affect potential interactions among different EO in mixes. The selection criteria to identify EO with positive effects on rumen microbial fermentation in these conditions were an increase or no change in total VFA, an increase in the proportion of propionate, and/or a decrease in the proportion of acetate, the A:P ratio and/or the ammonia-N concentration. These effects would improve the efficiency of energy and N utilization [1]. On the other hand, Calsamiglia et al. [1] suggested that, because of different mechanisms of action, the combination of EO may result in additive or synergistic effects. In fact, most EO products in the market are based on combination of different EO or their active molecules (i.e., CRINA-Ruminants®, DSM, Switzerland; AGOLIN®, Agolin Sa, Switzerland, XTRACT®, Pancosma, Switzerland), but scientific evidence of their additive or synergistic effects is lacking.

Monensin has been recognized as an effective in-feed antibiotic that results in an increase in propionate at the expenses of acetate, and reduces amino acid deamination [15,16] and was used as a positive control. The effects of monensin in Experiments 1 and 2 show these effects and confirm the reliability of the method to measure these changes in the selected EO. At similar concentrations as monensin, EO in Experiment 1 and EO mixes in Experiment 2 had no effects on rumen fermentation, even when considering the concentration of active components. This observation indicates that the activity of EO is weaker compared with monensin, and higher doses are required to provoke a change in rumen microbial fermentation.

4.1. *Experiment 1 (Individual Essential Oils)*

Ammonia-N was only reduced in Thy and Sag at the highest dose, and the magnitude of the reduction was much lower than that observed for monensin. The effect of Thy on ammonia-N concentration agrees with previous reports where Thy was tested in similar conditions and doses [17]. Because Sag had negative effects on VFA and Thy did not affect VFA at the high dose, the effects of EO on ammonia-N could not be considered as a criterion to select EO for Experiment 2.

Thyme and Ore contain high concentrations of thymol and carvacrol as major active components, respectively. Thymol and carvacrol are phenolic monoterpenes with a wide-spectrum antimicrobial activity due to its capacity to act as membrane permeabilizer [18–20]. Juven et al. [21] observed that their antibacterial activity was greater at pH 5.5 compared with the activity at pH 6.5. In the beef-type diet conditions of this experiment, Ore and Thy at 50 and 200 mg/L were affected in the same way as total and individual VFA, reflecting a similar mechanism of action, and are in agreement with previous results in similar conditions [5,17]. The lack or negative effect of these EO up to 500 mg/L also agrees with previous reports [22,23], but higher doses inhibit rumen microbial fermentation. These results indicate that Ore and Thy should be included at moderate doses to modify rumen fermentation profile without inhibiting microbial activity. Because phenolic compounds with similar structure may have additive effects [6], these two EO were selected for further exploration of possible interactions in Experiment 2.

Terpinen-4-ol is the main active component in TeTr and has a wide-spectrum antimicrobial activity [24]. The TeTr at 50 and 200 mg/L decreased the molar proportion of acetate and the A:P ratio and increased the molar proportion of propionate and the molar proportion of butyrate, a fermentation profile consistent with better energy utilization [1]. In fact, at 200 mg/L, the effect on the A:P ratio was the largest among all EO tested in Experiment 1. Results agree with Castillejos et al. [22] using similar fermentation conditions and doses. Although some reports have shown negative effects of TeTr on rumen microbial fermentation [25–27], with a significant reduction in total VFA and propionate concentrations, these trials were carried out with ruminal liquid and conditions for dairy cows and at pH > 6.0. Several authors have already recognized the role of media pH and type of diet on the antimicrobial activity of EO [5,21]. Therefore, results obtained in the present report with high concentrate beef-type fermentation conditions justify the selection of TeTr for further study in Experiment 2.

Clove bud oil at 200 mg/L increased the molar proportion of butyrate, and at 400 mg/l decreased the molar proportion of propionate and increased the molar proportion of butyrate, the A:P ratio and the pH compared with the control. Castillejos et al. [22] also reported that Clo at 500 mg/L reduced the molar proportion of propionate and increased the molar proportion of butyrate and the A:P ratio. This fermentation profile is not adequate to improve energy utilization in beef cattle diets. In the present trial, Clo at moderate dose (10 and 50 mg/L) increased the ammonia-N concentrations in 1% and 2% vs. control, but differences were not significant. In contrast, at lower doses (30 mg/L), Cardozo et al. [5] reported that eugenol increased total VFA by 20% and increased the concentration of ammonia-N, which could be desirable if ammonia-N concentration limits microbial protein synthesis in feedlot cattle fed high percentages of concentrate [28].

Rosemary oil only had effects at the highest dose but reduced total VFA concentrations and the proportion of butyrate, and increased the acetate proportion. This fermentation profile is not desirable in beef cattle conditions, where an increase in propionate is desired [1]. Sage oil at the highest dose decreased the A:P ratio, but at the same time total VFA was also reduced, suggesting that microbial fermentation was inhibited. It is interesting to observe that the active components of Ros and Sag are 1,8-cineole and thujone, respectively, and both are oxygenated phenols. Oxygenated phenols have been reported to have negative or even antagonistic effects when mixed with other phenolic compounds [6], and therefore, were discarded for Experiment 2.

4.2. Experiment 2 (Combination of Essential Oils)

In Experiment 1, TeTr, Ore and Thy at 50 and 200 mg/L had the most appropriate fermentation profile to improve energy utilization in beef cattle diets. The main active components of these EO are phenolic monoterpens. Bassole and Juliani [6] indicated that phenolic compounds tend to have additive effects, while synergistic or antagonistic effects would occur with other chemical compounds and vary depending on the microbial ecosystem. Clove bud oil was selected because previous research suggested that at pH 5.5 and at moderate doses stimulated deamination activity and inhibited peptide lyses [3,5]. Because of the different mechanisms of action involved, synergies may be expected. Furthermore, the main active component, eugenol, is a phenylpropanoid, which may have synergies or antagonistic effects with phenolic monoterpenes [6]. There are no previous reports designed to explore interactions between the selected EO on rumen fermentation profile in beef feedlot-type fermentation conditions.

In Experiment 1, TeTr was the EO with the strongest response at doses of 50 to 200 mg/L. For this reason, most of the mixes proposed contained TeTr at different effective concentrations. Treatments at 200 or 400 mg/L and with 50 or 75% of TeTr (corresponding to an effective dose of TeTr of 100 to 300 mg/L) reduced the proportion of acetate and the A:P ratio, and increased the proportion of propionate without affecting the total concentration of VFA (T1 to T5). The effective doses of TeTr in the mixes were within the range of those tested in Experiment 1, and the magnitude of the response was never higher that the effect observed when TeTr oils was supplemented alone in Experiment 1. Therefore, there is no evidence that the combination of different EO had an additive or synergistic effect neither with monoterpenes, nor with phenolpropanoids. It is also interesting to observe that whenever Ore, Thy or both were mixed with Clo (see T7 to T13, Table 1), the fermentation profile was not affected. In treatments T7 to T13 (Table 1), 100 mg of Thy, Ore or the combination of both resulted in no effect if mixed with 100, 200 or 300 mg of Clo. This is surprising because both Thy and Ore reduced acetate, increase propionate and reduced the A:P ratio when incubated alone in Experiment 1, but this effect was inhibited by the presence of Clo. Treatment T6 contained TeTr, and even in cases where the effective concentration of TeTr was 100 and 200 mg/L that resulted in improved fermentation profile in Experiment 1, there was only a small reduction in the acetate proportion (−5%), no effect on propionate and a non-significant reduction of the A:P ratio. There seem to be an antagonistic effect by which the combination of either Thy, Ore or both with Clo resulted in no effect on rumen fermentation profile.

None of the mixes in Experiment 2 affected ammonia-N concentration or BCVFA, suggesting that deamination was not affected. This is not surprising, because only Thy and Sag at 400 mg in Experiment 1 had effects, and in none of the treatments in Experiment 2 the effective dose of these EO was reached. Clove bud oil was selected because Busquet et al. [27] indicated that it reduced peptide degradation. We hypothesized that the combination of the inhibition of peptide degradation by eugenol in Clo may be synergistic with inhibition of deaminnation shown in Thy. However, the combination of different EO appear no to have this additive or synergistic effect in protein degradation. These results question the common practice of mixing different EO based on potential additive or synergistic effects.

5. Conclusions

Tea tree oil, oregano oil and thyme oil can be effective as rumen fermentation modifiers under feedlot beef-type fermentation conditions. The combinations of TeTr oil with Ore oil, Thy oil or Ore + Thy did not have additive effects over those observed with TeTr oil alone. When Thy, Ore or Thy + Ore where mixed with Clo, all effects on rumen fermentation profile disappeared even when TeTr was part of the mix, suggesting an antagonistic interaction of Clo with Thy and Ore. Results do not support the hypothesis of additivity among the EO tested, and antagonistic effects of Clo mixed with Thy or Ore were demonstrated at least in a low pH, beef-type fermentation conditions.

Author Contributions: Conceptualization, I.F., A.F., D.M., L.C. and S.C.; methodology, I.F., D.M. and S.C.; formal analysis, I.F. and D.M.; resources, S.C. and L.C.; data curation, I.F., G.F.-T. and S.C.; writing—Original draft preparation, I.F and G.F.-T.; writing—Review and editing, A.F., D.M., L.C. and S.C.; supervision, S.C.; project administration, S.C. All authors have read and agreed to the published version of the manuscript.

References

1. Calsamiglia, S.; Busquet, M.; Cardozo, P.W.; Castillejos, L.; Ferret, A. Invited Review: Essential oils as modifiers of rumen microbial fermentation. *J. Anim. Sci.* **2007**, *90*, 2580–2595. [CrossRef]
2. Benchaar, C.; McAllister, T.A.; Chouinard, P.Y. Digestion, ruminal fermentation, ciliate protozoal populations, and milk production from dairy cows fed cinnamaldehyde, quebracho condensed tannin, or *Yucca schidigera* saponin extracts. *J. Dairy Sci.* **2008**, *91*, 4765–4777. [CrossRef]
3. Busquet, M.; Calsamiglia, S.; Ferret, A.; Kamel, C. Screening for effects of plant extracts and active compounds of plants on dairy cattle rumen microbial fermentation in a continuous culture system. *Anim. Feed Sci. Technol.* **2005**, *123–124*, 597–613. [CrossRef]
4. Castillejos, L.; Calsamiglia, S.; Ferret, A.; Losa, R. Effects of a specific blend of essential oil compounds and the type of diet on rumen microbial fermentation and nutrient flow from a continuous culture system. *Anim. Feed Sci. Technol.* **2005**, *119*, 29–41. [CrossRef]
5. Cardozo, P.W.; Calsamiglia, S.; Ferret, A.; Kamel, C. Screening for the effects of natural plant extracts at different pH on *in vitro* rumen microbial fermentation of a high-concentrate diet for beef cattle. *J. Anim. Sci.* **2005**, *83*, 2572–2579. [CrossRef]
6. Bassolé, I.H.N.; Juliani, H.R. Essential oils in combination and their antimicrobial properties. *Molecules.* **2012**, *17*, 3989–4006. [CrossRef] [PubMed]
7. Hyldgaard, M.; Mygind, T.; Meyer, R.L. Essential oils in food preservation: Mode of action, synergies, and interactions with food matrix components. *Front. Microbiol.* **2012**, *3*, 1–24. [CrossRef]
8. Langeveld, W.T.; Veldhuizen, E.J.A.; Burt, S.A. Synergy between essential oil components and antibiotics: A review. *Crit. Rev. Microbiol.* **2013**, *40*, 76–94. [CrossRef] [PubMed]
9. NRC. *Nutrient Requirements of Beef Cattle*, 7th rev. ed; Nat. Acad. Press: AcadWashington DC, USA, 1996.
10. McDougall, E.I. Studies on ruminant saliva. 1. The composition and output of sheep's saliva. *Biochem. J.* **1948**, *43*, 99–109. [CrossRef] [PubMed]
11. Chaney, A.L.; Marbach, E.P. Modified reagents for determination of urea and ammonia. *Clin. Chem.* **1962**, *8*, 130–132. [PubMed]
12. Jouany, J.P. Volatile fatty acid and alcohol determination in digestive contents, silage juices, bacterial cultures and anaerobic fermentor contents. *Sci. Aliment.* **1982**, *2*, 131–144.
13. Benchaar, C.; Duynisveld, J.L.; Charmley, E. Effects of monensin and increasing dose levels of a mixture of essential oil compounds on intake, digestion and growth performance of beef cattle. *Can. J. Anim. Sci.* **2006**, *86*, 91–96. [CrossRef]
14. Nanon, A.; Suksombat, W.; Yang, W.Z. Effects of essential oils supplementation on *in vitro* and *in situ* feed digestion in beef cattle. *Anim. Feed Sci. Technol.* **2014**, *196*, 50–59. [CrossRef]
15. Wallace, R.J.; Czerkawski, J.W.; Breckenridge, G. Effect of monensin on the fermentation of basal rations in the Rumen Simulation Technique (Rusitec). *Br. J. Nutr.* **1981**, *46*, 131–148. [CrossRef] [PubMed]
16. Chalupa, W.; Corbett, W.; Brethour, J.R. Effects of monensin and amicloral on rumen fermentation. *J. Anim. Sci.* **1980**, *51*, 170–179. [CrossRef] [PubMed]
17. Castillejos, L.; Calsamiglia, S.; Ferret, A. Effect of essential oil active compounds on rumen microbial fermentation and nutrient flow in in vitro systems. *J. Dairy Sci.* **2006**, *89*, 2649–2658. [CrossRef]
18. Dorman, H.J.D.; Deans, S.G. Antimicrobial agents from plants: Antibacterial activity of plant volatile oils. *J. Appl. Microbiol.* **2000**, *88*, 308–316. [CrossRef] [PubMed]
19. Walsh, S.E.; Maillard, J.Y.; Russell, A.D.; Catrenich, C.E.; Charbonneau, D.L.; Bartolo, R.G. Activity and mechanisms of action of selected biocidal agents on Gram-positive and -negative bacteria. *J. Appl. Microbiol.* **2003**, *94*, 240–247. [CrossRef]
20. Lambert, R.J.W.; Skandamis, P.N.; Coote, P.J.; Nychas, G.J.E. A study of the minimum inhibitory concentration and mode of action of oregano essential oil, thymol and carvacrol. *J. Appl. Microbiol.* **2001**, *91*, 453–462. [CrossRef]

21. Juven, B.J.; Kanner, J.; Schved, F.; Weisslowicz, H. Factors that interact with the antibacterial action of thyme essential oil and its active constituents. *J. Appl. Bacteriol.* **1994**, *76*, 626–631. [CrossRef]
22. Castillejos, L.; Calsamiglia, S.; Martín-Tereso, J.; Ter Wijlen, H. *In vitro* evaluation of effects of ten essential oils at three doses on ruminal fermentation of high concentrate feedlot-type diets. *Anim. Feed Sci. Technol.* **2008**, *145*, 259–270. [CrossRef]
23. Macheboeuf, D.; Morgavi, D.P.; Papon, Y.; Mousset, J.L.; Arturo-Schaan, M. Dose–response effects of essential oils on *in vitro* fermentation activity of the rumen microbial population. *Anim. Feed Sci. Technol.* **2008**, *145*, 335–350. [CrossRef]
24. Davidson, P.M.; Naidu, A.S. Phyto-phenols. In *Natural Food Antimicrobial Systems*; CRC Press: Boca Raton, FL, USA, 2000; pp. 265–293.
25. Hristov, A.N.; Ropp, J.K.; Zaman, S.; Melgar, A. Effects of essential oils on *in vitro* ruminal fermentation and ammonia release. *Anim. Feed Sci. Technol.* **2008**, *144*, 55–64. [CrossRef]
26. Gunal, M.; Ishlak, A.; AbuGhazaleh, A.A.; Khattab, W. Essential oils effect on rumen fermentation and biohydrogenation under *in vitro* conditions. *Czech., J. Anim. Sci.* **2014**, *59*, 450–459. [CrossRef]
27. Busquet, M.; Calsamiglia, S.; Ferret, A.; Kamel, C. Plant Extracts Affect *In Vitro* Rumen Microbial Fermentation. *J. Dairy Sci.* **2006**, *89*, 761–771. [CrossRef]
28. Devant, M.; Ferret, A.; Gasa, J.; Calsamiglia, S.; Casals, R. Effects of protein concentration and degradability on performance, ruminal fermentation, and nitrogen metabolism in rapidly growing heifers fed high-concentrate diets from 100 to 230 kg body weight. *J. Anim. Sci.* **2000**, *78*, 1667–1676. [CrossRef]

17

Ruminal In Vitro Protein Degradation and Apparent Digestibility of Energy and Nutrients in Sheep Feed Native or Ensiled + Toasted Pea (*Pisum sativum*) Grains

Martin Bachmann [1], Christian Kuhnitzsch [1,2], Paul Okon [1], Siriwan D. Martens [2], Jörg M. Greef [3], Olaf Steinhöfel [1,2] and Annette Zeyner [1,*]

[1] Institute of Agricultural and Nutritional Sciences, Martin Luther University Halle-Wittenberg, Theodor-Lieser-Straße 11, 06120 Halle (Saale), Germany

[2] Saxon State Office for Environment, Agriculture and Geology, Am Park 3, 04886 Köllitsch, Germany

[3] Institute for Crop and Soil Science, Federal Research Centre for Cultivated Plants, Julius Kühn Institute (JKI), Bundesallee 58, 38116 Braunschweig, Germany

[*] Correspondence: annette.zeyner@landw.uni-halle.de

Simple Summary: Pea grains may partially replace soybean or rapeseed meals and cereals in ruminant diets, but this is limited by high solubility of pea protein in the rumen. Hydro-thermic treatments such as toasting may stabilize the protein and shift digestion from the rumen to the small intestine. The effect of toasting of ensiled pea grains on rumen-undegraded protein was tested in vitro and on apparent digestibility of organic matter, gross energy, and proximate nutrients in a digestion trial with sheep. Ensiling plus toasting increased rumen-undegraded protein from 20 to 62% of crude protein, but it also increased acid detergent insoluble protein, which is unavailable for digestive enzymes in the small intestine from 0.5 to 2.6% of crude protein. Ensiling plus toasting did not, however, affect total tract apparent digestibility of organic matter, energy, crude protein, or any other nutrient fraction, nor did it alter the concentration of metabolizable energy or net energy lactation in the peas. The technique can be implemented on farms and might have a positive impact on field pea production.

Abstract: Pea grains may partially replace soybean or rapeseed meals and cereals in ruminant diets, but substitution by unprocessed peas is limited by high ruminal protein solubility. The effect of combined ensiling and toasting of peas using a mobile toaster (100 kg/h throughput rate, 180 to 190 °C supplied air temperature) on rumen-undegraded protein (RUP) was tested in vitro using the *Streptomyces griseus* protease test. The effects of ensiling plus toasting on apparent digestibility of organic matter (OM), gross energy (GE), and proximate nutrients were examined in a digestion trial. Concentrations of metabolizable energy (ME) and net energy lactation (NEL) were calculated. Native peas had 38 g RUP/kg dry matter (DM), which was 20% of crude protein (CP). Rumen-undegraded protein increased three-fold after ensiling plus toasting ($p < 0.001$). Acid detergent insoluble protein increased five-fold. Apparent digestibility was 0.94 (OM), 0.90 (CP), and above 0.99 (nitrogen-free extract, starch, and sugars) and was not altered by the treatment. The ME (13.9 MJ/kg DM) or the NEL (8.9 MJ/kg DM) concentration was similar in native and ensiled plus toasted peas. This technique can easily be applied on farms and may increase RUP. However, it needs to be clarified under which conditions pea protein will be damaged.

Keywords: field peas; ensiling; hydro-thermic treatment; nutrient digestibility; rumen-undegraded protein; *Streptomyces griseus* protease test

1. Introduction

The production of non-genetically modified foods and animal feeds is increasingly prominent in the public and the political spotlights. In the livestock sector, soybean meal is, apart from rapeseed meal (after oil extraction), undoubtedly the most intensively used protein feed, but it nearly completely originates from genetically modified sources [1]. The European Union (EU-28) currently imports about 95% of the utilized soybeans [2]. National programs now intend to reduce soybean imports and partly replace the classical protein feeds by more cost-effective and sustainable indigenous protein plants such as lupines, faba beans, or field peas. This supports local feed production and nutrient cycles as well. Field pea grains combine high percentages of protein and starch—210 to 261 and 404 to 496 g/kg dry matter (DM), respectively [3–5]. Although post-extraction soybean and rapeseed meals have higher protein contents (508 to 564 and 365 to 411 g/kg DM, respectively [6,7]), they do not provide comparable amounts of starch or other non-structural carbohydrates. The one-to-one substitution of soybean meal and cereals by field pea grains was successful in dairy cows on moderate and high production levels and did not affect feed intake and (fat-corrected) milk yields as long as equal levels of rumen-undegraded protein (RUP) were considered [8,9]. However, continuously high and increasing levels of milk production require a better utilization of feed energy and nutrients. Starch, but especially protein, from field peas is highly degradable in the rumen when the grains are unprocessed, which is limiting small intestinal availability [10]. Increasing RUP might therefore enable the substitution of soybean or rapeseed meals in high-yielding animals [8].

Previous in vitro [11] and in vivo [12] studies demonstrated that the combination of ensiling and subsequent toasting of pea grains is suitable to obtain harvesting and storage stability regardless of the weather conditions. Contrary to our expectations, ensiling of pea grains still led to reduction of protein solubility and an increase of RUP in vitro [11]. Toasting of dry and unprocessed peas after harvesting did not decrease protein solubility, and did not increase RUP or post-ruminal crude protein (CP). Therefore, we further used the combination of ensiling and toasting. We hypothesized that hydro-thermic treatment (toasting) of ensiled field pea grains with 180 to 190 °C supplied air temperature (i.e., 85 to 90 °C grain temperature) and 100 kg/h throughput rate in the toaster would increase RUP without damaging the protein. Apparent digestibility of organic matter (OM), crude nutrient fractions, starch, sugars, and gross energy (GE) was expected to remain unaffected, which would (regarding pea protein) increase the amount that is available for digestion in the small intestine. The effect of ensiling plus toasting on RUP contents of field pea grains was tested in vitro using the *Streptomyces griseus* protease test. Treatment effects on apparent digestibility of energy, OM, and proximate nutrients were proven in a standard digestion trial with sheep. Additionally, the data were used to calculate GE, metabolizable energy (ME), and net energy lactation (NEL) of the pea treatments, applying official equations of the Society of Nutrition Physiology (GfE) [13,14] and the National Research Council (NRC) [15].

2. Materials and Methods

2.1. Pea Treatments

The field pea cultivar Alvesta (KWS SAAT SE, Einbeck, Germany) was used. It was grown and harvested in 2017 in Köllitsch, Saxony. A total quantity of 27.3 t DM of the native grains (i.e., as harvested; not ensiled or toasted) were re-moistened from 779 to 749 g/kg by adding 120 g of a homo-fermentative lactic acid bacteria inoculant including *Lactobacillus plantarum* and *Pediococcus acidilactici* strains (LAB; together 6.8×10^6 colony forming units per g fresh matter; Josilac® classic, Josera GmbH & Co. KG, Kleinheubach, Germany) in 200 L water (i.e., 0.6 g LAB/L), crushed using a Murska 2000 S2x2 (Murska, Ylivieska, Finland), and ensiled in a silage plastic bag (BAG Budissa Agroservice GmbH, Malschwitz, Germany) for 9 months. The grain silage had 749 g DM/kg, 34 g crude ash (CA)/kg DM, 189 g CP/kg DM, 20 g acid ether extract (AEE)/kg DM, and 63 g crude fiber (CF)/kg DM. Further ensiling characteristics are specified in Table 1. The aerobic stability of the field pea grain silage was tested following the procedure of H. Honig [16]. Subsequently, the ensiled grains were toasted using a mobile toaster (EcoToast 100, Agrel GmbH, Arnstorf, Germany) at atmospheric pressure with a throughput rate of 100 kg/h, 180 to 190 °C supplied air temperature, and 85 to 90 °C grain temperature.

Table 1. Fermentation characteristics of the field pea grain silage.

pH [1]	6.1
pH [2]	6.3
Lactic acid	2.3
Acetic acid	0.3
Propionic acid	<0.2
iso-Butyric acid	<0.3
n-Butyric acid	<0.1
iso-Valeric acid	<0.1
n-Valeric acid	<0.1
Ethanol	9.4
1,2-Propanediol	<0.3
1-Propanol	<0.4
Aerobic stability	≥7

The fatty acid and alcohol concentrations are given in g/kg dry matter, and the aerobic stability is given in days until the temperature difference between material and ambient exceeds 3 °C. [1] After ensiling. [2] After 7 days of aerobic storage.

2.2. *In Vitro Estimation of RUP*

The concentrations of RUP in native (i.e., not ensiled or toasted) and ensiled plus toasted pea grains were estimated in vitro using the *Streptomyces griseus* protease test [17] with 10 analytical replicates each. Briefly, 0.5 g of material was weighed into the 136 mL glass bottles, 40 mL of borate phosphate buffer consisting of 12.20 g/L $NaH_2PO_4 \times H_2O$ + 8.91 g/L $Na_2B_4O_7 \times 10H_2O$ (pH 6.7 to 6.8) was added, and the solution was incubated for 1 h at 39 °C in a shaking water bath (80 rpm). A solution of nonspecific type XIV *Streptomyces griseus* protease (Sigma-Aldrich Chemie GmbH, Munich, Germany; ≥3.5 U/mg) with an activity of 0.58 U/mL was made following Licitra et al. [18]. The protease preparation combined aminopeptidase and caseinolytic activities [19]. One unit was defined to hydrolyze casein producing color equivalent (Folin-Ciocalteu reagent) to 1.0 μmol (i.e., 181 μg) of tyrosine per min at pH 7.5 and 37 °C. After 1 h, 3.63 and 3.55 mL of protease solution were added to the bottles (i.e., 24 U/g true protein) based on 179 and 175 g true protein/kg DM in native and ensiled plus toasted peas, respectively (calculated relative to a soybean standard with 493 g true protein/kg DM, which requires 10 mL of the solution) [17]. True protein was calculated from protein fractionation according to Licitra et al. [20] as CP—the non-protein nitrogen (A). The incubation time was set to 24 h considering a lag time of approximately 2 h [18]. After incubation, the bottles' contents were filtered through Whatman #41 filter circles, and each was washed with 200 mL deionized water. The filters were air-dried, and nitrogen was determined in the residues and the blank filters using a FOSS Kjeltec™ 8400 (FOSS GmbH, Hamburg, Germany).

The RUP content of the pea treatments was calculated as follows considering a weighed portion of 0.5 g of each treatment:

$$\text{RUP, g/kg DM} = \left(\frac{((N_{residue} - N_{blank}) \times 6.25 \times 10)}{0.5 \times DM_{feed}}\right) \times 10 \quad (1)$$

where $N_{residue}$ is nitrogen measured in the filter residues (mg); N_{blank} is mean nitrogen measured in the blank filters (mg); and DM_{feed} is the DM content of the native or ensiled plus toasted field pea grains (%).

2.3. *In Vivo Determination of OM, GE, and Nutrient Digestibilities*

The sheep used in this study were kept and cared for by the Research Centre for Agricultural and Nutritional Sciences, Martin Luther University Halle-Wittenberg, Wettin/Löbejün, Saxony-Anhalt, Germany. The experiment was carried out with approval by the Saxony-Anhalt Federal Administration Authority (approval no. 2-1524 MLU).

Eight adult Pomeranian coarsewool wethers were used as model animals for the digestion trial. All animals were clinically healthy and under regular veterinary supervision.

The composition of the offered diets is given in Table 2. Tap water was offered ad libitum. The feeding level of the sheep was close to the energy maintenance level [13,21]. The recommended feed protein content for digestibility trials with sheep is a minimum of 120 g CP/kg diet DM [22]. This was met during the current experiment. The diets were offered in two equal meals per day. The chemical compositions of the diet components and the mixed diets are given in Table 3.

Table 2. Composition of mixed diets offered during the experiment.

Component (g/Day as Fed)	Control Diet	Test Diet (Native Peas)	Test Diet (Ensiled + Toasted Peas)
Lucerne (chopped)	450	225	225
Barley (crushed, Ø 3.5 mm)	450	225	225
Wheat straw (chopped, Ø 6.0 mm)	100	50	50
Native peas	0	500	0
Ensiled + toasted peas	0	0	500
Mineral feed [1]	10	10	10

[1] basu-kraft® Top-Mineral (BASU Heimtierspezialitäten GmbH, Bad Sulza, Germany).

The digestion trial was run as a difference test in two consecutive periods following the guidelines of GfE [22]. Each period consisted of 14 days adaptation to the diet followed by 6 days of total collection of feces. In total, 4 sheep received the native pea grains (i.e., not ensiled or toasted; 4 analytical replicates), and 5 sheep received the ensiled plus toasted peas (i.e., 5 analytical replicates). During the experiment, the sheep were individually housed in metabolic cages. All feed components were weighed for each mealtime and animal before the experiment started, and composited samples were taken. During total collection, the animals were fitted with harnesses for feces collection, which were emptied each morning prior the first meal. Daily defecation was quantified, and an aliquot of 0.2 was taken. Feed residuals were recorded. The fecal samples were composited individually per sheep per collection period. The feed and the fecal samples were stored dry or frozen at −20 °C, respectively.

The animals' body weights were recorded in each period before adaptation, before feces collection, and at the end of the experiment. The initial body weight was 78 ± 9.7 kg. After a light decline, it was kept constant during the experiment (76 ± 7.5 kg before the first collection, 74 ± 7.8 kg before the second adaptation, 73 ± 11 kg before the second collection, and 73 ± 10 kg at the end of the experiment).

Table 3. Chemical composition of the feeds and the diets used in the experiment.

	Lucerne	Wheat Straw	Barley	Native Peas	Ensiled + Toasted Peas	Mixed Diet (Control)	Mixed Diet (Including Native Peas) [1]	Mixed Diet (Including Ensiled + Toasted Peas) [1]
Dry matter	927	945	888	779	970	912	851	942
Crude ash	71	71	22	31	32	50	41	41
Organic matter	929	929	978	969	968	950	959	959
Crude protein	146	46	115	186	186	122	152	155
Acid ether extract	21	10	25	13	10	22	18	16
Starch	8	14	537	533	496	240	375	372
Sugars	47	12	42	77	38	41	58	40
Crude fiber	363	448	52	62	61	236	157	146
Neutral detergent fiber	524	843	210	128	197	420	285	305
Acid detergent fiber	389	506	64	80	79	259	176	166
Acid detergent lignin	85	59	8	5	15	49	29	31
Cellulose	56	447	304	75	64	210	147	135
Hemicellulose	135	337	146	48	118	161	109	139
Nitrogen-free extract	399	425	786	708	711	570	632	642
ESOM	555	344	880	953	944	675	803	814
Gross energy	19.1	17.9	18.6	18.4	18.3	18.8	18.6	18.5

ESOM = enzyme-soluble organic matter. [1] Pea treatments were included at an amount of 500 g dry matter per day. Dry matter (DM) is given in g/kg, gross energy is given in MJ/kg DM, and all other analytes are given in g/kg DM.

2.4. Analyses

Dry matter, CA, CP, AEE, CF, and Van Soest detergent fiber contents of feeds and feces were analyzed according to the German key book for feed analysis (VDLUFA methods no. 3.1, 4.1.1, 5.1.1 B, 6.1.1, 6.5.1, 6.5.2, 6.5.3, and 8.1) [23]. Neutral detergent fiber (NDF) was determined after 1 h treatment with heat stable amylase. Neutral detergent fiber and acid detergent fiber (ADF) were expressed exclusive of residual ash. Organic matter was calculated as OM = 1000 − CA, and the nitrogen-free extract (NFE) was calculated as NFE = OM − CP − AEE − CF. The amount of cellulose was calculated as cellulose = ADF − acid detergent lignin, and hemicellulose (HEM) was calculated as HEM = NDF − ADF. Neutral detergent insoluble CP (NDICP) was determined in the feeds according to VDLUFA method no. 4.13.1 [24]. Gross energy was determined in the feeds and the feces by bomb calorimetry using a C7000 Oxygen Bomb Calorimeter (IKA® Werke, Staufen, Germany). Starch was determined in the feeds and the feces according to the amyloglucosidase method (VDLUFA method no. 7.2.5) [23]. The sugar content of the feeds and the feces was analyzed using anthrone reagent [25]. Enzyme-soluble organic matter (ESOM) was analyzed according to VDLUFA method no. 6.6.1 [23]. The protein fractions A (i.e., non-protein nitrogen), B1 (i.e., true protein, which is soluble in borate phosphate buffer at pH 6.7 to 6.8 but precipitable), B2 (i.e., true protein, which is insoluble in the borate phosphate buffer minus true protein, which is insoluble in neutral detergent), B3 (i.e., true protein, which is insoluble in neutral detergent but soluble in acid detergent), and C (i.e., true protein, which is insoluble in acid detergent) were determined in native and ensiled plus toasted field peas according to Licitra et al. [21] (method no. LKS FMUAA 1402015-11). For each fraction, residual nitrogen was determined according to the Kjeldahl method. The protein fractions were used to calculate the true protein content (i.e., B1 + B2 + B3 + C) and the soluble protein (i.e., A + B1). The protein insoluble in pepsin (CP_{ip}) was also analyzed using the Kjeldahl method after 48 h of incubation in a pepsin-hydrochloric acid solution (method no. LKS FMUAA 1112014-07) [26]. The organic acids and the alcohols produced during the fermentation of the silage were determined by high performance liquid chromatography and refractive index detection (method no. LKS FMUAA 1662018-05) using a Shimadzu LC-20A Prominence (Shimadzu Corp., Kyoto, Japan) and a Hi-Plex H 8 μm column (300 × 7.7 mm; Agilent Technologies Inc., Santa Clara, CA, USA). All LKS methods were accredited according to DIN EN ISO/IEC 17025:2005.

2.5. Calculations and Statistical Analysis

Apparent digestibility coefficients of OM, CA, CP, AEE, CF, NDF, ADF, NFE, starch, sugars, and GE were calculated as the difference between intake and fecal output divided by the intake on a daily basis for each individual. Feed residuals were consistently lower than 2% of what was offered to the sheep. In accordance with the method prescription [22], feed residuals were thus not considered for digestibility calculation.

The current measured GE data and apparent digestibility coefficients were used to calculate GE, ME, and NEL according to GfE [13] as follows: GE, MJ/kg DM = 0.0239 × CP + 0.0398 × AEE + 0.0201 × CF + 0.0175 × NFE; ME, MJ/kg DM = 0.0312 × digestible AEE + 0.0136 × digestible CF + 0.0147 (digestible OM − digestible AEE − digestible CF) + 0.00234 × CP; and NEL, MJ/kg DM = 0.6 (1 + 0.004 (q − 57)) × ME, where q = ME/GE × 100. These equations were derived from the large data pool of total metabolic trials provided by the Institute for Animal Nutrition "Oskar Kellner" in Rostock (Germany) [27]. Moreover, GE, ME, and NEL were calculated on the basis of crude nutrient analyses using equations provided by GfE [13,14] and NRC [15].

Statistical analysis was performed using SAS 9.4 (SAS Institute Inc., Cary, NC, USA). The pooled *t*-test was used to compare RUP estimates and digestibility coefficients between the field pea treatments at a significance level of $p < 0.05$. Homogeneity of the sample variances was confirmed using the folded *F*-test, and the studentized residuals were confirmed to have Gaussian distribution using the UNIVARIATE procedure.

3. Results

Native field pea grains had 38 g RUP/kg DM, which was 20% of CP. The concentration of RUP was increased to a three-fold in the ensiled plus toasted grains (115 g/kg DM, i.e., 62% of CP; $p < 0.001$), as shown in Figure 1. The protein fractionation procedure revealed that, after toasting the ensiled grains, protein solubility was five-fold decreased, soluble protein (B1) was nine-fold decreased, and the insoluble fractions B2 and B3 were two-fold and 27-fold increased, respectively (Table 4). The CP_{ip} was slightly increased. However, the C fraction was also increased to a five-fold after ensiling plus toasting (Table 4).

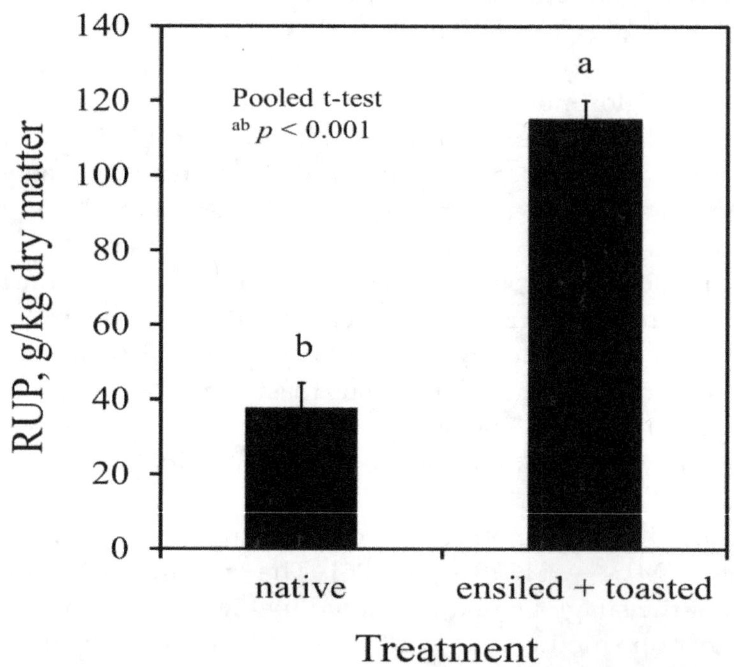

Figure 1. In vitro estimation of rumen-undegraded protein (RUP) in native and ensiled + toasted field pea grains using *Streptomyces griseus* protease (3.63 and 3.55 mL of 0.58 U/mL protease solution, respectively, and 24 h incubation time); native peas had 779 g DM/kg, 186 g CP/kg DM, and 179 g TP/kg DM; ensiled + toasted peas had 970 g DM/kg, 186 g CP/kg DM, and 175 g TP/kg DM; CP = crude protein, DM = dry matter, TP = true protein.

Table 4. Crude protein composition of the field pea treatments.

	Native Peas	Ensiled + Toasted Peas
Crude protein (CP)	186	186
True protein [1]	179	175
Protein solubility [2]	74	16
Protein fraction A	6.5	9.0
Protein fraction B1	67.7	7.2
Protein fraction B2	24.5	56.7
Protein fraction B3	0.9	24.5
Protein fraction C	0.5	2.6
CP_{ip}	5.2	7.2
NDICP	1.1	61.6

A = non-protein nitrogen, B1 = buffer-soluble true protein, B2 = buffer-insoluble true protein—true protein that is insoluble in neutral detergent, B3 = true protein that is insoluble in neutral detergent, but soluble in acid detergent, C = true protein that is insoluble in acid detergent, CP_{ip} = protein insoluble in pepsin, NDICP = neutral detergent-insoluble CP. [1] True protein was calculated as CP − A. [2] Protein solubility was calculated as A + B1. CP, true protein, and NDICP are given in g/kg DM, and protein solubility, the protein fractions (A, B1, B2, B3, and C), and CP_{ip} are given in % of CP.

The apparent digestibility coefficients determined for OM, CA, CP, CF, ADF, NFE, starch, sugars, and GE did not differ between the treatments (Table 5). Native and ensiled plus toasted field peas did differ in AEE digestibility by tendency (0.49 vs. 0.61; $p = 0.0548$). They significantly differed in apparent digestibility of the NDF fraction (0.69 vs. 0.81; $p < 0.05$; Table 5).

Table 5. Total tract apparent digestibility of energy and proximate nutrients in native and ensiled + toasted field pea grains.

	Native Peas	Ensiled + Toasted Peas
Organic matter	0.94 (0.019)	0.94 (0.026)
Crude ash	0.38 (0.200)	0.39 (0.140)
Crude protein	0.90 (0.033)	0.89 (0.042)
Acid ether extract	0.49 (0.049)	0.61 (0.088)
Crude fiber	0.61 (0.055)	0.65 (0.078)
Neutral detergent fiber	0.69 (0.059) ^{b}	0.81 (0.055) ^{a}
Acid detergent fiber	0.65 (0.069)	0.66 (0.074)
Nitrogen-free extract	0.99 (0.008)	0.99 (0.015)
Starch	1.00 (0.0007)	1.00 (0.001)
Sugars	1.00 (0.003)	0.99 (0.006)
Gross energy	0.91 (0.022)	0.91 (0.027)

Standard deviations are given in parentheses. ^{ab} Different superscripts mark significant differences between the treatments ($p < 0.05$).

Gross energy contents measured by bomb calorimetry and calculated according to GfE [13] were similar in the native and the ensiled plus toasted field peas (18.4 and 18.3 vs. 18.6 MJ/kg DM; Table 6). The native field peas had a digestible energy concentration of 16.7 MJ/kg DM, and the ensiled plus toasted peas had a digestible energy concentration of 16.6 MJ/kg DM, which was calculated based on measured GE and GE digestibility. The ME contents calculated on the basis of measured GE and nutrient digestibilities were 13.8 and 13.9 MJ/kg DM in native and ensiled plus toasted peas, respectively. A similar ME content was calculated in the native peas using GfE and NRC equations. In ensiled plus toasted peas, ME was slightly underestimated by 0.2 MJ/kg DM using GfE and by 0.4 MJ/kg DM using NRC equations (Table 6). This was similar with the NEL contents (Table 6).

Table 6. Gross energy (GE), metabolizable energy (ME), and net energy lactation (NEL) in native and ensiled + toasted field pea grains.

	Native Peas			Ensiled + Toasted Peas		
	Measured/Calculated [1]	GfE [2]	NRC [3]	Measured/Calculated [1]	GfE [2]	NRC [3]
GE	18.4	18.6	n.a.	18.3	18.6	n.a.
ME	13.8 [4]	13.9	13.9	13.9 [4]	13.7	13.5
NEL	8.9 [4]	8.9	9.0	8.9 [4]	8.8	8.7

n.a. = not applicable. [1] Calculated on the basis of measured GE concentrations and apparent digestibility of crude nutrients according to GfE [13]. [2] Calculated according to GfE [13,14]. [3] Calculated according to NRC [15]. [4] Given as the mean of 4 and 5 measurements in native and ensiled + toasted peas, respectively. All items are given in MJ/kg dry matter.

4. Discussion

The social, the political, and the scientific interest in using regionally grown legumes as protein feeds for farm animals is increasing. Faba beans, field peas, lupines, clover, or lucerne can offer an alternative to soybean meal and rapeseed meal in Europe. Using field peas as at least a partial replacement of soybean or rapeseed meals and/or cereals was possible without addressed digestibility reduction, performance depression, or impaired quality of the end products in growing and finishing pigs [28], broiler chickens [29], laying hens [30], growing lambs and beef cattle [31,32], and dairy cows [8,9]. However, a complete replacement was not always successful in either monogastric [33,34] or ruminant livestock [8,32] when the peas remained unprocessed. A main factor that is limiting the substitution efficiency of pea grains is the higher ruminal protein solubility compared to soybean [35] or rapeseed meal. The RUP content of unprocessed peas is around 20% of CP, but in post-extraction soybean and rapeseed meals, 25 to 69% of CP was reported [36–39]. Moreover, ruminal starch degradation of unprocessed peas is approximately 56%, while it is 44% in maize and 83% in barley [40].

Ensiling may generally lead to proteolysis and carbohydrate degradation [41–43] but should not affect ruminal degradation of protein, starches, and other carbohydrates as long as no perishable fermentation (e.g., yeast fermentation) and concomitant heating occurs. Despite a good sensorial quality, the current pea grain silage was not reduced in pH (6.1), and ethanoic acid and ethanol contents were increased compared to previously used silages (by 1.4 and 6.8 g/kg DM) [11]. However, the proportion of true protein was comparable [11,12].

In native pea grains, storage proteins are mainly composed of the globulins 7S vicilin and 11S legumin, which are soluble to more than 70% in the digestive tract (74% in this study) [11,44]. Heat or heat plus pressure treatments such as dry roasting, steam flaking, autoclaving, toasting, extrusion, or expander treatment clearly decrease protein degradability in the rumen [4–6,10,45–48]. Feed protein structures are stabilized by complex Maillard reactions [46,47,49–53]. Next, proteolysis in the rumen can be inhibited or slowed down, but total tract digestibility of OM or protein should not be affected [4,54]. These effects largely depend on the type and the conditions of the treatment (e.g., temperature, throughput rate/duration of heat exposure, feed quantity, and moisture content) [11,53]. The native field pea grains in our study had a RUP content of about 20% of CP, which was similar to what Masoero et al. [5] measured. Then, RUP tripled in the ensiled plus toasted pea grains. Protein solubility decreased to one-fifth, B1 to one-ninth, and especially B3 increased by 27 times. We, however, also found that the C fraction increased five-fold after ensiling plus toasting, which might point to some protein damage.

The determination of total tract apparent digestibility of OM and crude nutrients is the official method to obtain ME and NEL of feedstuffs in Germany [13,21]. In native peas, total tract apparent digestibility coefficients of OM, GE, and proximate nutrients were similar to those reported by the German Agricultural Society (DLG) [39]. In our case, sheep were fed slightly below the maintenance requirement, which might have caused a minor overestimation of digestibility. Total tract apparent digestibility of OM, CA, CP, CF, ADF, NFE, starch, sugars, and GE was not affected by ensiling plus toasting, whereas AEE and NDF digestibility increased by approximately 10%. Total tract apparent digestibility does not allow conclusions on partial (e.g., ruminal) nutrient digestibility. The latter, however, might be altered by ensiling and toasting of peas, even if total tract digestibility remains unaffected. The increase in AEE and NDF digestibility was probably more an analytical issue and not a physiological response. Increased AEE digestibility was probably caused by the low AEE contents in both feed and feces, which needs to be set into relation to the inherent analytical error. Following ensiling plus toasting, NDICP was increased by approximately 30% of NDF. This probably led to an overestimation of the NDF intakes by the sheep and thus to an overestimation of NDF digestibility. When we reduced the NDF contents in the feed by the respective NDICP amounts, apparent digestibilities of NDF were 0.68 ± 0.063 and 0.74 ± 0.063 in native and ensiled plus toasted peas, respectively ($p = 0.22$). Although the samples were treated with amylase during NDF analysis, it is still possible that parts of starch contributed to an overestimation of the NDF intakes, at least when parts of the starch from the feed were altered by the feed treatment. Fiber, ash, and AEE digestibility coefficients had a large variation among the animals (standard deviation ranged from 0.05 to 0.20). Thus, the apparent increase of AEE and NDF digestibility was due to analytical uncertainties.

No effect of ensiling plus toasting on ME or NEL estimations was found. The available equations for ME and NEL estimation provided by the GfE [14] and the NRC [15] had a high conformity with calculations based on measured GE and ME calculated on the basis of measured GE and nutrient digestibilities.

5. Conclusions

The combination of ensiling and toasting of field pea grains on farms using a mobile toaster with 100 kg/hour throughput rate and 180 to 190 °C temperature of the supplied air (i.e., 85 to 90 °C grain temperature) led to a three-fold increase of RUP. However, boundary conditions for heat damage of proteins have to be clarified. Protein fractions that are fully insoluble to digestive enzymes (e.g., CPip or acid detergent insoluble protein) also increased after ensiling plus toasting, which should be avoided. Total tract apparent digestibility of OM, CA, CP, CF, ADF, NFE, starch, sugars, and GE was not affected by ensiling plus toasting. The apparent increase in AEE digestibility was probably an analytical issue. The NDF apparent digestibility probably increased due to increasing amounts of NDICP after toasting. Contents of GE, ME, and NEL remained unaffected. A high conformity was found among GE, ME, and NEL, which were measured, calculated, or estimated using GfE or NRC equations, respectively.

Author Contributions: Conceptualization, M.B.; data curation, M.B.; formal analysis, M.B., C.K., P.O., and J.M.G.; funding acquisition, O.S. and A.Z.; investigation, M.B., C.K., P.O., and S.D.M.; methodology, M.B., C.K., and J.M.G.; project administration, O.S. and A.Z.; resources, J.M.G.; supervision, O.S. and A.Z.; validation, J.M.G.; visualization, M.B.; writing—original draft, M.B.; writing—review & editing, M.B., C.K., P.O., J.M.G., S.D.M., O.S., and A.Z., M.B. and C.K. contributed equally to the work.

References

1. ISAAA. *Global Status of Commercialized Biotech/GM Crops: 2016*; ISAAA Brief No. 52; ISAAA: Ithaca, NY, USA, 2016.
2. Tillie, P.; Rodríguez-Crezo, E. Markets for Non-Genetically Modified, Identity-Preserved Soybean in the EU. 2015, pp. 1–72. Available online: http://publications.jrc.ec.europa.eu/repository/handle/JRC95457 (accessed on 6 March 2019).
3. Abreu, J.M.F.; Bruno-Soares, A.M. Chemical composition, organic matter digestibility and gas production of nine legume grains. *Anim. Feed Sci. Technol.* **1998**, *70*, 49–57. [CrossRef]

4. Goelema, J.O.; Spreeuwenberg, M.A.M.; Hof, G.; van der Poel, A.F.B.; Tamminga, S. Effect of pressure toasting on the rumen degradability and intestinal digestibility of whole and broken peas, lupins and faba beans and a mixture of these feedstuffs. *Anim. Feed Sci. Technol.* **1998**, *76*, 35–50. [CrossRef]
5. Masoero, F.; Pulimeno, A.M.; Rossi, F. Effect of extrusion, expansion and toasting on the nutritional value of peas, faba beans and lupins. *Ital. J. Anim. Sci.* **2005**, *4*, 177–189. [CrossRef]
6. Konishi, C.; Matsui, T.; Park, W.; Yano, H.; Yano, F. Heat treatment of soybean meal and rapeseed meal suppresses rumen degradation of phytate phosphorus in sheep. *Anim. Feed Sci. Technol.* **1999**, *80*, 115–122. [CrossRef]
7. Getachew, G.; Robinson, P.H.; DePeters, E.J.; Taylor, S.J. Relationships between chemical composition, dry matter degradation and in vitro gas production of several ruminant feeds. *Anim. Feed Sci. Technol.* **2004**, *111*, 57–71. [CrossRef]
8. Corbett, R.R.; Okine, E.K.; Goonewardene, L.A. Effects of feeding peas to high-producing dairy cows. *Can. J. Anim. Sci.* **1995**, *75*, 625–629. [CrossRef]
9. Khorasani, G.R.; Okine, E.K.; Corbett, R.R.; Kennelly, J.J. Nutritive value of peas for lactating dairy cattle. *Can. J. Anim. Sci.* **2001**, *81*, 541–551. [CrossRef]
10. Yu, P.; Holmes, J.H.G.; Leury, B.J.; Egan, A.R. Influence of dry roasting on rumen protein degradation characteristics of whole faba bean (*Vicia faba*) in dairy cows. *Asia. Austral. J. Anim. Sci.* **1998**, *11*, 35–42. [CrossRef]
11. Bachmann, M.; Kuhnitzsch, C.; Thierbach, A.; Michel, S.; Bochnia, M.; Greef, J.M.; Martens, S.D.; Steinhöfel, O.; Zeyner, A. Effects of toasting temperature and time on ruminal gas production kinetics and post-ruminal crude protein from field pea (*Pisum sativum*) grain silages measured *in vitro*. In Proceedings of the 73rd Conference of the Society of Nutrition Physiology, Göttingen, Germany, 13–15 March 2019; DLG-Verlag: Frankfurt (Main), Germany, 2019; p. 112.
12. Kuhnitzsch, C.; Martens, S.D.; Bachmann, M.; Zeyner, A.; Hofmann, T.; Steinhöfel, O. Vergleichende Untersuchungen zum Einsatz siliert und getoasteter Erbsen bzw. Erbsenschröpfschnitt-GPS in der Fütterung hochleistender Milchkühe. In Proceedings of the Forum angewandte Forschung in der Rinder- und Schweinefütterung, Fulda, Germany, 2–3 April 2019; Self-Publishing: Bad Sassendorf, Germany, 2019; pp. 61–64.
13. Society of Nutrition Physiology (GfE). *Empfehlungen zur Energie- und Nährstoffversorgung der Milchkühe und Aufzuchtrinder*; DLG-Verlag: Frankfurt (Main), Germany, 2001.
14. Society of Nutrition Physiology (GfE). New equations for predicting metabolisable energy of compound feeds for cattle. In Proceedings of the 63rd Conference of the Society of Nutrition Physiology, Göttingen, Germany, 10–12 March 2009; DLG-Verlag: Frankfurt (Main), Germany, 2009; pp. 143–146.
15. National Research Council (NRC). *Nutrient Requirements of Dairy Cattle*, 7th ed.; National Academy Press: Washington, DC, USA, 2001.
16. Honig, H. Evaluation of aerobic stability. In Proceedings of the EUROBAC Conference, Uppsala, Sweden, 12–16 August 1986; Lindgren, S., Pettersson, K., Eds.; Swedish University of Agricultural Sciences: Uppsala, Sweden, 1990; pp. 72–78.
17. Licitra, G.; Lauria, F.; Carpino, S.; Schadt, I.; Sniffen, C.J.; Van Soest, P.J. Improvement of the *Streptomyces griseus* method for degradable protein in ruminant feeds. *Anim. Feed Sci. Technol.* **1998**, *72*, 1–10. [CrossRef]
18. Licitra, G.; Van Soest, P.J.; Schadt, I.; Carpino, S.; Sniffen, C.J. Influence of the concentration of the protease from *Streptomyces griseus* relative to ruminal protein degradability. *Anim. Feed Sci. Technol.* **1999**, *77*, 99–113. [CrossRef]
19. Jurášek, L.; Johnson, P.; Olafson, R.W.; Smillie, L.B. An improved fractionation system for pronase on CM-Sephadex. *Can. J. Biochem.* **1971**, *49*, 1195–1201. [CrossRef] [PubMed]
20. Licitra, G.; Hernandez, T.M.; Van Soest, P.J. Standardization of procedures for nitrogen fractionation of ruminant feeds. *Anim. Feed Sci. Technol.* **1996**, *57*, 347–358. [CrossRef]
21. Society of Nutrition Physiology (GfE). Formeln zur Schätzung des Gehaltes an Umsetzbarer Energie und Nettoenergie-Laktation in Mischfuttern. In *Proceedings of the 50th Conference of the Society of Nutrition Physiology*; DLG-Verlag: Frankfurt (Main), Germany, 1996; pp. 153–155.
22. Society of Nutrition Physiology (GfE). Leitlinien für die Bestimmung der Verdaulichkeit von Rohnährstoffen an Wiederkäuern. *J. Anim. Physiol. Anim. Nutr.* **1991**, *65*, 229–234. [CrossRef]

23. VDLUFA. *Die Chemische Untersuchung Von Futtermitteln Methodenbuch*, 3rd ed.; VDLUFA-Verlag: Darmstadt, Germany, 2012.
24. VDLUFA. *Methode 4.13.1, Bestimmung des Neutral-Detergenzien-Löslichen Rohproteins (NDLXP)*; VDLUFA-Verlag: Darmstadt, Germany, 2016.
25. Yemm, E.W.; Willis, A.J. The estimation of carbohydrates in plant extracts by anthrone. *Biochem. J.* **1954**, *56*, 508–514. [CrossRef] [PubMed]
26. Weissbach, F.; Prym, R.; Peters, G.; Lengerken, J.V. Pepsin-insoluble crude protein-Criterion of green forage silage quality. *Tierzucht* **1985**, *39*, 346–349.
27. Schiemann, R.; Nehring, K.; Hoffmann, L.; Jentsch, W.; Chudy, A. *Energetische Futterbewertung und Energienormen*; VEB Deutscher Landwirtschaftsverlag: Berlin, Germany, 1972.
28. White, G.A.; Smith, L.A.; Houdijk, J.G.M.; Homer, D.; Kyriazakis, I.; Wiseman, J. Replacement of soya bean meal with peas and faba beans in growing/finishing pig diets: Effect on performance, carcass composition and nutrient excretion. *Anim. Feed Sci. Technol.* **2015**, *209*, 202–210. [CrossRef]
29. Dotas, V.; Bampidis, V.A.; Sinapis, E.; Hatzipanagiotou, A.; Papanikolaou, K. Effect of dietary field pea (*Pisum sativum* L.) supplementation on growth performance, and carcass and meat quality of broiler chickens. *Livest. Sci.* **2014**, *164*, 135–143. [CrossRef]
30. Frau-Nji, F.; Niess, E.; Pfeffer, E. Effect of graded replacement of soybean meal by faba beans (*Vicia faba* L.) or field peas (*Pisum sativum* L.) in rations for laying hens on egg production and quality. *J. Poult. Sci.* **2007**, *44*, 34–41. [CrossRef]
31. Loe, E.R.; Bauer, M.L.; Lardy, G.P.; Caton, J.S.; Berg, P.T. Field pea (*Pisum sativum*) inclusion in corn-based lamb finishing diets. *Small Rum. Res.* **2004**, *53*, 39–45. [CrossRef]
32. Reed, J.J.; Lardy, G.P.; Bauer, M.L.; Gilbery, T.C.; Caton, J.S. Effect of field pea level on intake, digestion, microbial efficiency, ruminal fermentation, and in situ disappearance in beef steers fed growing diets. *J. Anim. Sci.* **2004**, *82*, 2123–2130. [CrossRef]
33. Valencia, D.G.; Serrano, M.P.; Centeno, C.; Lázaro, R.; Mateos, G.G. Pea protein as a substitute of soya bean protein in diets for young pigs: Effects on productivity and digestive traits. *Livest. Sci.* **2008**, *118*, 1–10. [CrossRef]
34. Valencia, D.G.; Serrano, M.P.; Jiménez-Moreno, E.; Lázaro, R.; Mateos, G.G. Ileal digestibility of amino acids of pea protein concentrate and soya protein sources in broiler chicks. *Livest. Sci.* **2009**, *121*, 21–27. [CrossRef]
35. Vander Pol, M.; Hristov, A.N.; Zaman, S.; Delano, N.; Schneider, C. Effect of inclusion of peas in dairy cow diets on ruminal fermentation, digestibility, and nitrogen losses. *Anim. Feed Sci. Technol.* **2009**, *150*, 95–105.
36. Cone, J.W.; Van Gelder, A.H.; Steg, A.; Van Vuuren, A.M. Prediction of in situ rumen escape protein from in vitro incubation with protease from *Streptomyces griseus*. *J. Sci. Food Agric.* **1996**, *72*, 120–126. [CrossRef]
37. Can, A.; Hummel, J.; Denek, N.; Südekum, K.-H. Effects of non-enzymatic browning reaction intensity on in vitro ruminal protein degradation and intestinal protein digestion of soybean and cottonseed meals. *Anim. Feed Sci. Technol.* **2011**, *163*, 255–259. [CrossRef]
38. Böttger, C.; Weber, T.; Südekum, K.-H. Proteinwert von Rapsextraktionsschroten für Wiederkäuer—Schätzung mittels chemischer und in vitro-Verfahren. In Proceedings of the 129th VDLUFA-Kongress, Freising, Germany, 12–15 September 2017; VDLUFA-Verlag: Darmstadt, Germany, 2017; p. 107.
39. Deutsche Landwirtschafts-Gesellschaft (DLG). *DLG-Futterwerttabellen Wiederkäuer*, 7th ed.; DLG-Verlag: Frankfurt (Main), Germany, 1997.
40. Ljøkjel, K.; Skrede, A.; Harstad, O.M. Effects of pelleting and expanding of vegetable feeds on in situ protein and starch digestion in dairy cows. *J. Anim. Feed Sci.* **2003**, *12*, 435–449. [CrossRef]
41. Mustafa, A.F.; Seguin, P. Characteristics and in situ degradability of whole crop faba bean, pea, and soybean silages. *Can. J. Anim. Sci.* **2003**, *83*, 793–799. [CrossRef]
42. Hoedtke, S.; Gabel, M.; Zeyner, A. Protein degradation in feedstuffs during ensilage and changes in the composition of the crude protein fraction. *Übers. Tierernährg.* **2010**, *38*, 157–179.
43. Gefrom, A.; Ott, E.M.; Hoedtke, S.; Zeyner, A. Effect of ensiling moist field bean (*Vicia faba*), pea (*Pisum sativum*) and lupine (*Lupinus* spp.) grains on the contents of alkaloids, oligosaccharides and tannins. *J. Anim. Physiol. Anim. Nutr.* **2013**, *97*, 1152–1160. [CrossRef] [PubMed]
44. Duranti, M. Grain legume proteins and nutraceutical properties. *Fitoterapia* **2006**, *77*, 67–82. [CrossRef] [PubMed]

45. Focant, M.; Van Hoecke, A.; Vanbelle, M. The effect of two heat treatments (steam flaking and extrusion) on the digestion of *Pisum sativum* in the stomachs of heifers. *Anim. Feed Sci. Technol.* **1990**, *28*, 303–313. [CrossRef]
46. Aufrère, J.; Graviou, D.; Melcion, J.P.; Demarquilly, C. Degradation in the rumen of lupin (*Lupinus albus* L.) and pea (*Pisum sativum* L.) seed proteins. Effect of heat treatment. *Anim. Feed Sci. Technol.* **2001**, *92*, 215–236. [CrossRef]
47. Azarfar, A.; Tamminga, S.; Pellikaan, W.F.; van der Poel, A.F.B. In vitro gas production profiles and fermentation end-products in processed peas, lupins and faba beans. *J. Sci. Food Agric.* **2008**, *88*, 1997–2010. [CrossRef]
48. Vaga, M.; Hetta, M.; Huhtanen, P. Effects of heat treatment on protein feeds evaluated in vitro by the method of estimating utilisable crude protein at the duodenum. *J. Anim. Physiol. Anim. Nutr.* **2017**, *101*, 1259–1272. [CrossRef] [PubMed]
49. Hellwig, M.; Henle, T. Baking, ageing, diabetes: A short history of the Maillard reaction. *Angew. Chem. Int. Ed.* **2014**, *53*, 10316–10329. [CrossRef] [PubMed]
50. Holum, J.R. *Fundamentals of General, Organic, and Biological Chemistry*, 2nd ed.; Wiley: New York, NY, USA, 1982.
51. Hurrell, R.E.; Finot, R.A. Effect of food processing on protein digestibility and amino acid availability. In *Digestibility and Amino Acid Availability in Cereals and Oilseeds*; Finley, J.W., Hopkins, D.T., Eds.; American Association of Cereal Chemists: St. Paul, MN, USA, 1985; pp. 516–527.
52. Broderick, G.A.; Wallace, R.J.; Ørskov, E.R. Control of Rate and Extent of Protein Degradation. In *Physiological Aspects of Digestion and Metabolism in Ruminants, Proceedings of the 7th International Symposium on Ruminant Physiology, Clermont—Ferrand, France, 3–7 September 1979*; Tsuda, T., Sasaki, Y., Kawashima, R., Eds.; Academic Press: San Diego, CA, USA, 1991; pp. 541–592.
53. Yu, P.; Goelema, J.O.; Leury, B.J.; Tamminga, S.; Egan, A.R. An analysis of the nutritive value of heat processed legume seeds for animal production using the DVE/OEB model: A review. *Anim. Feed Sci. Technol.* **2002**, *99*, 141–176. [CrossRef]
54. Poncet, C.; Rémond, D. Rumen digestion and intestinal nutrient flows in sheep consuming pea seeds: The effect of extrusion or chestnut tannin addition. *Anim. Res.* **2002**, *51*, 201–216. [CrossRef]

Permissions

All chapters in this book were first published by MDPI; hereby published with permission under the Creative Commons Attribution License or equivalent. Every chapter published in this book has been scrutinized by our experts. Their significance has been extensively debated. The topics covered herein carry significant findings which will fuel the growth of the discipline. They may even be implemented as practical applications or may be referred to as a beginning point for another development.

The contributors of this book come from diverse backgrounds, making this book a truly international effort. This book will bring forth new frontiers with its revolutionizing research information and detailed analysis of the nascent developments around the world.

We would like to thank all the contributing authors for lending their expertise to make the book truly unique. They have played a crucial role in the development of this book. Without their invaluable contributions this book wouldn't have been possible. They have made vital efforts to compile up to date information on the varied aspects of this subject to make this book a valuable addition to the collection of many professionals and students.

This book was conceptualized with the vision of imparting up-to-date information and advanced data in this field. To ensure the same, a matchless editorial board was set up. Every individual on the board went through rigorous rounds of assessment to prove their worth. After which they invested a large part of their time researching and compiling the most relevant data for our readers.

The editorial board has been involved in producing this book since its inception. They have spent rigorous hours researching and exploring the diverse topics which have resulted in the successful publishing of this book. They have passed on their knowledge of decades through this book. To expedite this challenging task, the publisher supported the team at every step. A small team of assistant editors was also appointed to further simplify the editing procedure and attain best results for the readers.

Apart from the editorial board, the designing team has also invested a significant amount of their time in understanding the subject and creating the most relevant covers. They scrutinized every image to scout for the most suitable representation of the subject and create an appropriate cover for the book.

The publishing team has been an ardent support to the editorial, designing and production team. Their endless efforts to recruit the best for this project, has resulted in the accomplishment of this book. They are a veteran in the field of academics and their pool of knowledge is as vast as their experience in printing. Their expertise and guidance has proved useful at every step. Their uncompromising quality standards have made this book an exceptional effort. Their encouragement from time to time has been an inspiration for everyone.

The publisher and the editorial board hope that this book will prove to be a valuable piece of knowledge for researchers, students, practitioners and scholars across the globe.

List of Contributors

Zhenguo Yang, Tianle He, Gifty Ziema Bumbie, Hong Hu, Qingju Chen, Changwen Lu and Zhiru Tang
Laboratory for Bio-feed and Molecular Nutrition, College of Animal Science and Technology, Southwest University, Chongqing 400715, China

Ridha Ibidhi
Department of Eco-Friendly Livestock Science, Institute of Green Bio Science and Technology, Seoul National University, Pyeongchang, Gangwon 08826, Korea
National Institute of Agronomic Recherche of Tunisia (INRAT), Laboratory of Animal Production and Forages, Université de Carthage, rue Hédi Karray, Ariana 2049, Tunisia

Sonia Tassone, Riccardo Fortina and Silvia Boggero
Department of Agriculture, Forest, and Food Sciences, University of Turin, 10095 Grugliasco, TO, Italy

Emanuela Valle, Federica Raspa, Domenico Bergero and Manuela Renna
Department of Veterinary Sciences, University of Turin, 10095 Grugliasco, TO, Italy

Laura Cavallarin
Institute of Sciences of Food Production, National Research Council, 10095 Grugliasco, TO, Italy

Mauro Giammarino
Department of Prevention, Asl TO3, Veterinary Service, Area Animal Sanity, Via Torino 62, 10045 Piossasco, TO, Italy

Hèctor Salas, Lorena Castillejos, Montserrat López-Suárez and Alfred Ferret
Animal Nutrition and Welfare Service (SNIBA), Department of Animal and Food Science, Universitat Autonòma de Barcelona, 08193 Bellaterra, Spain

Ives C. S. Bueno, Roberta A. Brandi and Gabriela Benetel
Department of Animal Science, University of São Paulo, Pirassununga, São Paulo 1365-900, Brazil

Gisele M. Fagundes
Department of Animal Science, Federal University of Roraima – UFRR, BR 174, km 12, Boa Vista, Roraima 69300-000, Brazil

James Pierre Muir
Texas A&M AgriLife Research, Texas A&M University, Stephenville, TX 76401, USA

Maria Chiaravalli, Luca Rapetti, Andrea Rota Graziosi, Gianluca Galassi, Gianni Matteo Crovetto and Stefania Colombini
Dipartimento di Scienze Agrarie e Ambientali, Università degli Studi di Milano, 20133 Milano, Italy

Sihem Dabbou
Center Agriculture Food Environment (C3A), University of Trento, Via E. Mach 1, 38010 San Michele all'Adige, Italy
Research and Innovation Centre, Fondazione Edmund Mach, 38010 San Michele all'Adige, Italy

Ilario Ferrocino, Laura Gasco, Ilaria Biasato, Luca Cocolin and Daniele Michele Nucera
Department of Agricultural, Forest and Food Sciences, University of Torino, Largo Paolo Braccini 2, 10095 Grugliasco, Italy

Achille Schiavone, Sandra Maione and Dominga Soglia
Department of Veterinary Science, University of Turin, Largo P. Braccini 2, 10095 Grugliasco, Italy

Angela Trocino
Department of Comparative Biomedicine and Food Science, University of Padova, Viale dell'Universita 16, Legnaro, 35020 Padova, Italy

Gerolamo Xiccato
Department of Agronomy, Food, Natural Resources, Animal, and Environment, University of Padova, Viale dell'Universita 16, 35020 Legnaro, Padova, Italy

Ana C. Barroeta
Nutrition and Animal Welfare Service, Department of Animal and Food Science, Faculty of Veterinary Medicine, Autonomous University of Barcelona, Bellaterra, 08193 Barcelona, Spain

Francesco Gai
Institute of Sciences of Food Production, National Research Council, Largo Paolo Braccini 2, 10095 Grugliasco, Italy

Yafeng Huang, Fangfang Zhou and Zhibiao Nan
State Key Laboratory of Grassland Agro-ecosystems, Key Laboratory of Grassland Livestock Industry Innovation of Ministry of Agriculture and Rural Affairs, College of Pastoral Agriculture Science and Technology, Lanzhou University, Lanzhou 730020, China

Jae-Cheol Jang, Zhikai Zeng and Gerald C. Shurson
Department of Animal Science, University of Minnesota, St. Paul, MN 55108, USA

Pedro E. Urriola
Department of Animal Science, University of Minnesota, St. Paul, MN 55108, USA
Department of Veterinary Population Medicine, University of Minnesota, St. Paul, MN 55108, USA

Juana C. Chagas, Mohammad Ramin and Sophie J. Krizsan
Department of Agricultural Research for Northern Sweden, Swedish University of Agricultural Sciences (SLU), Skogsmarksgränd, 90183 Umeå, Sweden

Kristina Bacova, Andrea Laukova, Michaela Madarova and Iveta Placha
Centre of Biosciences, Slovak Academy of Sciences, Institute of Animal Physiology, Soltesovej 4-6, 040 01 Kosice, Slovakia

Karin Zitterl-Eglseer
Institute of Animal Nutrition and Functional Plant Compounds, University of Veterinary Medicine Vienna, Veterinarplatz 1, A-1210 Wien, Austria

Lubica Chrastinova
National Agricultural and Food Centre, Hlohovecka 2, 951 41 Nitra-Luzianky, Slovakia

Sona Gancarcikova
Laboratory of Gnotobiology, Department of Microbiology and Immunology, University of Veterinary Medicine and Pharmacy, Komenskeho 73, 041 81 Kosice, Slovakia

Drahomira Sopkova and Zuzana Andrejcakova
Department of Anatomy, Histology and Physiology, University of Veterinary Medicine and Pharmacy, Komenskeho 73, 041 81 Kosice, Slovakia

Jennifer L. Ellis, Christopher D. Powell and James France
Centre for Nutrition Modelling, Department of Animal Biosciences, University of Guelph, 50 Stone Road East, Guelph, ON N1G 2W1, Canada

Héctor Alaiz-Moretón
Departamento de Ingeniería Eléctrica de Sistemas y Automática, Escuela de Ingeniería Industrial e Informática, Universidad de León, Campus Universitario de Vegazana, 24071 León, Spain

Alberto Navarro-Villa
Animal & Grassland Research and Innovation Centre, Teagasc, Grange, Dunsany, Co. Meath C15 PW93, Ireland
Trouw Nutrition R&D, Ctra. CM-4004 km 10.5, 45950 El Viso de San Juan, Spain

Emma J. McGeough
Animal & Grassland Research and Innovation Centre, Teagasc, Grange, Dunsany, Co. Meath C15 PW93, Ireland
Department of Animal Science, University of Manitoba, Winnipeg, MB R3T 2N2, Canada

Peter Purcell and Padraig O'Kiely
Animal & Grassland Research and Innovation Centre, Teagasc, Grange, Dunsany, Co. Meath C15 PW93, Ireland

Secundino López
Instituto de Ganadería de Montaña (IGM), CSIC-Universidad de León, Departamento de Producción Animal, Universidad de León, 24007 León, Spain

Ana de la Moneda and Eduarda Molina-Alcaide
Estación Experimental del Zaidin (Consejo Superior de Investigaciones Cientificas), Profesor Albareda, 1, 18008 Granada, Spain

Maria Dolores Carro
Departamento de Producción Agraria. Escuela Técnica Superior de Ingeniería Agronómica, Alimentaria y de Biosistemas, Universidad Politécnica de Madrid, Ciudad Universitaria, 28040 Madrid, Spain

Martin R. Weisbjerg
Aarhus University, AU Foulum, Blichers Allé 20, 8830 Tjele, Denmark

Vibeke Lind
Norwegian Institute of Bioeconomy Research (NIBIO), PB 115, 1431 Ås, Norway

Michael Y. Roleda
Norwegian Institute of Bioeconomy Research (NIBIO), PB 115, 1431 Ås, Norway
The Marine Science Institute, College of Science, University of the Philippines, Diliman, Quezon City 1101, Philippines

Margarita Novoa-Garrido
Norwegian Institute of Bioeconomy Research (NIBIO), PB 115, 1431 Ås, Norway
Faculty of Biosciences and Aquaculture, Nord University, 8049 Bodø, Nordland, Norway

Panthiphaporn Chankuang, Achira Linlawan, Kawisara Junda, Chittikan Kuditthalerd and Tuksaorn Suwanprateep
Faculty of Veterinary Medicine, Kasetsart University, Bangkok 10900, Thailand

Attawit Kovitvadhi and Pipatpong Chundang
Department of Physiology, Faculty of Veterinary Medicine, Kasetsart University, Bangkok 10900, Thailand

Pornchai Sanyathitiseree
Department of Large Animal and Wildlife Clinical Science, Faculty of Veterinary Medicine, Kasetsart University, Nakorn Pathom 73140, Thailand

Chaowaphan Yinharnmingmongkol
Animal Space Pet Hospital, Bangkok 10170, Thailand

Pier Giorgio Peiretti
Institute of Sciences of Food Production, National Research Council, 10095 Grugliasco, Italy

Ignacio Fandiño, Diego Moya and Sergio Calsamiglia
Departamento de Ciència Animal i dels Aliments, Universitat Autònoma de Barcelona, Servicio de Nutrición y Bienestar Animal (SNiBA), 08193 Bellaterra, Spain

Gonzalo Fernandez-Turren
Departamento de Producción Animal, Instituto de Producción Animal, Facultad de Veterinaria, Universidad de la República, Ruta 1 km 42, CP 80100 San José, Uruguay

Martin Bachmann, Paul Okon and Annette Zeyner
Institute of Agricultural and Nutritional Sciences, Martin Luther University Halle-Wittenberg, Theodor-Lieser-Straße 11, 06120 Halle (Saale), Germany

Christian Kuhnitzsch and Olaf Steinhöfel
Institute of Agricultural and Nutritional Sciences, Martin Luther University Halle-Wittenberg, Theodor-Lieser-Straße 11, 06120 Halle (Saale), Germany
Saxon State Office for Environment, Agriculture and Geology, Am Park 3, 04886 Köllitsch, Germany

Siriwan D. Martens
Saxon State Office for Environment, Agriculture and Geology, Am Park 3, 04886 Köllitsch, Germany

Jörg M. Greef
Institute for Crop and Soil Science, Federal Research Centre for Cultivated Plants, Julius Kühn Institute (JKI), Bundesallee 58, 38116 Braunschweig, Germany

Index

www.ingramcontent.com/pod-product-compliance
Lightning Source LLC
LaVergne TN
LVHW081015210926
848485LV00004B/871
* 9 7 8 1 6 4 1 1 6 7 7 3 4 *